Antisense Therapeutics

METHODS IN MOLECULAR MEDICINE™

John M. Walker, SERIES EDITOR

Human Cell Culture Protocols, edited by *Gareth E. Jones*, 1996

Antisense Therapeutics, edited by *Sudhir Agrawal,* 1996

Vaccine Protocols, edited by *Andrew Robinson, Graham H. Farrar, and Christopher N. Wiblin,* 1996

Prion Diseases, edited by *Harry F. Baker and Rosalind M. Ridley,* 1996

Molecular Diagnosis of Cancer, edited by *Finbarr Cotter,* 1996

Molecular Diagnosis of Genetic Diseases, edited by *Rob Elles,* 1996

Herpes Simplex Virus Protocols, edited by *Moira S. Brown and Alasdair MacLean,* 1996

***Helicobacter Pylori* Protocols,** edited by *Christopher L. Clayton and Harry T. Mobley,* 1996

Lectins in Medical Research, edited by *Jonathan M. Rhodes and Jeremy D. Milton,* 1996

Gene Therapy Protocols, edited by *Paul Robbins,* 1996

METHODS IN MOLECULAR MEDICINE™

Antisense Therapeutics

Edited by

Sudhir Agrawal

Hybridon, Inc., Worcester, MA

Humana Press ✳ **Totowa, New Jersey**

Cover illustration: Fig. 9 from Chapter 7, "c-*Myb* in Smooth Muscle Cells," by Michael Simons and Robert D. Rosenberg.

Printed in the United States of America. 10 9 8 7 6 5 4 3 2 1

Library of Congress Cataloging in Publication Data

Main entry under title:

Methods in molecular medicine™.

Antisense therapeutics/edited by Sudhir Agrawal.
    p.   cm.—(Methods in molecular medicine™)
    Includes index.
    ISBN 0-89603-305-8 (alk. paper)
    1. Antisense nucleic acids—Therapeutic use. I. Agrawal, Sudhir.   II. Series
    [DNLM: 1. Oligonucleotides, Antisense—therapeutic use. 2. Oligonucleotides, Antisense—pharmacology. QU 57 A633 1996]
    RM666.A456A585 1996
    615'.31—dc20
    DNLM/DLC
    for Library of Congress   96-3441
                                                                                CIP

Preface

Antisense nucleic acids (antisense therapeutics) have attracted much interest as a novel class of therapeutic agents for the treatment of viral infections, cancers, and genetic disorders because of their ability to inhibit the expression of disease-associated genes. Antisense therapeutics inhibit gene expression in a sequence-specific manner through hybridizing to the target gene through Watson-Crick base pairing. Antisense therapeutics have two characteristics required for successful drug design—specificity and affinity. Theoretically, an antisense oligonucleotide comprising 17 nucleotides is unique for its target in the human genome and its affinity for the target is much higher than the 37°C body temperature.

The laboratory of Paul Zamecnik published the first report of antisense oligonucleotides in 1978 in two papers in the Proceedings of the National Academy of Sciences. He and his colleague Mary Stephenson showed that a synthetic oligonucleotide (13-mer) complementary to the Rous sarcoma virus genome inhibited the viral replication when added exogenously to the infected cell culture. Based on these findings, he remarked in his paper on the general chemotherapeutic potential of this observation. I am very fortunate to have worked in Paul Zamecnik's laboratory and have learned a great deal from him about various aspects of antisense therapeutics.

Research efforts exploring the possibility of using oligonucleotides as therapeutic agents started in the mid-1980s. The focus shifted to identification of oligonucleotide analogs that have specificity and affinity for the targeted gene and can be administered locally or systemically, are stable in vivo, and are safe in animals. These analogs thereby provide a wide therapeutic index. Enormous progress has been made in the last 10 years: Many analogs of oligonucleotides have been synthesized and studied as antisense agents. The most widely studied analog of oligonucleotides is phosphorothioate. The results of preclinical studies using oligodeoxynucleotide phosphorothioates have shown that antisense oligonucleotides have good biological activity, pharmacology, pharmacokinetics, and safety both in vitro and in vivo, and they are currently being evaluated in human clinical trials for the treatment of viral infections and cancers.

This volume will provide applications of antisense therapeutics in various design models with contributions from leading experts in the antisense field. These experts will describe their findings, ranging from design of oligonucleotides in the laboratory to in vivo applications, including state-of-the-art strategies currently being used. The book provides extensive knowledge of various aspects of antisense therapeutics in 14 chapters. The topics that have been covered in this book include oligonucleotide synthesis, purification, selection of oligonucleotide for a particular gene target, various cell culture systems, use of various cell lines, cellular uptake, in vitro biological activity, in vivo biological activity in various animal model systems, pharmacology, and pharmacokinetics of oligonucleotides. The book also provides extensive knowledge of the application of oligonucleotides in relation to CNS application and touches on various routes of administration (e.g., intravenous, subcutaneous, intraperitoneal, intravitreal, intraocular, intracerebral, osmotic pumps, and so forth) of oligonucleotides using various animal models.

In the first chapter, Paul Zamecnik describes the historical aspects of antisense oligonucleotides, clearly depicting how the idea of antisense therapeutics became reality. He also describes the application of antisense oligonucleotides in various disease models. In the second chapter, by C. F. Bennett and colleagues, the pharmacology and pharmacokinetics of antisense oligonucleotides have been discussed in respect to their application in cancer and inflammation models. Leonard Necker and colleagues discuss the additional challenge of delivery across the blood–brain barrier to the central nervous system in the third chapter. In the fourth chapter, Marina Catsicas describes the use of oligonucleotides in a CNS-related target in vivo. In Chapter 5, Eric Wickstrom describes various characteristics of antisense oligonucleotides, including uptake, stability, and efficacy, using a tumor model. Chapter 6, by Richard Fine and colleagues, describes the intravitreal administration of antisense oligonucleotides and their effects in an animal model. Chapter 7, by Michael Simons and Robert Rosenberg, describes the use of antisense oligonucleotide in control of smooth muscle cell growth and proliferation following balloon angioplasty.

The use of antisense oligonucleotides as antiviral agents in a duck hepatitis B virus model is described by Wolf-Bernard Offensperger and colleagues in Chapter 8. In Chapter 9, Wolfgang Brysch and colleagues describe the pharmacokinetics, organ uptake, and efficacy following intracerebral administration. Ramaswamy Narayanan (Chapter 10) describes the use of antisense oligonucleotides as a tool to identify a disease target gene and the subsequent use of the same oligonucleotide to inhibit the gene's function and control the tumor growth.

Patrick Iverson (Chapter 11) discusses in detail the pharmacokinetics of oligonucleotide phosphorothioates administered to mice, rats, monkeys, and humans. Chapter 12, by Yoon Cho-Chung and Marina Nesterova, describes the arrest of tumor growth in nude mice by single dose administration of antisense oligonucleotides targeting protein kinase A in tumor-bearing mice. The application of antisense therapeutics in the CNS following infusion of oligonucleotides into the striatum is described by Harold Robertson and colleagues in Chapter 13. The last chapter, written by me and my colleague Jamal Temsamani, describes the impact of chemical modification of oligonucleotides on pharmacokinetics, tissue disposition, and in vivo stability.

I am, of course, deeply indebted to the contributors of this book for their hard work and their patience during its preparation; they deserve the credit for its utility and interest. I would also like to congratulate each of them for the fine work carried out in their laboratories. I would like to thank Dr. John Walker for guidance and encouragement during the preparation of the book. I would also like to thank Mr. Tom Lanigan and Ms. Fran Lipton of Humana Press for constant editorial support and guidance, and the production team of Humana Press, who have worked diligently on this book. Finally, I would like to thank Ms. Kathy Anthony for expert secretarial assistance in putting together this book.

Sudhir Agrawal

Contents

Contributors

SUDHIR AGRAWAL • *Hybridon, Inc., Worcester, MA*

ANIL P. AMARATUNGA • *Department of Biochemistry, Boston University School of Medicine, Boston, MA*

JOHN N. ARMSTRONG • *Departments of Pharmacology, Anatomy, and Neurobiology, Laboratory of Molecular Neurobiology, Faculty of Medicine, Dalhousie University, Halifax, Canada*

GUIDON AYALA • *Glaxo Institute for Molecular Biology, Geneva, Switzerland*

ELIEL BAYEVER • *Department of Hematology and Oncology, Children's National Medical Center, Washington, DC*

C. FRANK BENNETT • *ISIS Pharmaceuticals, Carlsbad, CA*

RAYMOND BERGAN • *Clinical Pharmacology Branch, National Cancer Institute, Bethesda, MD*

MICHAEL R. BISHOP • *Department of Internal Medicine, and The Eppley Institute for Cancer Research, University of Nebraska Medical Center, Omaha, NE*

HUBERT E. BLUM • *Department of Medicine, University Hospital, Zurich, Switzerland*

WOLFGANG BRYSCH • *Max-Planck Institute for Biophysical Chemistry, Gottingen, Germany*

MARINA CATSICAS • *Glaxo Institute for Molecular Biology, Geneva, Switzerland. Present Address: Department of Physiology, University College, London, UK*

STEFAN CATSICAS • *Glaxo Institute for Molecular Biology, Geneva, Switzerland*

CHRISTINE CHAVANY • *Clinical Pharmacology Branch, National Cancer Institute, Bethesda, MD*

BERNARD J. CHIASSON • *Department of Pharmacology, Laboratory of Molecular Neurobiology, Faculty of Medicine, Dalhousie University, Halifax, Canada*

YOON S. CHO-CHUNG • *Cellular Biochemistry Section, Laboratory of Tumor Immunology and Biology, National Cancer Institute, Bethesda, MD*

BRYAN L. COPPLE • *Department of Pharmacology and The Eppley Institute for Cancer Research, University of Nebraska Medical Center, Omaha, NE*

NICHOLAS DEAN • *ISIS Pharmaceuticals, Carlsbad, CA*

DAVID J. ECKER • *ISIS Pharmaceuticals, Carlsbad, CA*

RICHARD E. FINE • *Department of Biochemistry, Boston University School of Medicine; ENR VA Hospital, Boston, MA*

DANIEL GESELOWITZ • *Clinical Pharmacology Branch, National Cancer Institute, Bethesda, MD*

GABRIELE GRENNINGLOH • *Glaxo Institute for Molecular Biology, Geneva, Switzerland*

KIMBERLY A. HIGGINS-SOCHASKI • *Division of Oncology, Roche Research Center, Hoffmann-La Roche, Nutley, NJ*

MURRAY HONG • *Department of Pharmacology, Laboratory of Molecular Neurobiology, Faculty of Medicine, Dalhousie University, Halifax, Canada*

MICHELE L. HOOPER • *Department of Pharmacology, Laboratory of Molecular Neurobiology, Faculty of Medicine, Dalhousie University, Halifax, Canada*

PATRICK L. IVERSEN • *Departments of Pharmacology and Pharmaceutical Sciences, and The Eppley Institute for Cancer Research, University of Nebraska Medical Center, Omaha, NE*

KENNETH A. JONES • *Glaxo Institute for Molecular Biology, Geneva, Switzerland*

JONATHON KNOWLES • *Glaxo Institute for Molecular Biology, Geneva, Switzerland*

KENNETH S. KOSIK • *Department of Neurology, Harvard Medical School, Boston, MA*

SUSAN E. LEEMAN • *Department of Pharmacology, Boston University School of Medicine, Boston, MA*

JEAN-YVES MALTESE • *Division of Oncology, Roche Research Center, Hoffmann-La Roche, Nutley, NJ*

BRETT P. MONIA • *ISIS Pharmaceuticals, Carlsbad, CA*

PAUL R. MURPHY • *Department of Physiology and Biophysics, Laboratory of Molecular Neurobiology, Faculty of Medicine, Dalhousie University, Halifax, Canada*

RAMASWAMY NARAYANAN • *Division of Oncology, Roche Research Center, Hoffmann-La Roche, Nutley, NJ*

LEONARD M. NECKERS • *Clinical Pharmacology Branch, National Cancer Institute, Bethesda, MD*

MARIA NESTEROVA • *Cellular Biochemistry Section, Laboratory of Tumor Immunology and Biology, National Cancer Institute, Bethesda, MD*

SILKE OFFENSPERGER • *Department of Medicine, University of Freiburg, Germany*

WOLF-BERNHARD OFFENSPERGER • *Department of Medicine, University of Freiburg, Germany*

ASTRID OSEN-SAND • *Glaxo Institute for Molecular Biology, Geneva, Switzerland*

JOSE R. PEREZ • *Division of Oncology, Roche Research Center, Hoffmann-La Roche, Nutley, NJ*

EMILIO MERLO PICH • *Glaxo Institute for Molecular Biology, Geneva, Switzerland*

ABDALLA RIFAI • *Max-Planck Institute for Biophysical Chemistry, Gottingen, Germany*

PETER A. RITTENHOUSE • *Department of Pharmacology, Boston University School of Medicine, Boston, MA*

HAROLD A. ROBERTSON • *Department of Pharmacology, Laboratory of Molecular Neurobiology, Faculty of Medicine, Dalhousie University, Halifax, Canada*

ROBERT D. ROSENBERG • *Department of Biology, Massachusetts Institute of Technology, Cambridge, MA*

KARL-HERMANN SCHLINGENSIEPEN • *Max-Planck Institute for Biophysical Chemistry, Gottingen, Germany*

MICHAEL SIMONS • *Department of Biology, Massachusetts Institute of Technology, Cambridge, MA*

JULIE K. STAPLE • *Glaxo Institute for Molecular Biology, Geneva, Switzerland*

JAMAL TEMSAMANI • *Hybridon, Inc., Worcester, MA*

HEMANT K. TEWARY • *Department of Pharmacology, University of Nebraska Medical Center, Omaha, NE*

WOLFGANG TISCHMEYER • *Max-Planck Institute for Biophysical Chemistry, Gottingen, Germany*

LUKE WHITESELL • *Clinical Pharmacology Branch, National Cancer Institute, Bethesda, MD*

ERIC WICKSTROM • *Department of Pharmacology, Thomas Jefferson University, Philadelphia, PA*

PAUL C. ZAMECNIK • *Worcester Foundation for Biomedical Research, Shrewsbury, MA*

1

History of Antisense Oligonucleotides

Paul C. Zamecnik

1. Introduction and Early Studies

Biological science is a rapidly flowing experimental stream, at times encountering a dam that impedes further progress. At such a point, a single crack may induce a major breakthrough. Discovery of the double helical structure of DNA in 1953 *(1)* caused such an event, with flooding of new information into the area now known as molecular biology.

At this same time, our laboratory *(2,3)* developed a cell-free system for the study of protein synthesis, a domain separate from the DNA world. In 1954, James Watson and this author examined his wire model of DNA and puzzled about how the information from the gene became translated into the sequence of a protein *(4)*. The histochemical studies of Brachet *(5)* and Caspersson *(6)* had shown that in the pancreas, an organ very actively synthesizing proteins for export, the cytoplasm was rich in what became known as ribonucleic acid. But how the DNA of the nucleus unwound its double strand and transcribed the RNA, the apparent intermediate in protein synthesis found in the cytoplasm, was unknown *(7)*.

The first example of the versatility of nucleic acid base pairing in the flow of information from DNA to protein was the discovery of transfer RNA *(8–11)* and perception of its role in translating the language of the gene into the sequence of protein *(12)*. Deciphering of the genetic code *(13,14)* next brought to light the precision of the tRNA–mRNA hybridization steps in protein translation. tRNA (an antisense or negatively stranded RNA) acts in four distinguishable ways, as follows:

1. By base pairing with messenger RNA to initiate translation of the message;
2. By base pairing with messenger RNA to propagate translation of the message;

From: *Methods in Molecular Medicine: Antisense Therapeutics*
Edited by: S. Agrawal Humana Press Inc., Totowa, NJ

3. By base pairing with ribosomal RNA to position the trinucleotide anticodon region for optimal hybridization with the messenger RNA codon; and

4. By presenting a terminating anticodon (which is an antisense trinucleotide) to end the nascent protein sequence.

Puromycin, a natural nucleotide analog, provided an early example of antisense inhibition of protein synthesis *(15)*.

Further experimentation supported the hypothesis *(16)* that hybridization of synthetic exogenously added oligonucleotides can influence cellular metabolism at three distinct levels: replication *(17)*, transcription *(18,19)*, and translation *(20,21)*. The variety and importance of these steps invited the thought that natural oligonucleotides might play such roles in living cells *(22)*. In the double helix, the DNA strand that carries the genetic message has been designated the sense strand. Its complementary mate, necessary as a template for the synthesis of a new sense strand, has become known as the antisense strand. Antisense polynucleotides have, in fact, for over two decades been known to occur naturally in prokaryotes *(23)*, and have recently been found in eukaryotes *(24)*. For some years synthetic oligonucleotides have also been reported to be capable of playing varied antisense inhibitory roles *(25–28)*.

Quite separate from the synthetic oligonucleotide field were independent developments from splicing larger segments of negatively stranded DNA into the genomes of cells, with the help of plasmids and viruses. These antisense strands of DNA were successfully integrated into the genomes of relatively few host cells. Nevertheless, by selection processes these antisense sequences were picked out and found to be replicated along with the recipient's genomic material. Thus, they were capable of blocking or altering the expression of cellular genes in a hereditary way. An example of the success of this technique is the permanent alteration of the color of petunias by antisense interference with synthesis of the flavonoid genes *(29)*. An important difference between these approaches is that the synthetic, relatively short oligomers enter virtually all cells in a tissue culture or living animal *(30–32)* whereas the plasmids carrying much longer antisense polynucleotides generally enter a small percentage of cells, but may nevertheless have a dramatic genetic therapeutic effect *(33)*.

In the early 1960s the limits of DNA and RNA synthesis were in general trinucleotides, but by the end of the decade skilled scientists were able to construct oligomers 10–15 U in length, and to ligate such segments together *(34)*. The message encoded in DNA or RNA, however, remained difficult to decipher. In 1965, Holley and colleagues *(35)* sequenced the primary structure of tRNA$_{ala}$, using the new method for scale-up and isolation of the tRNA family of molecules worked out in our laboratory by Roger Monier *(36)*. In this highly competitive quest, several groups accomplished the sequencing of other particular tRNAs shortly thereafter *(37–39)*, getting little credit for their efforts.

In 1970, the discovery of reverse transcriptase *(40,41)* made it more feasible to sequence oligonucleotides by synthesizing the primary structure of DNA enzymatically, and then sequencing the nascent DNA so formed. The wandering spot-analysis technique of Sanger and Coulsen *(42)* at that time made it possible to determine the sequence of approx 15–25 monomer units at the 3' end of a polynucleotide.

The present discussion focuses on the role of the synthetic antisense oligonucleotides (20–30-mers) as chemotherapeutic agents, and omits the splicing insertion into genomes of the larger (1,000-2,000 monomer unit) biologically synthesized polynucleotides. Those of us raised on the principle of Occum's razor, which advises making explanations as simple as possible, continue to be surprised at the unfolding complexity of this synthetic oligonucleotide approach to chemotherapy. One was prepared for the nuclease sensitivity of unmodified oligodeoxynucleotides in a living cell system *(16,43)*, and the enhancement of therapeutic efficacy by blocking both ends of the oligodeoxynucleotide *(16)*. The effect of ribonuclease H *(44)* was generally unexpected, however, particularly the major role it plays in some antisense inhibitions. As is now known, a stoichiometrically acting oligodeoxynucleotide inhibitor may activate RNase H during its complementary hybridization with mRNA, then dissociate from its complement when the mRNA is hydrolyzed at the double-stranded area, and hybridize with another molecule of mRNA, this repetitive action resulting in a catalytic effect.

By 1976, we were able to sequence 21 nucleotides inside the 3'-polyA tail of the Rous sarcoma virus, a terminus similar to that we had previously found on the avian myelobastosis virus *(45)*. Rous sarcoma virus was the only purified virus for which a sufficient quantity was available to make a sequencing effort feasible. At this time, we learned that Maxam and Gilbert *(46)* had invented a revolutionary new DNA sequencing technique, and had, unbeknown to us, begun to decipher the 5'- end of the same Rous sarcoma virus. Astonishingly, both ends of this linear viral genome bore the same primary sequence, and were in the same polarity *(47,48)*.

It occurred to us that the new piece of DNA synthesized by reverse transcription at the 5'- end of this retrovirus might circularize and hybridize with the 3' end, like a dog biting its tail. Electron microscopic studies had suggested the presence of a circularized intermediate in the replicative process of this virus. Thus, we considered the possibility of inhibiting viral replication by adding to the replication system a synthetic piece of DNA to block the circularization step (or alternatively some other step essential for replication), in the former case by hybridizing specifically with the 3' end of the viral RNA in a competitive way.

It was at this time generally believed that oligonucleotides did not penetrate the external membrane of eukaryotic cells, to enter the cytosol and nucleus

(49). Clearly, neither did ATP, except under unusual circumstances, nor Ap_4A *(50)*. Segments of cellular genomes were currently coaxed into cell entry by an inefficient calcium phosphate precipitation procedure. The negative charge of the oligonucleotide was regarded as presenting a major impediment to traverse of an oligonucleotide through the eukaryotic external cell wall. Nevertheless, an experiment testing the possibility of synthetic oligonucleotide cell wall penetration was performed. We added a 13-mer synthetic oligodeoxynucleotide, complementary to the 3' end of the virus, to the medium of chick fibroblasts in tissue culture, along with Rous sarcoma virus itself. It inhibited the formation of new virus, and also prevented transformation of chick fibroblasts into sarcoma cells—both of these startling observations *(16)*. In a cell-free system, translation of the Rous sarcoma viral message was also dramatically impaired *(20)*.

Until 1985, little further progress occurred, for three intertwined reasons: first, there was still widespread disbelief that oligonucleotides could enter eukaryotic cells; second, it was difficult to synthesize an oligomer of sufficient length to hybridize well at 37°C and of specificity requisite to target a chosen segment of genome; and third, there was very little DNA (or RNA) genome sequence available for targeting in this way. This latter reason determined the choice of the Rous sarcoma virus, which Haseltine et al. *(47)* and our laboratory *(48)* were sequencing contemporaneously and whose results were published in tandem.

2. Independent Complementary Developments

Two important developments in the late 1970s and early 1980s increased the feasibility of the synthetic oligonucleotide hybridization inhibition approach. The first were the dramatic improvements in DNA sequencing that came from the Maxam-Gilbert chemical degradation procedure *(46)*, and the more convenient dideoxy enzymatic sequencing technique of Sanger's laboratory *(51)*. The second was the solid-phase oligonucleotide synthetic approach introduced successfully by Letsinger and Lunsford *(52)* and Caruthers *(53)*. At this time, as well, there developed a growing acceptance that oligonucleotides could pass through the eukaryotic cell membrane and enter the cell, and that they could readily be synthesized and purified. Finally, an abundance of potential DNA sequence targets began to appear like fireworks in a previously darkened genetic sky.

The unmodified antisense oligodeoxynucleotide has proven to be the best RNase H activator, provided there are at least four or more contiguous hybridizing base pairs. The phosphorothioate modified oligodeoxynucleotides, although not so effective, still activate Rnase H, and are quite nuclease resistant *(54)*. These two properties account for the early general preference of the latter in synthetic oligodeoxynucleotide experiments. In contrast, other varied modifications at the internucleotide bridging phosphate site result in inability to activate RNase H, and thus present a disadvantage. Included in this category are methyl phosphonates, α-oligonucleotides, internucleotide peptide bonds, and others. Modifications on the ribosyl moiety, such as the 2'-0 methyl group, also fail to activate RNase H. Hybrid and chimeric oligonucleotides are coming into increasing usage, since they combine terminal nuclease-resistant segments of an oligonucleotide with a central RNase-sensitive portion (*see* Chapter 14).

Furthermore, if only the central portion of the oligomer is phosphorothioate modified, whereas the peripheral 3' and 5' segments are, for example, 2'-0 methyl-modified oligomer moieties *(55)*, the nonspecific effects of the totally phosphorothioate oligomer *(56)* are to a considerable extent minimized. The selfstabilized snap-back oligomer is particularly advantageous in providing enhanced nuclease resistance with other desirable properties *(57)*.

In addition to the promise of the antisense approach documented in these pages and elsewhere, there are reasons why the competitive oligonucleotide hybridization technique may not be successful in attempts to inhibit noxious genes, wherever they may exist in the animal and plant kingdoms. A central reason is failure to find a single-stranded segment of genome that is highly conserved and accessible. Secondary and tertiary structure of the genome may prevent hybridization. Proteins that have a high association constant with the area of genome targeted, i.e., promoters, enhancers, and modulators, for example, may prove to be barriers to oligonucleotide hybridizations. The aggregation effect on oligonucleotides of a G-quartet motif *(58,59)* is also an impediment. Zon *(60)* touches on some of these aspects in a historical review.

It would be advantageous if hybridization inhibition could be achieved at the transcription level. This would block the amplification step, which results in numerous copies of mRNA for translation. A favorable site for transcription hybridization is the transcription bubble, consisting of 12–17 nucleotides of unwound double-helical DNA. As an example of this approach, we have found in an in vitro transcription system that specific hybridization inhibition can be induced using a linearized plasmid segment of HIV for a template. With T7RNA polymerase, a gag RNA of about 640 nucleotides can be synthesized. This synthesis can be inhibited by a complementary 14-mer unmodified oligodeoxynucleotide, just downstream of the T7 promoter *(19)*. The most effective inhibitor is a plus-sense oligonucleotide complementary to the negative DNA strand that serves as a template for pre-mRNA synthesis.

3. Examples of Current Disease Targets

Let me mention a few current medically related investigations involving our own laboratory that appear to show promise: HIV, influenza, and malaria. A study on HIV *(61)* shows that target selection in the HIV genome is important for prevention of development of escape mutants. Whereas escape mutants appeared after 20 d treatment of chronically infected Molt-3 cells with an antisense phosphorothioate oligomer pinpointing a splice acceptor site, continued inhibition without escape over an 84-d experimental period occurred when *rev1–28* or *gag-28* were the targets.

A second example is the use of antisense oligomers to inhibit influenza viral replication *(62)*. At 10 μ*M* concentration, replication of influenza C virus was inhibited 90% in tissue cultures of MDCK cells by a sense oligophosphorothioate targeted against the replicase gene of the negatively stranded virus. In 10-d-old embryonated chick eggs, phosphorothioate oligomer injection also induced marked inhibition of virus production *(63)*.

Another example is inhibition of replication of *Plasmodium falciparum* malaria by a phosphorothioate oligodeoxynucleotide targeted against the dihydrofolate reductase-thymidylate synthase gene of the parasite *(64)*. This enzyme is essential as a donor of a methyl group in the conversion of deoxyuridine monophosphate to thymidine monophosphate in the parasite, which must synthesize its own pyrimidines, being unable to use exogenous thymidine for synthesis of DNA. The adult erythrocyte is one of the rare eukaryotic cells that oligodeoxynucleotides do not penetrate. Fortunately, however, when a malarial parasite pushes its way into a red cell, it creates a permeabilized erythrocyte membrane plus a parasitophorous duct *(65)*, through either or both of which the oligodeoxynucleotide reaches the parasite inside its protective erythrocyte envelope. A fluorescently labeled oligodeoxynucleotide lights up a circular area inside an erythrocyte in which the *P. falciparum* parasite resides, surrounded by its own membrane, whereas the uninfected red cells fail to show evidence of cell entry *(66)*. The above-mentioned antisense oligomer shows a sequence-specific ID_{50} for replication of the parasite at $2–5 \times 10^{-8} M$ concentration *(67)*.

Thus, in summary, the synthetic antisense oligonucleotide technology has potential application to human diseases and displays promising results in cell-free systems, tissue cultures, and animal models. It is also at early trial points *(68,69)* in human testing against HIV, leukemia, Herpes virus, and other diseases, whose outcome remain for the future. The current status of these varied approaches is presented in later chapters in this book.

References

1. Watson, J. D. and Crick, F. H. C. (1953) Molecular structure of nucleic acids: a structure for deoxyribonucleic acids. *Nature (Lond.)* **171**, 737–738.
2. Siekevitz, P. and Zamecnik, P. C. (1951) In vitro incorporation of 1-C[14]-DL-alanine into proteins of rat liver granular fractions. *Fed. Proc.* **10**, 246 (abstract).

3. Zamecnik, P. C. and Keller, E. B. (1954) Relationship between phosphate energy donors and incorporation of labeled amino acids into proteins. *J. Biol. Chem.* **209,** 337–354.
4. Zamecnik, P. C. (1979) Historical aspects of protein synthesis. *Ann. NY Acad. Sci.* **325,** 269–301.
5. Brachet, J. (1950) *Chemical Embryology.* Interscience, New York.
6. Caspersson, T. O. (1950) *Growth and Cell Function.* Norton, New York.
7. Keller, E. B., Zamecnik, P. C., and Loftfield, R. B. (1954) The role of microsomes in the incorporation of amino acids into proteins. *J. Histochem. Cytochem.* **2,** 378–386.
8. Zamecnik, P. C., Hoagland, M. B., and Stephenson, M. L. (1957) Synthesis of protein in the cell nucleus. *NY Acad. Sci.* **V,** 273–274.
9. Zamecnik, P. C., Stephenson, M. L., Scott, J. F., and Hoagland, M. B. (1957) Incorporation of C^{14}-ATP into soluble RNA isolated from $105,000 \times g$ supernatant of rat liver. *Fed. Proc.* **16,** 275.
10. Zamecnik, P. C., Stephenson, M. L. and Hecht, L. I. (1958) Intermediate reactions in amino acid incorporation. *Proc. Natl. Acad. Sci. USA* **44,** 73–78.
11. Hoagland, M. B., Stephenson, M. L., Scott, J. F., Hecht, L. I., and Zamecnik, P. C. (1958) A soluble ribonucleic acid intermediate in protein synthesis. *J. Biol. Chem.* **231,** 241–256.
12. Hoagland, M. B., Zamecnik, P. C., and Stephenson, M. L. (1959) A hypothesis concerning the roles of particulate and soluble ribonucleic acids in protein synthesis, in *A Symposium on Molecular Biology* (Zirkle, R. E., ed.), University of Chicago Press, Chicago, IL, pp. 105–114.
13. Nirenberg, M. W. and Matthaei, J. H. (1961) The dependence of cell-free protein synthesis in E. coli upon naturally occurring or synthetic polyribonucleotides. *Proc. Natl. Acad. Sci. USA* **47,** 1588–1602.
14. Nishimura, S., Jones, D. S., and Khorana, H. G. (1965) The in vitro synthesis of a copolypeptide containing two amino acids in alternating sequence dependent upon a DNA-like polymer containing two nucleotides in alternating sequence. *J. Mol. Biol.* **13,** 302–324.
15. Allen, D. W. and Zamecnik, P. C. (1962) The effect of puromycin on rabbit reticulocyte ribosome. *Biochim. Biophys. Acta.* **55,** 865–874.
16. Zamecnik, P. C. and Stephenson, M. L. (1978) Inhibition of Rous sarcoma virus replication and transformation by a specific oligodeoxynucleotide. *Proc. Natl. Acad. Sci. USA* **75,** 280–284.
17. Helene, C. (1993) Control of gene expression by triple-helix-forming oligonucleotides: the antigene strategy, in *Antisense Research and Applications* (Crooke, S. T. and Lebleu, B., eds.), CRC, Boca Raton, FL, pp. 375–385.
18. Maher, L. J., Dervan, P. B., and Wold, B. (1992) Analysis of promoter-specific repression by triple- helical DNA complexes in a eukaryotic cell-free transcription system. *Biochemistry* **31,** 70–81.
19. Temsamani, J., Metelev, V., Levina, A., Agrawal, S., and Zamecnik, P. (1994) Inhibition of in vitro transcription by oligodeoxynucleotides. *Antisense Res. Devel.* **4,** 279–284.

20. Stephenson, M. L. and Zamecnik, P. C. (1978) Inhibition of Rous sarcoma viral RNA translation by a specific oligodeoxynucleotide. *Proc. Natl. Acad. Sci. USA* **75,** 285–288.

21. Zamecnik, P. C., Goodchild, J., Taguchi, Y., and Sarin, P. S. (1986) Inhibition of replication and expression of human T-cell lymphotropic virus type III in cultured cells by exogenous synthetic oligonucleotides complementary to viral RNA. *Proc. Natl. Acad. Sci. USA* **83,** 4143–4146.

22. Plesner, P., Goodchild, J., Kalckar, H., and Zamecnik, P. C. (1987) Oligonucleotides with rapid turnover of the phosphate groups occur endogenously in eukaryotic cells. *Proc. Natl. Acad. Sci. USA* **84,** 1936–1939.

23. Inouye, M. (1988) Antisense RNA: its functions and applications in gene regulation—a review. *Gene* **72,** 25–34.

24. Kimelman, D. (1992) Regulation of eukaryotic gene expression by natural antisense transcripts, in *Gene Regulations: Biology of Antisense RNA and DNA* (Erickson, P. and Izant, J. G., eds.), Raven, New York, pp. 1–10.

25. Belikova, A. M., Zarytova, V. F., and Grineva, N. I. (1967) Synthesis of ribonucleosides and diribonucleoside phosphates containing 2-chloroethylamine and nitrogen mustard residues. *Tetrahedron Lett.* **37,** 3557–3562.

26. Miller, P. S., Braiterman, I. T., and Tso, P. O. P. (1977) Effects of a trinucleotide ethylphosphtriester, Gmp(Et)Gmp(Et)U, on mammalian cells in culture. *Biochemistry* **16,** 1988–1996.

27. Paterson, B. M., Roberts, B. E., and Kuff, E. L. (1977) Structural gene identification and mapping by DNA-mRNA hybrid-arrested cell-free translation. *Proc. Natl. Acad. Sci. USA* **74,** 4370–4374.

28. Hastie, N. D. and Held, W. A. (1978) Analyses of mRNA populations by cDNA mRNA hybrid-mediated inhibition of cell-free protein synthesis. *Proc. Natl. Acad. Sci. USA* **75,** 1217–1221.

29. van der Krol, A. R., Stuitje, A. R., and Mol, J. N. M. (1991) Modulation of floral pigmentation by antisense technology, in *Antisense Nucleic Acids and Proteins* (Mol, J. M. N. and van der Krol, A. R., eds.), Marcel Dekker, New York, pp. 125–140.

30. Zamecnik, P., Aghajanian, J., Zamecnik, M., Goodchild, J., and Witman, G. (1994) Electron micrographic studies of transport of oligodeoxynucleotides across eukaryotic cell membranes. *Proc. Natl. Acad. Sci. USA* **91,** 3156–3160.

31. Temsamani, J., Kubert, M., Tang, J., Padmapriya, A., and Agrawal, S. (1994) Cellular uptake of oligodeoxynucleotides and their analogs. *Antisense Res. Devel.* **4,** 35–42.

32. Agrawal, S., Temsamani, J., and Tang, J. Y. (1991) Pharmacokinetics, biodistribution and stability of oligodeoxynucleotide phosphorothioates in mice. *Proc. Natl. Acad. Sci USA* **88,** 7595–7599.

33. Anderson, W. F. (1992) Human gene therapy. *Science* **256,** 808–813.

34. Khorana, H. G., Buchi, H., Ghosh, H., Gupta, N., Jacob, T. M., Kossel, H., Morgan, R., Narang, S. A., Ohtsuka, E., and Wells, R. D. (1966) Polynucleotide synthesis and the genetic code. *Cold Spring Harbor Symp. Quant. Biol.* **31,** 39–49.

35. Holley, K. W., Apgar, J., Everett, G. A., Madison J. T., Marquisee, M., Merrill, S. H., Penswick, J. R., and Zamir, A. (1965) Structure of a ribonucleic acid. *Science* **147**, 1462–1465.

36. Monier, R., Stephenson, M. L., and Zamecnik, P. C. (1960) The preparation and some properties of a low molecular weight ribonucleic acid from baker's yeast. *Biochim. Biophys. Acta* **43**, 1–8.

37. Zachau, H. G., Dutting, D., Feldmann, H., Melchers, F., and Karan, W. (1966) Serine specific transfer ribonucleic acids. XIV. Comparison of nucleotide sequence and secondary structure models. *Cold Spring Harbor Symp. Quant. Biol.* **31**, 417–424.

38. Raj Bhandary, U. L., Stuart, A., Faulkner, R. D., Chang, S. H., and Khorana, H. G. (1966) Nucleotide sequence studies on yeast phenylalanyl sRNA. *Cold Spring Harbor Symp. Quant. Biol.* **31**, 425–434.

39. Ingram, V. M. and Sjoquist, J. A. (1963) Studies on the structure of purified alanine and valine transfer RNA from yeast. *Cold Spring Harbor Quant. Biol.* **28**, 133–138.

40. Temin, H. M. and Mizutani, S. (1970) RNA-dependent DNA polymerase in virions of Rous sarcoma virus. *Nature* **226**, 1211–1213.

41. Baltimore, D. (1970) Viral RNA-dependent DNA polymerase. *Nature (Lond.)* **226**, 1209–1210.

42. Sanger, F. and Coulsen, A. R. (1975) A rapid method for determining sequences in DNA by primed synthesis with DNA polymerase. *J. Mol. Biol.* **94**, 441–448.

43. Wickstrom, E. (1986) Oligodeoxynucleotide stability in subcellular extracts and culture media. *J. Biochem. Biophys. Methods* **13**, 97–102.

44. Walder, R. W. and Walder, J. A. (1988) Role of RNase H in hybrid-arrested translation by antisense oligonucleotides in current communications, in *Molecular Biology. Antisense RNA and DNA* (Melton, D. A., ed.), Cold Spring Harbor Laboratory, Cold Spring Harbor, NY, pp. 35–40.

45. Stephenson, M. L., Scott, J. F., and Zamecnik, P. C. (1973) Evidence that the polyladenylic acid segment of RNA of avian myeloblastosis virus is located at the "35S" 3'-OH terminus. *Biochem. Biophys. Res. Commun.* **55**, 8–16.

46. Maxam, A. M. and Gilbert, W. (1977) A new method of sequencing DNA. *Proc. Natl. Acad. Sci. USA* **74**, 560–564.

47. Haseltine, W. A., Maxam, A. M., and Gilbert, W. (1977) Rous sarcoma virus is terminally redundant: the 5' sequence. *Proc. Natl. Acad. Sci. USA* **74**, 989–993.

48. Schwartz, D., Zamecnik, P. C., and Weith, H. L. (1977) Rous sarcoma virus is terminally redundant: the 3' sequence. *Proc. Natl. Acad. Sci. USA* **74**, 994–998.

49. Pitha, P., and Pitha, J. (1980) Polynucleotide analogs as inhibitors of DNA and RNA polymerases, in *International Encyclopedia of Pharmacology and Therapeutics, Section 103. Inhibitors of DNA and RNA Polymerases* (Sarin, P. S. and Gallo, R. C., eds.), Pergamon, New York, pp. 235–247.

50. Zamecnik, P. C. and Stephenson, M. L. (1969) Nucleoside pyrophosphate compounds related to the first step in protein synthesis, in *The Role of Nucleotides for the Function and Conformation of Enzymes.* Alfred Benzon Symposium I. (Kalckar, H. M., Klenow, H., Munch-Petersen, A., Ottesen, M., and Thaysen, J. H., eds.), Munksgaard, Copenhagen, pp. 276–291.

51. Sanger, F., Nicklens, S., and Coulsen, A. R. (1977) DNA sequencing with chain-terminating inhibitors. *Proc. Natl. Acad. Sci. USA* **74,** 5463–5467.

52. Letsinger, R. L. and Lunsford, W. B. (1976) Synthesis of thymidine oligonucleotides by phosphate triester intermediates. *J. Am. Chem. Soc.* **98,** 3655–3661.

53. Caruthers, M. H. (1985) Gene synthesis machines: DNA chemistry and its uses. *Science* **230,** 281–285.

54. Agrawal, S., Mayrand, S. H., Zamecnik, P. C., and Pederson T. (1990) Site-specific excision from RNA by RNase H and mixed phosphate backbone oligodeoxynucleotides. *Proc. Natl. Acad. Sci. USA* **87,** 1401–1405.

55. Metelev, V., Lisziewicz, J., and Agrawal, S. (1994) Study of antisense oligonucleotide phosphorothioates containing segments of oligodeoxynucleotides and 2'-0-methyl oligoribonucleotides. *Bioorg. Med. Chem. Lett.* **4,** 2929–2934.

56. Stein, C. A. and Krieg, A. M. (1994) Editorial. Problems in interpretation of data derived from *in vitro* and *in vivo* use of antisense oligodeoxynucleotides. *Antisense Res. Devel.* **4,** 67–69.

57. Tang, J. Y., Temsamani, J., and Agrawal, S. (1993) Self-stabilized antisense oligonucleotide phosphorothioates: properties and anti-HIV activity. *Nucleic Acids Res.* **21(11),** 2729–2735.

58. Buckheit, R. W., Jr., Roberson, J. L., Lackman-Smith, C., Wyatt, J. R., Vickers, T. A., and Ecker, D. J. (1994) Potent and specific inhibition of HIV envelope-mediated cell fusion and virus binding by G-quartet-forming oligonucleotide (Isis - 5320). *AIDS Res. Hum. Retrovir.* **10(11),** 1497–1506.

59. Kandimalla, E. R. and Agrawal, S. (1995) Single strand targeted triplex-formation. Destabilization of guanine quadruplex structures by foldback triplex-forming oligonucleotides. *Nucleic Acids Res.* **23,** 1068–1074.

60. Zon, G. (1993) *History of Antisense Drug Discovery in Antisense Research and Applications* (Crooke, S. T. and Lebleu, B., eds.), CRC, Boca Raton, FL, pp. 1–5.

61. Lisziewicz, J., Sun, D., Metelev, V., Zamecnik, P., Gallo, R. C., and Agrawal, S. (1993) Long-term treatment of human immunodeficiency virus-infected cells with antisense oligonucleotide phosphorothioates. *Proc. Natl. Acad. Sci. USA* **90,** 3860–3864.

62. Leiter, J. M., Agrawal, S., Palese, P., and Zamecnik, P. C. (1990) Inhibition of influenza virus replication by phosphorothioate oligodeoxynucleotides. *Proc. Natl. Acad. Sci. USA* **87,** 3430–3434.

63. Zamecnik, P. C., Agrawal, S., and Palese, P., unpublished data.

64. Rapaport, E., Misiura, K., Agrawal, S., and Zamecnik, P. C. (1992) Antimalarial activities of oligodeoxynucleotide phosphorothioates in chloroquine-resistant *Plasmodium falciparum. Proc. Natl. Acad. Sci. USA* **89,** 8577–8580.

65. Dluzewski, A. R., Mitchell, G. H., Fryer, P. R., Griffiths, S., Wilson, R. J. M., and Gratzer, W. B. (1992) Origins of the parasitophorous vacuole membrane of the malaria parasite, *Plasmodium falciparum,* in human red blood cells. *J. Cell Sci.* **102,** 527–532.

66. Zamecnik, P. C., Rapaport, E., Metelev, V., and Barker, R. (1996) Inhibition of replication of drug resistant *P. falciparum in vitro* by specific antisense phosphorothioate oligodeoxynucleotides, in *Antisense Oligodeoxynucleotides: From Technology to Therapy* (Schlingensiepen, K. H., Schlingensiepen, R., and Brysch, W., eds.), Blackwell International/Blackwell Wissenschaft, Berlin, in press.

67. Barker, R. H., Jr., Metelev, V., Rapaport, E., and Zamecnik, P. (1996) Inhibition of *Plasmodium falciparum* malaria using antisense oligodeoxynucleotides. *Proc. Natl. Acad. Sci. USA,* in press.

68. Crooke, S. J. (1994) Editorial. Progress in evaluation of the potential of antisense technology. *Antisense Res. Devel.* **4,** 145–146.

69. Hawkins, J. W. (1995) Editorial. Oligonucleotide therapeutics: coming 'round the clubhouse turn. *Antisense Res. Devel.* **5,** 1.

2

Pharmacology of Antisense Therapeutic Agents

Cancer and Inflammation

C. Frank Bennett, Nicholas Dean, David J. Ecker, and Brett P. Monia

1. Introduction
1.1. The Promise of Antisense Therapeutics

Antisense oligonucleotides represent a new paradigm for drug discovery that holds great promise to deliver potent and specific drugs with fewer undesired side effects. The antisense paradigm offers the opportunity to identify rapidly lead compounds based on knowledge of the biology of a disease process, and a relevant target gene sequence. With this information, the practitioner of antisense drug discovery can rapidly design, synthesize, and test a series of compounds in cell culture and determine if the target gene is specifically inhibited. A compound thus identified can then be tested in an animal model, either to determine whether targeted gene expression can be inhibited in various animal tissues or to determine if there is activity in an animal model of a human disease. The length of time and the resources required to identify a lead compound by the antisense paradigm is much less than by any other drug discovery method.

Although the antisense paradigm holds great promise, the field is still in its early stages, and there are a number of key questions that need to be answered and technical hurdles that must be overcome. Antisense technology focuses on a class of chemicals, oligonucleotide analogs, that have not been extensively explored as therapeutic agents. The key issues concerning this class of chemicals center on whether these compounds have acceptable properties as drugs. These include pharmacokinetic, pharmacological, and toxicological properties.

The first generation of antisense oligonucleotide analogs to be broadly examined for their properties as drugs are the phosphorothioates, where one of

From: *Methods in Molecular Medicine: Antisense Therapeutics*
Edited by: S. Agrawal Humana Press Inc., Totowa, NJ

the nonbridging phosphoryl oxygens of DNA is substituted with a sulfur. This relatively simple modification results in dramatic improvements in nuclease stability and the in vitro and in vivo pharmacokinetics. In this chapter, we will briefly review the recent advances in understanding the pharmacokinetic properties of phosphorothioate antisense oligonucleotides and then focus on the pharmacological properties of these compounds in animal models. We describe our current understanding of the specificity of these compounds in inhibiting gene expression in animal tissues and providing therapeutic activity in animal models of disease.

1.2. Pharmacokinetics

With the recent focus on phosphorothioates as the first class of oligonucleotide analogs to be broadly explored as drug candidates, we now have a fairly detailed understanding of their pharmacokinetic properties. Several groups have recently published in-depth pharmacokinetic studies *(1–6)*. Phosphorothioates are highly water-soluble compounds. On parenteral administration, phosphorothioate oligonucleotides become associated with serum proteins, which have a high capacity and low-affinity binding capability. Binding to serum proteins is saturable, but only at concentrations that are anticipated to exceed the amount that would be given therapeutically. The association with serum proteins provides an initial reservoir of compound and prevents rapid clearance by the kidneys.

Distribution of material from the blood to the tissues occurs very rapidly, with plasma half-lives on the order of <1 h *(2)*. Phosphorothioates distribute broadly into all tissues with the highest percentage of the dose in the kidneys, liver, and bone marrow. The only exception is the brain, which excludes phosphorothioates via the blood–brain barrier. After accumulation in the tissues, phosphorothioates are metabolized slowly. The rate of metabolism is dependent to some extent on the tissue. The liver, for example, eliminates phosphorothioates more rapidly ($t_{1/2}$ = 58 h) than other tissues, such as the kidney cortex ($t_{1/2}$ = 156 h) or the bone marrow ($t_{1/2}$ = 157 h) *(2)*. Phosphorothioates are extensively metabolized through a combination of nucleases and other metabolic enzymes. Phosphorothioates are not excreted intact unless administered at a high dose that exceeds the buffering capacity of the serum proteins. Overall, metabolism and elimination of phosphorothioates occur in a time frame consistent with once a day or every other day dosing.

2. Pharmacology of Oligonucleotides in Tumor Models
2.1. Antisense Oligonucleotides in Oncology

The ability to use antisense oligonucleotides to target selectively the genetic processes involved in cancer has raised the exciting possibility that this class of compounds could be developed as novel chemotherapeutic agents. There have been a large number of published studies in which antisense oligonucleotides have been used to inhibit the expression of gene products thought to be involved in the oncogenic process (reviewed in refs. *7–12*). Recently, several publications have documented that oligonucleotides identified in cellular based assays as inhibitors of gene expression are also effective in animal models in inhibiting growth of tumor cells in mice. One of the first published studies that demonstrated in vivo activity of oligonucleotides was a study by Whitesell et al. *(13)*, in which a phosphodiester oligonucleotide directed toward N-*myc* was infused, using an Alzet miniosmotic pump, in the vicinity of a subcutaneous transplanted neuroepithelioma cell line. The authors demonstrated a loss of N-*myc* protein, a change in cellular morphology, and a decrease in tumor mass by the antisense oligonucleotide, but not the sense oligonucleotide *(13)*. These results are somewhat surprising considering how unstable phosphodiester oligonucleotides are when administered to mice *(5)*. More recently, several studies have demonstrated that antisense oligonucleotides targeting the p65 subunit of NF-κB decreased growth of fibrosarcomas and melanomas *(14)*. Oligonucleotides targeting GAPDH and *jun*-D were without effect on tumor growth. The expression of p65 mRNA in the tumors was measured by RT-PCR and found to be decreased by the antisense oligonucleotide, but not the nontargeted oligonucleotides.

In a detailed study, Skorski and colleagues *(15)* have investigated the effects of a 26-mer phosphorothioate oligodeoxynucleotide targeting the *BCR-ABL* transcript on the in vivo growth of a Philadelphia chromosome-positive chronic myeloid leukemia (BV173) cell line. The oligonucleotide was an effective inhibitor of leukemia progression in vivo and enhanced the life-span of mice given the antisense oligonucleotide. A control (sense) oligonucleotide was without effect on the progression of the disease. The level of *BCR-ABL* transcript, as a result of infiltrating leukemia cells determined in tissues isolated from mice, was determined by RT-PCR and found to decrease with oligonucleotide treatment. The authors conclude that their results support the hypothesis that the oligonucleotide is working through an antisense mechanism of action. However, based on the data presented, the authors could not discriminate between reduction of *BCL-ABL* transcripts in the leukemia cells (an antisense mechanism) or decreased infiltration of leukemia cells into murine tissues (a cytotoxic effect of the oligonucleotide or an inhibition of tumor invasion).

Finally, in a series of reports, Gewirtz and colleagues *(16,17)* describe the antineoplastic effects of phosphorothioate oligonucleotides targeting c-*myb* on the growth of K562 human leukemia cells and melanoma cells in *scid* mice. Tumor growth inhibition was dependent on the route of oligonucleotide administration with the best results seen when the drugs were given by infusions with Alzet miniosmotic pumps. Some inhibition of tumor growth was seen with a control oligonucleotide; however, this inhibition was described as being statistically insignificant. The authors raise an important point that should be considered in studies with human xenografts. When attempting to measure oligonucleotide-dependent changes in expression of a human tumor gene, it may be necessary to discriminate between human and murine gene expression. Samples of mRNA obtained from xenografted human tumors can contain up to 50% "host" (murine) mRNA, since the tumors will contain "host" fibroblasts and other connective tissue. If the oligonucleotide given to the animal is specific for human sequences (as is often the case), then it should only reduce the expression of the human target gene. However, if the method used for analyzing expression of the gene does not discriminate between the two species, then this reduction in expression may be masked by high levels of expression of the "host" gene.

We have been interested in using antisense oligonucleotides to target specific members of multigene families that could play roles in cellular proliferation and transformation, in particular, proteins that are involved in cell signaling and response to mitogenic signals. Because of their unique specificity, an attractive feature of antisense oligonucleotides is that it is relatively easy to design a compound that specifically inhibits a member of a multigene family. Other approaches that design inhibitors to bind to enzyme-active sites or mimic natural ligands for receptors often fail to demonstrate specificity for the isoenzyme or receptor subtype of interest. The lack of specificity of conventional drugs often leads to undesirable side effects, which could be avoided by practicing isotypic pharmacology. Examples of this application of antisense oligonucleotides are described in the following sections.

2.2. ras Oncogenes

ras Gene products are plasma-membrane-associated, guanine nucleotide binding proteins, which are involved in transducing signals controlling cellular growth and differentiation *(18–20)*. In normal cells, the proportion of *ras* protein in the active (GTP-bound) state is tightly controlled by its own intrinsic GTPase activity as well as by a family of *ras* interacting proteins that stimulate *ras* GTP exchange *(18–20)*. Receptor-mediated stimulation of *ras* causes an increase in the proportion of cellular *ras* in the GTP-bound state relative to the GDP-bound state. This leads to the formation of a complex between *ras* and *raf* kinase, an event that stimulates *raf* kinase activity, initiating a multistep phosphorylation cascade, which ultimately leads to the activation of specific transcription factors whose activity is required for mitogenesis *(21)*. This phosphorylation pathway is commonly referred to as the mitogen-activated protein (MAP) kinase signaling pathway *(21)*.

To date, three different *ras* genes (Ki-*ras*, Ha-*ras*, and N-*ras*) and at least 40 closely related small GTP binding proteins have been identified and characterized in mammalian tissues. *ras* Genes acquire transforming potential by single base point mutations in their coding regions, resulting in single amino acid substitutions in the critical GTPase regulatory domain of the protein *(22–24)*. These mutations increase the proportion of *ras* in the GTP-bound state relative to the GDP-bound state, abrogating the normal function of *ras*, thereby converting a normally regulated cell protein to one that is constitutively active. Such deregulation of normal *ras* protein function is believed to be responsible for the transforming activity of *ras* oncogene products.

Naturally occurring mutations in *ras* oncogenes associated with human neoplasms are most commonly localized to codons 12 and 61 *(22,23)*. More than 20% of all human tumors contain mutations in at least one of the three *ras* genes, and these mutations occur at a relatively early stage of tumor progression *(23–25)*. However, the incidence of *ras* mutations varies substantially between different types of cancers *(22,25)*. Furthermore, for a particular type of cancer, a strong association often exists with the DNA sequence, the codon and the *ras* isotype mutated *(22–25)*.

2.2.1. Antisense Inhibition of ras Gene Expression in Cell Culture

To identify antisense oligonucleotides capable of inhibiting expression of Ha-*ras* and Ki-*ras* mRNA expression, a series of phosphorothioate oligodeoxynucleotides were designed and tested for inhibition of the appropriate *ras* isotype. In both cases, oligonucleotides 20 bases in length were targeted to mRNA sequences comprising the 5'-untranslated regions, coding regions (including codons 12 and 61), and the 3'-untranslated regions. Two cell lines were chosen for these studies: the T24 bladder carcinoma cell line, which expresses a mutation-bearing Ha-*ras* mRNA (codon-12, GGC → GTC), and the SW480 colon carcinoma cell line, which expresses a mutant Ki-*ras* mRNA (codon-12, GGT → GTT) *(26,27)*. Cells were treated with oligonucleotides at a concentration up to 20 μM in the absence of cationic lipid formulation or at a concentration up to 0.5 μM in the presence of cationic lipid. Inhibition of Ha-*ras* and Ki-*ras* mRNA expression was observed for select oligonucleotides administered in the presence of cationic lipid (Fig. 1). The oligonucleotides failed to inhibit expression of the respective *ras* gene products significantly in the absence of cationic lipids in the two cell lines studied. The degree of inhibition of the two different *ras* gene products varied depending on the mRNA target site and the particular *ras* message. For example, the 5'-untranslated region, including the AUG site of Ha-*ras* mRNA, was very sensitive to inhibition with antisense oligonucleotides, whereas oligonucleotides targeted to the 3'-untranslated region of this message were without effect. In contrast, oligo-

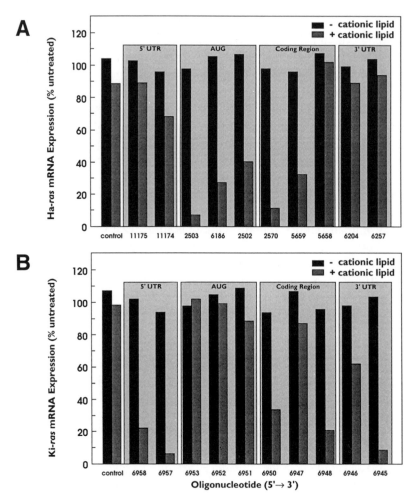

Fig. 1. Inhibition of *ras* mRNA expression by antisense oligodeoxynucleotides targeted to Ha-*ras* and Ki-*ras* in cell culture. Phosphorothioate oligodeoxynucleotides targeted to the 5'-untranslated region, translation start site (AUG), coding region, or 3'-untranslated region were administered to cultured cells in the presence (light shade) or absence (dark shade) of cationic lipid for a 4-h period and targeted mRNA levels were analyzed by Northern blot 24 h following oligodeoxynucleotide administration. Oligodeoxynucleotide concentrations were 200 nM in the presence of cationic lipid and 10 μM in the absence of cationic lipid. **(A)** Treatment of human T24 bladder carcinoma cells with oligodeoxynucleotides targeted to Ha-*ras* and analysis of Ha-*ras* mRNA levels. **(B)** Treatment of human SW480 colon carcinoma cells with oligodeoxynucleotides targeted to Ki-*ras* and analysis of Ki-*ras* mRNA levels. mRNA levels were quantitated by phosphorimage analysis and normalized to G3PDH mRNA levels.

nucleotides targeted to the AUG site of Ki-*ras* mRNA were poor inhibitors of Ki-*ras* expression whereas the 5'-untranslated region was very sensitive to antisense inhibition. Interestingly, for both target mRNAs, oligonucleotides designed to hybridize with codons 12 and 61 were effective in inhibiting expression of the respective mRNA targets, suggesting that mutant-specific inhibition of *ras* mRNA expression is feasible (Fig. 1). Oligodeoxynucleotides displaying activity against either Ki-*ras* or Ha-*ras* mRNA were also tested as unmodified phosphodiester oligodeoxynucleotides in the presence and absence of cationic lipid, and were found to be completely without effect on target message (B. Monia, unpublished results). This result is most likely explained by the susceptibility of unmodified DNA to nucleolytic degradation. However, novel modifications have been identified that increase target affinity and/or resistance to nucleolytic degradation. Some of these modifications have been incorporated into active *ras* oligonucleotides and found to increase antisense activity up to 15-fold *(28)*. However, despite the fact that these novel oligonucleotide modifications greatly enhance antisense activity in the presence of cationic lipid, they are still without effect when administered to cultured cells in the absence of cationic lipid.

2.2.2. Isotype-Specific Inhibition of ras Gene Expression

The structures of the three *ras* isotypes (Ha-, Ki-, N-) at the protein level are virtually identical throughout the protein, except for a small region at the carboxy terminus *(18–20)*. Thus, protein-targeting drugs, which selectively target the different *ras* isozymes, have not been described. However, because of the redundancy of the genetic code and the presence of noncoding (untranslated) sequences, highly related proteins are often encoded by highly diverged mRNA sequences. Therefore, it should be possible to design inhibitors to block expression of one particular isotype with minimal consequence to related isotypes. In fact, such "isotype specificity" has been demonstrated through the use of antisense inhibitors targeted against protein kinase C *(29)*.

To demonstrate isotype-specific inhibition of *ras* gene expression by antisense oligonucleotides, oligonucleotides that were specifically designed to hybridize with either the Ha-*ras* mRNA or the Ki-*ras* mRNA were evaluated for specificity for their targeted mRNA. ISIS 2503, an active 20 base deoxy phosphorothioate targeted to the Ha-*ras* mRNA AUG region *(30)*, is complementary to the AUG region of the Ki-*ras* message in only 9 of 20 bases and, therefore, would not be expected to bind efficiently to Ki-*ras* mRNA. Similarly, ISIS 6957, an active 20-base phosphorothioate oligodeoxynucleotide targeted to the 5'-untranslated region of Ki-*ras* mRNA, is complementary to the 5'-untranslated region of Ha-*ras* mRNA in only 4 of 20 bases, and therefore should not affect Ha-*ras* mRNA expression. In addition to these two oligonucleotides, a third 20 base phosphorothioate oligodeoxynucleotide (ISIS

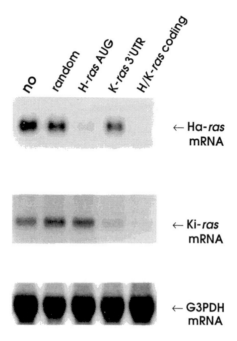

Fig. 2. Isotype-specific inhibition of *ras* gene expression. Human A549 lung carcinoma cells were treated with phosphorothioate 20-mer oligodeoxynucleotides at a concentration of 200 n*M* in the presence of cationic lipid, and *ras* mRNA levels were
determined as described in Fig. 1. Gel lanes are as follows: no, no oligodeoxynucleotide treatment; random, random 20-mer phosphorothioate control; H-*ras* AUG,
ISIS-2503 targeted to the AUG site of human Ha-*ras*; K-*ras* 3'UTR, ISIS-6957 targeted to the 5'-untranslated region of human Ki-*ras* mRNA; H/K-*ras* coding, ISIS-
9827 targeted to the coding region of both Ha-*ras* and Ki-*ras* mRNA.

9827) was tested that targeted a conserved sequence within the coding region
of both *ras* isotypes. Cells treated with each of these oligonucleotides were
analyzed for Ha-*ras* and Ki-*ras* mRNA expression by Northern analysis. ISIS
2503 reduced Ha-*ras* mRNA to undetectable levels without affecting Ki-*ras*
mRNA levels, whereas ISIS 6957 inhibited Ki-*ras* mRNA expression without
affecting Ha-*ras* mRNA levels (Fig. 2). Furthermore, ISIS 9827, targeted to
the coding regions of both *ras* isotypes, inhibited expression of both targets.
These studies demonstrate that isotype-selective inhibition of *ras* gene expression is possible through the use of properly designed antisense inhibitors.

2.2.3. Point Mutation-Specific Inhibition of ras Gene Expression

ras Genes acquire their tumor-promoting potential by single base point
mutations in their coding regions. Since the function of normal *ras* isotypes is

presumably important for cell survival, inhibiting expression of the mutated *ras* gene in tumors is preferred without affecting expression of the nonmutated *ras* isotypes.

Helene and coworkers have demonstrated inhibition of the mutant form of Ha-*ras* mRNA expression using a 9-base phosphodiester linked to an acridine intercalating agent *(31)*. Chang and coworkers have also demonstrated selective targeting of a mutant Ha-*ras* message in which a mutation at codon 61 was targeted and methylphosphonate oligodeoxynucleotides were employed *(32)*. Studies from our laboratory have demonstrated similar antisense specificity targeting the Ha-*ras* point mutation (GGC → GTC) at codon 12 using phosphorothioate oligodeoxynucleotides *(30)*. In our study, we demonstrated that mutation-specific inhibition can be achieved with phosphorothioate oligodeoxynucleotides, but that oligonucleotide affinity and concentration were critical to maintaining the selectivity. Oligonucleotides targeted to codon 12 ranging in length between 5 and 25 bases targeted to Ha-*ras* codon 12 were tested for overall activity and point mutation selectivity. Oligonucleotides <15 bases in length were inactive, whereas all oligonucleotides greater in length displayed good activity with potency correlating directly with oligonucleotide chain length (affinity). However, selective inhibition of mutant Ha-*ras* expression did not increase with oligonucleotide chain length, but required a specific length between 15 and 19 bases. The maximum selectivity observed for inhibition of mutant Ha-*ras* expression relative to normal Ha-*ras* was achieved with a 17-mer oligonucleotide (ISIS 2570) and was approximately fivefold.

We have now extended these studies to include point mutation-selective targeting of the Ki-*ras* isotype. In this case, a 15-base phosphorothioate was identified (ISIS 7453) that specifically inhibited a codon 12 mutation (GGT → GTT) of the Ki-*ras* gene product. To demonstrate selectivity, T24 bladder carcinoma cells, which express normal Ki-*ras* and mutant Ha-*ras* mRNA, and SW480 colon carcinoma cells, which express normal Ha-*ras* and mutant Ki-*ras* mRNA, were treated with oligonucleotides specific for each of these mutations, and Ha-*ras* and Ki-*ras* mRNA expression was analyzed by Northern blot. In addition, oligonucleotides targeted to the AUG of Ha-*ras* mRNA and 5'-untranslated region of Ki-*ras* mRNA were tested as controls. These oligonucleotides would not be expected to display point mutation selectivity.

ISIS 7453, targeted to the Ki-*ras* codon 12 point mutation in SW480 cells, reduced Ki-*ras* expression to undetectable levels in these cells without affecting normal Ki-*ras* expression in T24 cells (Fig. 3). Similarly, ISIS 2570 targeted to the Ha-*ras* codon-12 mutation in T24 cells reduced expression of Ha-*ras* mRNA in these cells without affecting expression of normal Ha-*ras* mRNA in SW480 cells (Fig. 3). Furthermore, oligonucleotides targeted to either the 5'-untranslated region of Ki-*ras* or AUG of Ha-*ras* blocked expres-

A T24 Cells
Ha-*ras* (GGC→ GTC)

B SW480 Cells
Ki-*ras* (GGT→ GTT)

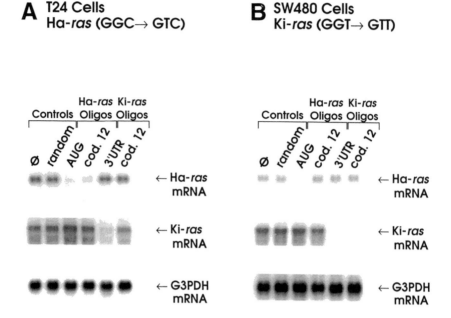

Fig. 3. Point mutation-specific inhibition of *ras* gene expression. Cells were treated with phosphorothioate oligodeoxynucleotides at a concentration of 100 n*M* in the presence of cationic lipid, and *ras* mRNA levels were analyzed as described in Fig. 1. **(A)** Inhibition of *ras* mRNA expression in human T24 bladder carcinoma cells. **(B)** Inhibition of *ras* mRNA expression in human SW480 colon carcinoma cells. Gel lanes for both panels are as follows: 0, no oligodeoxynucleotide treatment; random, control phosphorothioate sequence; AUG, ISIS 2503 (20-mer) targeted to the translation start site of Ha-*ras*; 3'UTR, (20-mer) targeted to the 5'-UTR of Ki-*ras*; cod. 12, oligodeoxynucleotides targeted to either the codon-12 point mutation of Ha-*ras* in T24 cells (17-mer) or the codon-12 point mutation of Ki-*ras* in SW480 cells (15-mer), as indicated in the figure.

sion of the appropriate isotype in both cell lines and in no case did an oligonucleotide specifically targeted to a particular isotype affect expression of the other isotype. These studies demonstrate that point mutation-specific inhibition of both Ha-*ras* and Ki-*ras* mRNA expression in tissue culture is possible through the use of properly designed antisense oligonucleotides.

2.2.4. Antiproliferative Effects of ras Antisense Oligonucleotides

Antisense oligonucleotides targeted to *ras* gene products have potent antiproliferative effects against tumor cells in culture, with different tumor lines exhibiting differential sensitivity to the antiproliferative effects of the *ras*

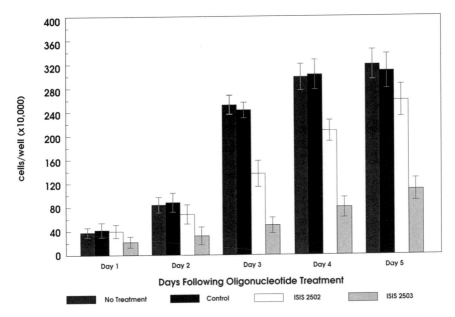

Fig. 4. Inhibition of human T24 cell proliferation with antisense oligodeoxy-nucleotides targeted to the translation initiation start site of Ha-*ras*. Oligodeoxy-nucleotides were administered at a concentration of 200 n*M* in the presence of cationic lipid for 4 h at d 0 followed by removal of lipid/oligodeoxynucleotide mixture and replacement with normal media. Following oligodeoxynucleotide treatment, cell number was determined at d 1, 2, 3, 4, and 5 by direct counting. Treatment conditions were as follows; no treatment, cationic lipid alone without oligodeoxynucleotide; control, randomized phosphorothioate control; ISIS-2502, 20-mer phosphorothioate oligo-deoxynucleotide targeted to the AUG site of human Ha-*ras* mRNA; ISIS-2503, phosphorothioate oligodeoxynucleotide also targeted to the AUG site of human Ha-*ras*, but shifted slightly to the 3' side of the AUG site as compared with ISIS-2502.

antisense oligonucleotides. These effects correlate well with the degree of inhibition of the *ras* gene product when analyzed by Northern blots. For example, although ISIS 2503 and 2502 are both targeted to the AUG region of Ha-*ras* mRNA, ISIS 2503 is approximately fivefold more potent as an inhibitor of Ha-*ras* expression. ISIS 2503 is also a more potent inhibitor of T24 cell proliferation as compared with ISIS 2502 (Fig. 4). Similar observations have been made with oligonucleotides containing novel modifications that increase potency; modifications that increase the degree of target mRNA inhibition also augment the antiproliferative effects observed in cell culture (B. Monia, unpublished results). Furthermore, in cell lines expressing a particular point mutation, cell proliferation is inhibited by oligonucleotides that selectively target mutant *ras* mRNA expression. For example, oligonucleotides targeted to a point mutation at codon-12 of either Ha-*ras* or Ki-*ras* mRNA selectively inhibit the proliferation of cells expressing the appropriate *ras* mutation, but have only modest effects on cells expressing the wild-type *ras* gene (B. Monia, unpublished data). Finally, as is the case for oligonucleotide-mediated inhibition of *ras* mRNA expression, the antiproliferative effects observed with *ras* oligonucleotides required the use of cationic lipid formulation in cell culture.

2.2.5. Antitumor Effects of ras Antisense Oligonucleotides in Nude Mouse Xenografts

There are several reports in which antisense approaches targeting *ras* gene products have been tested in animal models. For example, treatment of nude mice bearing human lung carcinoma cells expressing a Ki-*ras* mutation with a retroviral antisense Ki-*ras* vector resulted in complete elimination of the cancer cells in 87% of the treated mice after 30 d *(33)*. In another study, antisense oligonucleotides targeting the Ha-*ras* oncogene adsorbed onto polyalkyl-cyanoacrylate nanoparticles, which helped protect the oligonucleotides against nucleolytic degradation, inhibited the growth of an Ha-*ras* transformed cell line in vitro and in nude mice *(34)*. The effects of both Ha-*ras*- and Ki-*ras*-directed phosphorothioate antisense oligodeoxynucleotides in vivo against a spectrum of human tumor types have been evaluated (Table 1). Potent inhibition of human tumor cell growth in nude mice with *ras* antisense molecules was observed, which was oligonucleotide sequence-specific, isotype-specific, and tumor type-specific (B. Monia, manuscript in preparation). One important observation that has come out of these studies, as well as studies with other oligonucleotides in different in vivo models, is that although cationic lipids were required for inhibition of the targeted gene product in cell-based assays, cationic liposomes were not required for in vivo effects of the oligonucleotides. This observation is illustrated in Fig. 5 in which subcutaneously implanted A549 lung adenocarcinoma tumor cells were treated three times a week with ISIS 2503, which targets the AUG region of Ha-*ras*, in the presence or absence of a cationic liposome formulation (DMRIE:DOPE::50:50) at doses of either 20 mg/kg (without formulation) or 10 mg/kg (with formulation) over a 28-d period. Inhibition of tumor cell growth was observed for ISIS 2503 in both the presence and absence of cationic lipid. No effects on tumor growth were observed with a nonspecific phosphorothioate sequence formulated with or without cationic liposomes. These studies have been extended to include a wide range of human tumor types in the absence of cationic liposomes (Table 1).

The timing of initiation of oligonucleotide treatment following tumor implantation, route of oligonucleotide administration, and the effects of oligonucleotide chemistry on the in vivo antitumor effects observed for *ras* antisense

Table 1
Relative Antitumor Activity of Antisense Oligonucleotides Targeted to *ras* Isotypes in Human Nude Mouse Xenografts[a]

Oligonucleotide	Target	Description	Relative Anti-tumor Activity			
			T24 bladder	A549 lung	MDA.MB breast	SW480 colon
ISIS 2570	Ha-ras codon 12	Deoxy/P=S	++++	ND	+	ND
ISIS 3985	Ha-ras codon 12	2'OMe/Deoxy GAP P=S	>>+++++	ND	ND	ND
ISIS 2503	Ha-ras AUG	Deoxy/P=S	++++	++++	+	--
ISIS 6957	Ki-ras 5'UTR	Deoxy/P=S	ND	++++	ND	++
5FU	−	−	ND	+++	ND	---

[a]Oligonucleotides were administered either by ip injection or iv infusion at doses ranging between 0.06 and 20 mg/kg in the absence of cationic lipid. +, significant antitumor activity with the degree of antitumor activity proportional to the number of + symbols present; −, no significant antitumor activity observed; ND, not determined. 5-fluorouracil was administered to human A549 lung and SW480 colon carcinoma xenografts at a maximally tolerated dose of 60 mg/kg (ip). Abbreviations: UTR, untranslated region; deoxy, oligodeoxynucleotide (unmodified at the 2' sugar position); 2'-O-Me, 2'-O-methyl modified at the 2-sugar position; 2'-O-Me/Deoxy GAP, 2'-O-methyl/deoxy chimeric oligonucleotide; P=S, phosphorothioate.

25

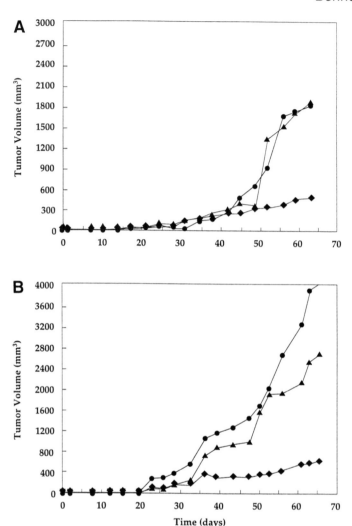

Fig. 5. Antitumor activity of Ha-*ras* targeted phosphorothioate oligodeoxynucleotides in nude mice containing subcutaneously implanted human A549 tumors. ISIS-2503, a 20-mer phosphorothioate oligodeoxynucleotide targeted to the AUG site of human Ha-*ras* mRNA, was administered by ip injection either in the presence of cationic lipid at a oligodeoxynucleotide dose of 10 mg/kg **(A)**, or in the absence of cationic lipid at a oligodeoxynucleotide dose of 20 mg/kg **(B)**. Oligodeoxynucleotide administration was initiated 25 d following tumor cell implantation and continued three times weekly over a 4-wk period. Symbols are as follows; solid circles, no oligodeoxynucleotide treatment (cationic lipid vehicle alone [A] or saline vehicle alone [B]); solid triangles, randomized 20-mer phosphorothioate oligodeoxynucleotide control; solid diamonds, ISIS-2503.

oligonucleotides have also been examined. Depending on the route and frequency of oligonucleotide administration, potent antitumor activity has been observed at oligonucleotide doses ranging from 0.1–25 mg/kg. Furthermore, we have demonstrated that novel modifications that greatly increase antisense potency in cell culture in the presence of cationic lipid *(28)* substantially increase antitumor activity in vivo without the need for cationic lipid (B. Monia et al., manuscript in preparation). These studies demonstrate that observations made from antisense experiments performed in cell culture, such as the need for cationic lipid, are not necessarily observed in vivo, whereas other findings made from cell culture experiments, such as an increase in antisense potency conferred by novel chemical modification, may accurately predict observations made during in vivo studies. The reasons for the apparent discrepancy in cationic liposome requirements for in vitro and in vivo studies are currently being explored.

2.3. Protein Kinase C-α

Protein kinase C (PKC) was originally identified as a serine/threonine kinase involved in mediating intracellular responses to a variety of growth factors, hormones, and neurotransmitters. Genetic analysis has revealed that PKC exists as a family of proteins consisting of at least 12 closely related and heterogeneously distributed isozymes *(35)*. PKC is also the major intracellular receptor for the tumor-promoting phorbol-ester class of compounds, and is therefore thought to be involved in the process of tumor promotion. Inhibitors of PKC function have been suggested to be of value as anticancer drugs *(36)*. We have previously identified a series of 20-mer phosphorothioate oligodeoxynucleotides that inhibit the expression of one isotype of PKC-α in human cells *(29)*. One of these oligonucleotides, ISIS 3521, specifically reduced the expression of PKC-α mRNA and protein with an IC_{50} of 200 nM (in the presence of cationic liposomes). The reduction in PKC-α expression promoted by ISIS 3521 was also highly specific. Oligonucleotides that were not complementary to PKC-α were without effect on the expression of either PKC-α mRNA or protein. Similarly, ISIS 3521 was without effect on the expression of the other closely related members of the PKC family.

The effects of ISIS 3521 and control oligodeoxynucleotides on the growth of subcutaneously transplanted human A549 cells in nude mice were examined. Oligonucleotides were administered ip at a dose of 20 mg/kg three times a week for 40 d. A significant reduction in the growth of these tumors in mice treated with ISIS 3521 was observed, with the control oligonucleotide having no effect (N. Dean, unpublished data). Furthermore, ISIS 3521 has demonstrated potent antiproliferative activity for several other tumor types with IC_{50} values <0.6 mg/kg/d when administered iv (N. Dean, manuscript submitted).

Expectedly, we have observed differential sensitivity of tumor types to the PKC-α oligonucleotide, suggesting that this isoenzyme may be important for the growth of some tumor types, but less so for others. These studies are currently being pursued to define further the mechanism of inhibition of tumor growth by PKC-α oligonucleotides and to demonstrate a reduction in PKC-α expression in the tumors.

2.4. Effect of ICAM-1 Antisense Oligonucleotides on Melanoma Metastasis

Intercellular adhesion molecule 1 (ICAM-1), a member of the immunoglobulin gene superfamily (described in more detail in Section 3.) at one time was considered to be a melanoma progression antigen *(37)*. Several studies have shown a correlation between ICAM-1 expression on the melanoma cells and stage of the disease, with the more advanced tumors exhibiting higher expression of ICAM-1 *(38)*. ICAM-1 expression also correlates with metastatic behavior of melanomas *(38,39)*. Treatment of human melanoma cells *ex vivo* with TNF-α or interferon-γ increased the number of metastatic lesions in lungs of nude mice when the treated cells were subsequently injected into the tail vein. Both cytokines induced a marked increase in expression of ICAM-1 on the melanoma cells *(40)*. To investigate whether ICAM-1 may be involved in the metastatic process, the melanoma cells were treated *ex vivo* with ICAM-1 antisense oligonucleotides, followed by a treatment with TNF-α, washed, and injected into mice. The ICAM-1 antisense oligonucleotides significantly reduced the number of metastases induced by TNF-α treatment, whereas an unrelated oligonucleotide failed to do so *(40)*. There was a good correlation between in vitro potency and in vivo activity, in that the more potent ICAM-1 antisense oligonucleotide in inhibiting ICAM-1 expression in cell culture was also more effective in reducing the number of metastatic lesions to the lung. The mechanism by which ICAM-1 contributes to enhanced metastasis is unclear at this time, but one possibility is that ICAM-1 on the melanoma cells could provide signals to adhering neutrophils or monocytes activating the leukocytes, resulting in increased release of proteases and other enzymes that break endothelial barriers aiding in the invasion into the tissue.

3. Pharmacology of Antisense Oligonucleotides in Inflammatory Models

As is evident by the number of reviews on the subject, much interest has been generated for the use of antisense oligonucleotides in viral infections and cancer *(8,9,12,41–43)*; however, antisense oligonucleotides could be useful for other applications. In particular, we have been intrigued about the possible use of oligonucleotides to inflammatory disorders. We will describe our work on cell adhesion molecules to exemplify the application of antisense oligonucleotide to inflammatory diseases. However, it should be noted that there are numerous other published reports in which antisense oligonucleotides have been used to inhibit the expression of a protein that is involved in the inflammatory process (reviewed in refs. *44,45*).

3.1. Cell Adhesion Molecules

The initiation and propagation of an immune response are dependent on a series of specific cell–cell interactions. Significant progress has been made in the identification of the molecules that mediate these cellular interactions and their downstream effectors resulting in the identification of a novel group of molecular targets. The endothelial cell adhesion molecules are representative of this class of proteins *(46–49)*. Under normal conditions, circulating leukocytes do not interact with the endothelial surfaces lining blood vessels. However, in response to inflammatory mediators, leukocytes adhere to and migrate through vascular endothelial cells utilizing a series of carefully orchestrated interactions.

These interactions with endothelium are owing to increased expression of adhesion molecules on the endothelium and activation of adhesion molecules on leukocytes. Under resting conditions, ICAM-1 is expressed at low levels on endothelial cells, but can be markedly unregulated on a variety of cell types in response to cytokines, such as tumor necrosis factor (TNF), interleukin 1 (IL-1), and interferon-γ *(50,51)*. ICAM-1 has been implicated in both firm adhesion and transmigration. In addition to its role in leukocyte emigration, ICAM-1 may also play a role in leukocyte activation by providing costimulatory signals *(52,53)*. Like ICAM-1, expression of VCAM-1 is induced on multiple cell types in response to specific cytokines *(54,55)*. E-selectin is not expressed on resting endothelium, but is markedly, although transiently, unregulated on endothelial cells in response to IL-1 or TNF *(56)*.

The induction of ICAM-1, VCAM-1, and E-selectin on the surface of endothelial cells requires *de novo* mRNA and protein synthesis. We have successfully identified oligonucleotides that inhibit expression of human ICAM-1, VCAM-1, and E-selectin by screening a large number of different oligodeoxynucleotides against each target *(57,58)*. The results demonstrated that in the presence of cationic liposomes, numerous phosphorothioate oligodeoxynucleotides are capable of inhibiting expression of the targeted protein; however, there are specific regions within a given mRNA that appear to be more sensitive to the effects of the oligonucleotide. Oligonucleotides targeting these sensitive regions were found to be severalfold more active than oligonucleotides targeting other regions of the mRNA. Both published *(57,58)* and unpublished data strongly suggest that the more active oligonucleotides inhibit expression of their respective protein by an antisense mechanism of action. This conclusion is based on the following data:

1. Inhibition of expression was dependent on the sequence of the oligonucleotide used. Sense and randomized oligonucleotides failed to inhibit expression of targeted gene product *(57,58)*.
2. A series of mismatched oligonucleotides to ICAM-1 were prepared in which 2, 4, 6, 8, or 10 bases were randomized. A 2-bp mismatch resulted in significant loss of activity, whereas oligonucleotides with four or more mismatches failed to inhibit ICAM-1 expression *(58,59)*.
3. Incubation of cells with a sense oligonucleotide in the presence of the antisense oligonucleotide reversed the inhibitory effects of the antisense oligonucleotide *(57)*.
4. Active oligonucleotides failed to inhibit constitutively expressed molecules, such as actin, G3PDH, or MHC molecules *(57)*.
5. At concentrations that inhibit expression of the targeted adhesion molecule, expression of the other adhesion molecules was not affected *(58)*.
6. The ICAM-1, VCAM-1, and E-selectin antisense oligonucleotides, but not control oligonucleotides, inhibit adhesion of HL-60 cells to cytokine-activated endothelial cells, demonstrating expected pharmacological activity.

The lead phosphorothioate antisense oligodeoxynucleotide from the ICAM-1 program, ISIS 2302, is currently in human clinical trials. This oligonucleotide is being investigated for efficacy in several inflammatory disorders, including acute renal allograft rejection, psoriasis, rheumatoid arthritis, and inflammatory bowel disease.

3.2. Effect of ICAM-1 Antisense Oligonucleotides in Inflammatory Models

3.2.1. Carrageenan-Induced Inflammation

Comparison of the nucleotide sequence of human and murine ICAM-1 cDNAs revealed the sequences to be only 65% conserved. In the 3'-untranslated region, where our most active human antisense oligonucleotides were identified, the degree of conservation is significantly less *(60)*. Thus, the human ICAM-1 antisense oligonucleotides would not be predicted to be effective in murine systems. We, therefore, designed and tested a number of oligonucleotides that target different regions on the murine ICAM-1 mRNA *(61)*. Similar to what was observed for human ICAM-1, the most effective oligonucleotide, ISIS 3082, targeted a specific sequence in the 3'-untranslated region of murine ICAM-1 mRNA *(61)*. It should be noted that although ISIS 3082 was identified in cell-culture-based assays using cationic liposome formulation, all the in vivo experiments performed with this oligonucleotide were done without the use of cationic liposomes, demonstrating in vivo pharmacological activity in the absence of cationic liposome formulations.

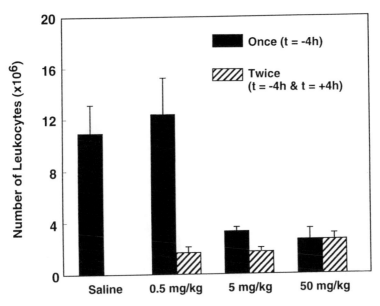

Fig. 6. Inhibition of carrageenan-mediated leukocyte migration by a murine ICAM-1 antisense oligodeoxynucleotide. ISIS 3082, a 20-mer phosphorothioate oligodeoxynucleotide that selectively inhibits ICAM-1 expression in murine tissues, was administered by iv injection at the indicated doses 4 h prior or 4 h prior and 4 h after sc implantation of a carrageenan-soaked sponge into CD-1 mice. Twenty-four hours after implantation, the sponge was removed, and the number of infiltrating leukocytes quantitated. Data represent mean ± SEM ($n = 5$).

To test whether the murine-specific ICAM-1 phosphorothioate antisense oligonucleotide, ISIS 3082, would inhibit leukocyte migration into inflamed tissues, we implanted carrageenan-soaked polyester sponges into the subcutaneous space on the back of mice. Carrageenan, an algal polysaccharide, has been widely used to induce inflammation at the site of administration and is a standard test for anti-inflammatory compounds *(62)*. Hallmarks of carrageenan-induced inflammation include a marked edematous reaction and leukocyte influx. The number of leukocytes infiltrating the sponge were determined 24 h after implantation. The oligonucleotides were either administered 4 h prior to the surgical implantation of the sponge or 4 h before and 4 h after implantation. Given as a single dose, 4 h prior to administration of carrageenan, ISIS 3082 significantly inhibited ($p > 0.05$) leukocyte migration at a dose of 5 and 50 mg/kg, but not 0.5 mg/kg (Fig. 6). No significant difference was observed between 5 and 50 mg/kg. When administered as two doses, before and after carrageenan administration, the oligonucleotide maximally inhibited leukocyte migration at

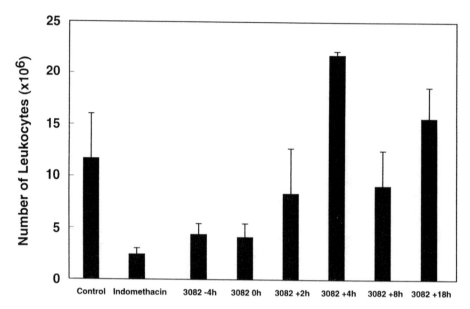

Fig. 7. Inhibition of leukocyte migration into area of inflammation. ISIS 3082, 5 mg/kg, was administered as a single iv injection at the indicated times relative to implantation of a carrageenan-soaked sponge into the sc space of mice. Twenty-four hours after implantation, the sponge was removed, and the number of infiltrating leukocytes was determined. Indomethacin was administered orally in three doses of 3 mg/kg, 0, 8, and 21 h after implantation of the sponge. Data represent the mean (SEM ($n = 5$).

a dose of 0.5 mg/kg. As expected, the human-specific ICAM-1 antisense oligonucleotide failed to inhibit leukocyte migration (data not shown). A second series of experiments was performed to determine the optimal time of administration of ISIS 3082 relative to challenge with carrageenan. ISIS 3082 at a dose of 5 mg/kg was administered as a single iv dose either 4 h prior to carrageenan administration, at the time of carrageenan administration, or 2, 4, 8, or 18 h after carrageenan. The antisense oligonucleotide significantly inhibited leukocyte influx when given 4 h prior to or at the time of carrageenan administration (Fig. 7). When administered 2 h after carrageenan, there was a slight decrease in leukocyte influx. If the oligonucleotide was administered 4 h or later after administration of carrageenan, there was no decrease in leukocyte influx; in fact, at 4 h there was actually an increase in leukocyte migration. Indomethacin was administered every 8 h as a positive control. These results are consistent with what would be predicted for such a model, i.e., once the inflammatory events are triggered and ICAM-1 is induced, the oligonucleotide would not be expected to block leukocyte migration.

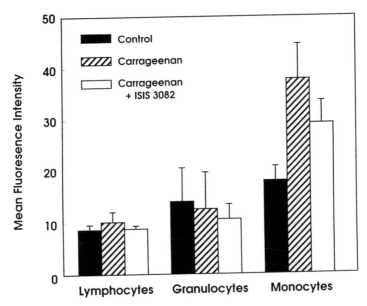

Fig. 8. Inhibition of ICAM-1 expression on peripheral blood leukocytes by ISIS 3082. CD-1 mice were treated with 5 mg/kg of ISIS 3082 by iv injection 2 h prior to surgical implantation of a carrageenan-soaked sponge. Twenty-four hours after implantation peripheral blood was collected. ICAM-1 expression on peripheral blood leukocytes was determined by flow cytometry using PE-labeled murine ICAM-1 antibody. The different leukocyte populations were identified by gating on specific populations using dual light scatter and immunostaining with CD3 antibodies (T-lymphocytes) and MAC-1 antibodies (granulocytes and monocytes). Results represent the mean ± SD ($N = 5$).

Several experiments were performed to address the mechanism by which the ICAM-1 antisense oligonucleotide inhibited leukocyte migration. There was no significant effect of the oligonucleotide on hematology parameters, including leukocyte number or differential leukocyte counts. Therefore, the decrease in leukocyte emigration was not the result of the oligonucleotide producing a neutropenic state. Cells that infiltrated the graft were predominately neutrophils and monocytes (data not shown). Preliminary data suggest that the ICAM-1 antisense oligonucleotide decreased mononuclear cell infiltration more than neutrophils. Examination of ICAM-1 expression on circulating leukocytes demonstrated that ISIS 3082 significantly attenuated the increase in ICAM-1 expression on circulating monocytes induced by the carrageenan treatment (Fig. 8). Circulating lymphocytes and granulocytes expressed a low level of ICAM-1, which was not significantly affected by carrageenan or the antisense oligonucleotide treatments (Fig. 8). The expression of ICAM-1 in vessels surrounding the carrageenan-soaked sponge was not examined.

3.2.2. Cardiac Allograft Rejection

Previous studies have demonstrated that treatment of mice with ICAM-1 MAb significantly prolongs cardiac and renal allograft survival *(63–65)*. The effects of the murine ICAM-1 antisense oligonucleotide in a heterotopic cardiac allograft model were evaluated *(61)*. Treatment of C3H recipients of C57BL/10 hearts with the ICAM-1 antisense oligonucleotide resulted in a dose-dependent increase in allograft survival *(61)*. Treatment by continuous iv infusion for 7 d resulted in an increase in allograft survival from 11.0 d (mean survival time) at a dose of 1.25 mg/kg/d to 15.3 d at a dose of 10 mg/kg/d compared to 7.7 d for untreated animals. Neither a scrambled control nor unrelated phosphorothioate oligodeoxynucleotide significantly affected allograft survival, demonstrating that the effects were sequence-specific. Increasing the duration of treatment from 7 to 14 d further increased allograft survival to a mean survival time of 23.0 d. Based on a 24–48 h elimination half-life of this oligonucleotide (Cooke et al., manuscript submitted) and other phosphorothioate oligonucleotides *(1,3,5)*, these data are consistent with the expected response, in that the oligonucleotide protects the graft from being rejected as long as therapeutic concentrations are delivered.

The murine ICAM-1 antisense oligonucleotide also prolonged allograft survival in other strain combinations. Histological examination of the hearts 6 d following transplantation documented protection of the allograft. Hearts from untreated recipients demonstrated marked infiltration with mononuclear cells and neutrophils. This infiltration was associated with severe necrosis and mineralization in 60% of the heart tissue. In contrast, recipients treated with the ICAM-1 antisense oligonucleotide showed scattered infiltration with mononuclear cells in 20% of the myocardium and no signs of necrosis. Similar infiltration was observed in syngeneic grafts. Thus, the oligonucleotide markedly reduced leukocyte infiltration and damage in the allografts *(61)*. In this particular model, inhibition of ICAM-1 mRNA induction could be a direct effect of the oligonucleotide or the result of decreased cellular infiltrate and release of proinflammatory cytokines. Thus, in this example, demonstration of ICAM-1 mRNA reduction is not direct proof that the oligonucleotide is working by an antisense mechanism. However, the data in aggregate strongly suggest that this is the mechanism of action. Furthermore, we have been able to show selective reduction of basal ICAM-1 mRNA and inhibited induction of ICAM-1 mRNA by LPS (*see* Section 4.), demonstrating that the oligonucleotide is capable of reducing ICAM-1 mRNA in animal tissues.

The ICAM-1 antisense oligonucleotide was found to enhance synergistically cardiac allograft survival when administered with antilymphocyte serum, rapamycin, brequinar, and anti-LFA-1 MAb, but not cyclosporin A. Administration of the oligonucleotide allowed reduction of the dose of rapamycin or brequinar, both of which can produce severe toxic side effects. The reason for lack of synergy with cyclosporin A is unknown; however, it should be pointed out that cyclosporin is not a very effective immunosuppressive agent in mice and not predictive of efficacy in humans. The combination of ISIS 3082 with anti-LFA-1 MAb was particularly effective, in that a 7-d treatment of both agents immediately after transplantation produced indefinite survival of the allograft *(61)*. Similar results were previously reported using an ICAM-1 MAb with anti-LFA-1 MAb *(64)*. If this particular combination of agents is as effective in treating human allografts, it could dramatically change transplantation therapy.

3.2.3. Other Inflammatory Models

The ICAM-1 antisense oligonucleotide has been tested to a lesser extent in two additional inflammatory models, inflammatory bowel disease and collagen-induced arthritis. In dextran sulfate sodium (DSS)-induced inflammatory bowel disease *(66,67)*, the ICAM-1 antisense oligonucleotide markedly attenuates the development of colitis as well as reverses pre-existing colitis (Bennett et al., manuscript submitted). The ICAM-1 antisense oligonucleotide exhibited an IC_{50} value of approx 0.3 mg/kg when administered sc for 7 d. Optimal activity was observed at a dose of 1 mg/kg/d. Two scrambled control oligodeoxynucleotides and a herpes virus phosphorothioate oligodeoxynucleotide failed to produce a significant effect in the colitis model. The antisense oligonucleotide was at least as effective as either transforming growth factor β (TGF-β) or cyclosporin A, both of which have been reported to be active in this model *(67)*.

Type II collagen-induced arthritis in rodents is regarded as one of the better models for rheumatoid arthritis in humans *(68,69)*. The effects of the murine ICAM-1 antisense oligonucleotide were evaluated in this model. Mice treated three times per week with 5 or 10 mg/kg/d produced a dose-dependent, although incomplete, reduction in the percent arthritic animals and in X-ray scores (T. Geiger, unpublished data). A scrambled control oligonucleotide failed to affect significantly either parameter. Increasing the frequency of dosing to five times per week resulted in more dramatic effects. Mice treated with a dose of 10 mg/kg/d administered ip five times per week failed to develop arthritis, whereas a dose of 5 mg/kg reduced the percent of animals developing arthritis from 80 to <20% (T. Geiger, unpublished data). The ICAM-1 antisense oligonucleotide did not significantly reduce the level of serum amyloid protein in animals. This result suggests that the oligonucleotide did not affect the production of cytokines, such as IL-1 or IL-6, again consistent with the expected mode of action.

4. Inhibition of Gene Expression in Animal Tissues
4.1. Oligonucleotide Effects on Gene Expression in Animals

The utility of oligonucleotides as inhibitors of gene expression in animal tissue has also been examined recently. The focus of many of these studies has been to determine whether particular oligonucleotides can influence the outcome of a disease model, and has been limited to localized administration of the oligonucleotides. For example, inhibition of restenosis has been reported by oligonucleotides targeting c-*myb*, *cdc*2, and PCNA kinase. Oligonucleotides were administered in a pluronic gel to the outside of the carotid artery or intraluminally in the presence of liposomes *(70,71)*. In both instances, a reduction in expression in the targeted mRNA was reported; however, the methods used to quantitate these changes have been questioned *(72)*. Essentially, the criticism is that stringent quantitative data were not included in the analysis of protein and RNA levels in the targeted tissue *(72)*. Oligonucleotide-dependent reductions in gene expression have also been described subsequent to localized administration into the brain. c-*fos (73,74)*, D_2 receptors *(75)*, and neuropeptide Y-Y1 receptors *(76)* have all been targeted.

4.2. Inhibition of PKC-α Expression in Mice
After Systemic Oligonucleotide Administration

To understand better the in vivo utility of phosphorothioate oligonucleotides as inhibitors of gene expression, we have identified an oligonucleotide that inhibits expression of murine PKC-α (ISIS 4189). PKC-α was chosen as a target, since this is a member of a large multigene family, which would allow us to evaluate oligonucleotide specificity by comparing effects of this oligonucleotide on the other PKC isotypes.

ISIS 4189 is a 20-mer phosphorothioate oligodeoxynucleotide designed to hybridize to the AUG translation initiation codon of murine PKC-α. In the presence of cationic liposomes, ISIS 4189 reduced PKC-α mRNA and protein expression with an IC_{50} of 100–200 nM *(77)*. ISIS 4189 was administered ip to mice at doses ranging from 1–100 mg/kg daily for 1 wk. The oligonucleotide caused a dose-dependent reduction in PKC-α mRNA expression in liver, with an IC_{50} of 30–50 mg/kg, whereas a control oligodeoxynucleotide was without effect. PKC-α mRNA expression in other organs was not consistently reduced. Levels of the other PKC isozymes remained unchanged by the oligonucleotide treatment demonstrating isozyme-specific inhibition *(77)*. Gross

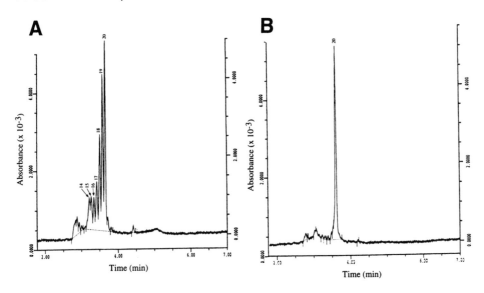

Fig. 9. Oligonucleotide metabolism in mouse liver. SK-1 hairless mice were given a single dose of 50 mg/kg oligonucleotides, either **(A)** ISIS 4189 (phosphorothioate oligodeoxynucleotide) or **(B)** ISIS 7817 (phosphorothioate 2'-*O*-propyl). Seventy-two hours later, animals were sacrificed, and oligonucleotides extracted from liver tissue and resolved by capillary gel electrophoresis. Traces show intact full-length (20 mo) material and n-1, n-2, and so forth, metabolites.

toxicity was not apparent in animals treated with oligonucleotides (both ISIS 4189 and the control phosphorothioate) at doses up to 100 mg/kg/d for up to 1 wk.

Our findings are consistent with the tissue distribution of phosphorothioate oligodeoxynucleotides in that the liver was the major organ of deposition *(2,3)*. It is also worth noting that reduction in PKC-α mRNA expression remained for at least 24 h after the final dose of oligodeoxynucleotide. This is consistent with the reported stability of phosphorothioate oligodeoxynucleotides *(3)* and our own observations on the stability of ISIS 4189 in vivo (Fig. 9A). To our knowledge, these results are the first reported example of a phosphorothioate oligodeoxynucleotide inhibiting expression of a normal, constitutively expressed host mRNA in vivo after systemic administration. It is also important to note that the inhibition of PKC-α seen with the oligonucleotide did not require the use of cationic liposome formulations, although ISIS 4189 was discovered in cell-based assays using cationic liposomes. These findings further demonstrate that a requirement for cationic liposomes in vitro to show oligonucleotide activity does not preclude systemic activities in vivo in the absence of cationic lipids or other delivery systems.

As part of our ongoing efforts to improve the pharmacokinetic and pharmacodynamic properties of oligonucleotides, we have synthesized and evaluated the properties in vivo of a number of analogs of the phosphorothioate oligodeoxynucleotide. Oligonucleotides are being examined for in vivo stability as well as their ability to reduce PKC-α expression in different organs. One of the most dramatic improvements in oligonucleotide activity found in tissue culture has been demonstrated by the synthesis of an analog of ISIS 4189 as a chimeric oligonucleotide containing six 2'-*O*-propyl residues at each of the 5'- and 3'-ends of the molecule (ISIS 7817). In this design, the centered region (eight nucleotides) is left unmodified at the 2' sugar position in order to serve as substrate for RNase H. A similar approach has previously been used by Monia and colleagues for oligonucleotides targeting Ha-*ras (28)*. The purpose behind 2'-*O*-propyl modification was to increase the nuclease stability of the oligonucleotide (McKay et al., unpublished observations). We have observed that, when incorporated into ISIS 4189, this modification causes a four- to fivefold increase in potency of the molecule in tissue culture compared with ISIS 4189. We have now begun examining the influence of this modification on oligonucleotide performance in vivo and have observed a substantial increase in the stability of the oligonucleotide. Capillary gel electrophoresis has been used to examine oligonucleotide integrity and distribution in different organs after systemic administration. Twenty-four hours after a single administration of ISIS 4189, substantial oligonucleotide degradation had occurred in mouse liver (Fig. 9A). Almost as many 19-mer oligonucleotides as full-length 20-mer oligonucleotides can be seen. Also n-2, n-3, and so forth, peaks can be seen, presumably representing the sequential removal of 3'- (or 5'-) residues by exonucleases. In contrast, the oligonucleotide containing 2'-*O*-propyl modifications (ISIS 7817) is almost completely intact at this time (Fig. 9B). These results confirm the utility of including 2'-*O*-propyl modifications into an oligonucleotide as a means of enhancing its nuclease stability in vivo. We are currently determining the effects of this modification on the ability of the oligonucleotide to inhibit PKC-α expression in different mouse tissues after systemic administration. It is anticipated that inclusion of modifications, such as 2'-*O*-propyl, into oligonucleotides will result in the development of compounds with considerably enhanced activity both in vitro and in vivo when compared to phosphorothioate oligodeoxynucleotides.

4.3. Inhibition of ICAM-1 Expression in Murine Tissues

The effects of the murine-specific ICAM-1 antisense oligonucleotide, ISIS 3082, on ICAM-1 expression in mouse tissues has been examined in two different models. ICAM-1 is expressed at low levels in many tissues and can be induced in multiple tissues by inflammatory mediators, such as TNF, IL-1, or

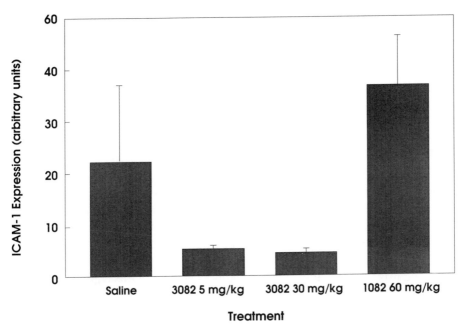

Fig. 10. Inhibition of ICAM-1 expression in murine lung tissue following treatment with ISIS 3082. ISIS 3082 or a control oligonucleotide, ISIS 1082, was administered by continuous sc infusion using Alzet pumps for 7 d at the indicated dose levels. Mice were sacrificed, and the amount of ICAM-1 expression in total lung tissue was determined by Northern blot analysis. ICAM-1 expression was normalized for loading by reprobing with G3PDH probes. Results represent the mean ±SD (*n* = 4).

bacterial endotoxin. The effect of ISIS 3082 on constitutive expression of ICAM-1 in murine lung was determined following continuous administration with Alzet miniosmotic pumps. In normal lung, ICAM-1 is moderately expressed on endothelial cells and more extensively expressed on type I Alzet pneumocytes *(78)*. Infusion of ISIS 3082, by implanting the Alzet pumps sc, at a dose of 5 mg/kg/d for 6 d, resulted in a 70% reduction of ICAM-1 expression (Fig. 10). Increasing the dose to 30 mg/kg /d did not lead to further reduction of ICAM-1 expression. In contrast, a control phosphorothioate oligonucleotide, ISIS 1082, slightly stimulated ICAM-1 expression. Thus, the ICAM-1 antisense oligonucleotide can selectively reduce constitutive ICAM-1 expression in mouse lung when administered by subcutaneous injection. Another model in which ISIS 3082 has demonstrated reduction of ICAM-1 expression is endotoxin-induced ICAM-1 expression in mouse lung. For this model, ISIS 3082 was administered by an iv injection 2 h prior to instillation of lipopolysaccharide into the lungs of mice. Instillation of bacterial lipopolysaccharide into the lung results in increased ICAM-1 expression on endothelial cells and type II pneumocytes *(78)*. Induction of ICAM-1 expression was markedly attenuated by a single dose of ISIS 3082 *(79)*. Maximal reduction occurred between 30 and 100 mg/kg (Doerschuk et al., manuscript submitted). Coincident with a reduction in ICAM-1 mRNA expression was a marked reduction in neutrophil influx *(80)*. These studies directly demonstrate that ISIS 3082 is capable of reducing ICAM-1 mRNA expression when administered by parenteral injection.

5. Conclusions and Future Directions

Based on the studies described here and by others in this volume, antisense oligonucleotides hold great promise as a novel class of therapeutic agents. Although we are just beginning to understand and exploit the technology, well-controlled experiments in animal models clearly show that antisense compounds have potent and specific pharmacological effects. Proof of mechanism for any drug in animals or in humans is very difficult to obtain experimentally. The standards for mechanistic proof that we describe in this chapter and those suggested by others *(72,81)* are substantially more stringent than those required for other drugs. There are numerous examples of clinically effective drugs on the shelves of pharmacies where the mechanism of action is poorly understood. Although rigorous proof of mechanism for an antisense compound has only been demonstrated in a few studies, the data strongly suggest that properly designed compounds work through antisense mechanisms in animals.

Future animal experiments should include efforts to obtain adequate proof of mechanism. Ideally, direct measurement of degradation of the target mRNA (in the case of an RNase H mechanism) or inhibition of expression of the target protein in a specific fashion is the best proof of mechanism. Where this is not possible, the use of composition-matched scrambled controls can serve as indirect evidence of an antisense mechanism. Even when a direct proof of mechanism is possible, scrambled controls assist in ruling out a chemical class effect of the compounds indirectly contributing to the activity.

Phosphorothioates have performed well as the first-generation antisense oligonucleotide analogs. The pharmacokinetics of these compounds demonstrates that once a day or every other day dosing is feasible. The potency observed in animal models of human disease suggests that treatment in the range of 0.05–5 mg/kg will produce significant effects. Given these data, and the recent dramatic reductions in cost of large-scale synthesis of phosphorothioates (A. Scozzari, unpublished results), this class of compounds holds great promise for commercial success.

Phosphorothioates represent only the first class of antisense compounds. There is substantial scope for improvement through medicinal chemistry, and many laboratories have demonstrated that novel antisense oligonucleotide analogs can have improved properties (reviewed in refs. *82,83*). In this chapter, we have described early experiments that demonstrate that improved analogs are more stable and more potent in animals. Although much remains to be learned, a great deal of progress has clearly been made in an area that is proving to be an important advance in therapeutics.

References

1. Cossum, P. A., Sasmor, H., Dellinger, D., Truong, L., Cummins, L., Owens, S. R., Markham, P. M., Shea, J. P., and Crooke, S. (1993) Disposition of the ^{14}C-labeled phosphorothioate oligonucleotide ISIS 2105 after intravenous administration to rats. *J. Pharmacol. Exp. Ther.* **267**, 1181–1190.
2. Cossum, P. A., Truong, L., Owens, S. R., Markham, P. M., Shea, J. P., and Crooke, S. T. (1994) Pharmacokinetics of a 14C-labeled phosphorothioate oligonucleotide, ISIS 2105, after intradermal administration to rats. *J. Pharmacol. Exp. Ther.* **269**, 89–94.
3. Agrawal, S., Temsamani, J., and Tang, J. Y. (1991) Pharmacokinetics, biodistribution, and stability of oligodeoxynucleotide phosphorothioates in mice. *Proc. Natl. Acad. Sci. USA* **88**, 7595–7599.
4. Temsamani, J., Tang, J., Padmapriya, A., Kubert, M., and Agrawal, S. (1993) Pharmacokinetics, biodistribution, and stability of capped oligodeoxynucleotide phosphorothioates in mice. *Antisense Res. Dev.* **3**, 277–284.
5. Sands, H., Gorey-Feret, L. J., Cocuzza, A. J., Hobbs, F. W., Chidester, D., and Trainor, G. L. (1994) Biodistribution and metabolism of internally ^{3}H-labeled oligonucleotides. I. Comparison of a phosphodiester and a phosphorothioate. *Mol. Pharmacol.* **45**, 932–943.
6. Saijo, Y., Perlaky, L., Wang, H., and Busch, H. (1994) Pharmacokinetics, tissue distribution, and stability of antisense oligodeoxynucleotide phosphorothioate ISIS 3466 in mice. *Oncol. Res.* **6**, 243–249.
7. Crooke, S. T. (1993) Therapeutic potential of oligonucleotides. *Curr. Opin. Invest. Drugs* **2**, 1045–1048.
8. Crooke, S. T. (1992) Therapeutic applications of oligonucleotides. *Annu. Rev. Pharmacol. Toxicol.* **32**, 329–376.
9. Calabretta, B. (1991) Inhibition of protooncogene expression by antisense oligodeoxynucleotides: biological and therapeutic implications. *Cancer Res.* **51**, 4505–4510.
10. Stein, C. A., Tonkinson, J. L., and Yakubov, L. (1991) Phosphorothioate oligodeoxynucleotides—antisense inhibitors of gene expression? *Pharmacol. Ther.* **52**, 365–384.
11. Milligan, J. F., Matteucci, M. D., and Martin, J. C. (1993) Current concepts in antisense drug design. *J. Med. Chem.* **36**, 1923–1937.
12. Stein, C. A. and Cheng, Y.-C. (1993) Antisense oligonucleotides as therapeutic agents—Is the bulllet really magical? *Science* **261**, 1004–1012.
13. Whitesell, L., Rosolen, A., and Neckers, L. M. (1991) In vivo modulation of N-myc expression by continous perfusion with an antisense oligonucleotide. *Antisense Res. Dev.* **1**, 343–350.

14. Kitajima, I., Shinohara, T., Bilakovics, J., Brown, D. A., Xu, X., and Nerenberg, M. (1992) Ablation of transplanted HTLV-1 tax-transformed tumors in mice by antisense inhibition of NF-kB. *Science* **258,** 1792–1795.

15. Skorski, T., Nieborowska-Skorska, M., Nicolaides, N. C., Szczylik, C., Iversen, P., Iozzo, R. V., Zon, G., and Calabretta, B. (1994) Suppression of Philadelphia leukemia cell growth in mice by *BCR-ABL* antisense oligodeoxynucleotide. *Proc. Natl. Acad. Sci. USA* **91,** 4504–4508.

16. Ratajczak, M. Z., Kant, J. A., Luger, S. M., Huiya, N., Zhang, J., Zon, G., and Gewirtz, A. M. (1992) In vivo treatment of human leukemia in a scid mouse model with c-myb antisense oligodeoxynucleotides. *Proc. Natl. Acad. Sci. USA* **89,** 11,823–11,827.

17. Hijiya, N., Zhang, J., Ratajczak, M. Z., Kant, J. A., DeRiel, K., Herlyn, M., Zon, G., and Gewirtz, A. M. (1994) Biologic and therapeutic significance of *MYB* expression in human melanoma. *Proc. Natl. Acad. Sci. USA* **91,** 4499–4503.

18. McCormick, F. (1989) *ras* GTPase activating protein: signal transmitter and signal terminator. *Cell* **56,** 5–8.

19. Hall, A. (1990) The cellular functions of small GTP-binding proteins. *Science* **249,** 635–649.

20. Bokoch, G. M. and Der, C. J. (1993) Emerging concepts in the Ras superfamily of GTP-binding proteins. *FASEB J.* **7,** 750–759.

21. Daum, G., Eisenmann-Tappe, I., Fries, H.-W., Troppmair, J., and Rapp, U. R. (1994) The ins and outs of Raf kinases. *TIBS* **19,** 474–480.

22. Bishop, J. M. (1987) The molecular genetics of cancer. *Science* **235,** 305–306.

23. Bos, J. L. (1989) *ras* Oncogenes in human cancer: a review. *Cancer Res.* **49,** 4682–4689.

24. Vogelstein, B., Fearon, E. R., Hamilton, S. R., Kein, S. E., Pressinger, A. C., Leppert, M., Nakamura, Y., White, R., Smits, A., and Bos, J. L. (1988) Genetic alterations during colorectal tumor development. *N. Engl. J. Med.* **319,** 525–532.

25. Lemoine, N. R., Mayall, E. S., Wyllie, F. S., Williams, E. D., Goyns, M., Stringer, B., and Wynford-Thomas, D. (1989) High frequency of *ras* oncogene activation in all stages of human thyroid tumorigenesis. *Oncogene* **4,** 159–164.

26. Reddy, E. P. (1983) Nucleotide sequence analysis of the T24 human bladder carcinoma oncogene. *Nature* **220,** 1061–1063.

27. Bos, J. L., Verlaan-de Vries, M., Marshall, C. J., Veeneman, G. H., van Boom, J. H., and van der Eb, A. J. (1986) A human gastric carcinoma contains a single mutated and an amplified normal allele of the Ki-*ras* oncogene. *Nucleic Acids Res.* **14,** 1209–1217.

28. Monia, B. P., Lesnik, E. A., Gonzalez, C., Lima, W. F., McGee, D., Guinosso, C., Kawasaki, A. M., Cook, P. D., and Freier, S. M. (1993) Evaluation of 2' modified oligonucleotides containing deoxy gaps as antisense inhibitors of gene expression. *J. Biol. Chem.* **268,** 14,514–14,522.

29. Dean, N. M., McKay, R., Condon, T. P., and Bennett, C. F. (1994) Inhibition of protein kinase C-a expression in human A549 cells by antisense oligonucleotides inhibits induction of intercellular adhesion molecule 1 (ICAM-1) mRNA by phorbol esters. *J. Biol. Chem.* **269,** 16,416–16,424.

30. Monia, B. P., Johnston, J. J., Ecker, D. J., Zounes, M. A., Lima, W. F., and Freier, S. M. (1992) Selective inhibition of mutant Ha-*ras* mRNA expression by antisense oligonucleotides. *J. Biol. Chem.* **267,** 19,954–19,962.

31. Saison-Behmoaras, T., Tocque, B., Rey, I., Chassignol, M., Thuong, N. T., and Helene, C. (1991) Short modified antisense oligonucleotides directed against Ha-Ras point mutation induce selective cleavage of the mRNA and inhibit T24 cells proliferation. *EMBO J.* **10,** 1111–1118.

32. Chang, E. H., Miller, P. S., Cushman, C., Devadas, K., Pirollo, K. F., Ts'o, P. O. P., and Yu, Z. P. (1991) Antisense inhibition of *ras* p21 expression that is sensitive to a point mutation. *Biochemistry* **30,** 8283–8286.

33. Georges, R. N., Mukhopadhyay, T., Zhang, Y., Yen, N., and Roth, J. A. (1993) Prevention of orthotopic human lung cancer growth by intratracheal instillation of a retroviral antisense K-*ras* construct. *Cancer Res.* **53,** 1743–1746.

34. Schwab, G., Chavany, C., Duroux, I., Goubin, G., Lebeau, J., Hélène, C., and Saison-Behmoaras, T. (1994) Antisense oligonucleotides adsorbed to polyalkyl-cyanoacrylate nanoparticles specifically inhibit mutated Ha-*ras*-mediated cell proliferation and tumorigenicity in nude mice. *Proc. Natl. Acad. Sci. USA* **91,** 10,460–10,464.

35. Nishizuka, Y. (1992) Intracellular signaling by hydrolysis of phospholipids and activation of protein kinase C. *Science* **258,** 607–614.

36. Basu, A. (1993) The potential of protein kinase C as a target for anticancer treatment. *Pharmacol. Ther.* **59,** 257–280.

37. Johnson, J. P., Stade, B. G., Hupke, U., Holzman, B., and Riethmuller, G. (1988) The melanoma progression-associated antigen P3. 58 is identical to the intercellular adhesion molecule, ICAM-1. *Immunobiol.* **178,** 275–284.

38. Natali, P., Nicotra, M. R., Cavaliere, R., Bigotti, A., Romano, G., Temponi, M., and Ferrone, S. (1990) Differential expression of intercellular adhesion molecule 1 in primary and metastatic melanoma lesions. *Cancer Res.* **50,** 1271–1278.

39. Johnson, J. P., Stade, B. G., Holzmann, B., Schwable, W., and Riethmuller, G. (1989) De novo expression of intercellular adhesion molecule-1 in melanoma correlates with increased risk of metastasis. *Proc. Natl. Acad. Sci. USA* **86,** 641–644.

40. Miele, M. E., Bennett, C. F., Miller, B. E., and Welch, D. R. (1994) Enhanced metastatic ability of TNF-a-treated malignant melanoma cells is reduced by intercellular adhesion molecule-1 (ICAM-1, CD54) antisense oligonucleotides. *Exp. Cell Res.* **214,** 231–241.

41. Cohen, J. S. (1991) Antisense oligodeoxynucleotides as antiviral agents. *Antiviral Res.* **16,** 121–133.

42. Dolnick, B. J. (1991) Antisense agents in cancer research and therapeutics. *Cancer Invest.* **9,** 185–194.

43. Agrawal, S. (1992) Antisense oligonucleotides as antiviral agents. *TIBTECH* **10,** 152–158.

44. Bennett, C. F. (1993) Antisense oligonucleotides in inflammation research and therapeutics, in *Antisense Research and Applications* (Crooke, S. T. and Lebleu, B., eds.), CRC, Boca Raton, pp. 547–562.

45. Bennett, C. F. and Crooke, S. T. (1995) Oligonucleotide based inhibitors of cytokine expression and function, in *Therapeutic Modulation of Cytokines* (Henderson, B. and Bodmer, M., eds.), CRC, Boca Raton, in press.

46. Springer, T. A. (1990) Adhesion receptors of the immune system. *Nature* **346,** 425–434.

47. Butcher, E. C. (1991) Leukocyte-endothelial cell recognition: three (or more) steps to specificity and diversity. *Cell* **67,** 1033–1036.

48. Albelda, S. M., Smith, C. W., and Ward, P. A. (1994) Adhesion molecules and inflammatory injury. *FASEB J.* **8,** 504–512.

49. Bevilacqua, M. P. (1993) Endothelial-leukocyte adhesion molecules. *Ann. Rev. Immunol.* **11,** 767–804.

50. Rothlein, R., Dustin, M. L., Marlin, S. D., and Springer, T. A. (1986) A human intercellular adhesion molecule (ICAM-1) distinct from LFA-1. *J. Immunol.* **137,** 1270–1274.

51. Rothlein, R., Czajkowski, M., O'Neill, M. M., Marlin, S. D., Mainolfi, E., and Merluzzi, V. J. (1988) Induction of intercellular adhesion molecule 1 on primary and continuous cell lines by pro-inflammatory cytokines. *J. Immunol.* **141,** 1665–1669.

52. Altmann, D. M., Hogg, N., Trowsdale, J., and Wilkinson, D. (1989) Cotransfection of ICAM-1 and HLA-DR reconstitutes human antigen-presenting cell function in mouse L cells. *Nature* **338,** 512–514.

53. Van Seventer, G. A., Shimizu, Y., Horgan, K. J., and Shaw, S. (1990) The LFA-1 ligand ICAM-1 provides an important costimulatory signal for T cell receptor-mediated activation of resting T cells. *J. Immunol.* **144,** 4579–4586.

54. Rice, G. E., Munro, J. M., and Bevilacqua, M. P. (1990) Inducible cell adhesion molecule 110 (INCAM-110) is an endothelial receptor for lymphocytes. *J. Exp. Med.* **171,** 1369–1374.

55. Osborn, L., Hession, C., Tizard, R., Vassallo, C., Luhowskyj, S., Chi-Rosso, G., and Lobb, R. (1989) Direct expression cloning of vascular cell adhesion molecule 1, a cytokine-induced endothelial protein that binds to lymphocytes. *Cell* **59,** 1203–1211.

56. Bevilacqua, M. P., Pober, J. S., Mendrick, D. L., Cotran, R. S., and Gimbrone, M. A. (1987) Identification of an inducible endothelial-leukocyte adhesion molecule. *Proc. Natl. Acad. Sci. USA* **84,** 9238–9242.

57. Chiang, M.-Y., Chan, H., Zounes, M. A., Freier, S. M., Lima, W. F., and Bennett, C. F. (1991) Antisense oligonucleotides inhibit intercellular adhesion molecule 1 expression by two distinct mechanisms. *J. Biol. Chem.* **266,** 18,162–18,171.

58. Bennett, C. F., Condon, T., Grimm, S., Chan, H., and Chiang, M.-Y. (1994) Inhibition of endothelial cell-leukocyte adhesion molecule expression with antisense oligonucleotides. *J. Immunol.* **152,** 3530–3540.

59. Nestle, F. O., Mitra, R. S., Bennett, C. F., Chan, H., and Nickoloff, B. J. (1994) Cationic lipid is not required for uptake and selective inhibitory activity of ICAM-1 phosphorothioate antisense oligonucleotides in keratinocytes. *J. Invest Dermatol.* **103,** 569–575.

60. Siu, G., Hedrick, S. M., and Brian, A. A. (1989) Isolation of the murine intercellular adhesion molecule 1 (ICAM-1) *Gene J. Immunol.* **143**, 3813–3820.

61. Stepkowski, S. M., Tu, Y., Condon, T. P., and Bennett, C. F. (1994) Blocking of heart allograft rejection by intercellular adhesion molecule-1 antisense oligonucleotides alone or in combination with other immunosuppressive modalities. *J. Immunol.* **153**, 5336–5346.

62. Vinegar, R., Truax, J. F., and Selph, J. L. (1976) Quantitative studies of the pathway to acute carrageenan inflammation. *Fed. Proc.* **35**, 2447–2456.

63. Cosimi, A. B., Conti, D., Delmonico, F. L., Preffer, F. I., Wee, S.-L., Rothlein, R., Faanes, R., and Colvin, R. B. (1990) In vivo effects of monoclonal antibody to ICAM-1 (CD54) in nonhuman primates with renal allografts. *J. Immunol.* **144**, 4604–4612.

64. Isobe, M., Yagita, H., Okumura, K., and Ihara, A. (1992) Specific acceptance of cardiac allograft after treatment with antibodies to ICAM-1 and LFA-1. *Science* **255**, 1125–1127.

65. Haug, C. E., Colvin, R. B., Delnonico, F. L., Auchincloss, H., Tolkoff-Rubin, N., Preffer, F. I., Rothlein, R., Norris, S., Scharschmidt, L., and Cosimi, A. B. (1993) A phase I trial of immunosuppression with anti-ICAM-1 (CD54) mAb in renal allograft recipients. *Transplantation* **55**, 766–773.

66. Okayasu, I., Hatakeyama, S., Yamada, M., Ohkusa, T., Inagaki, Y., and Nakaya, R. (1990) A novel method in the induction of reliable experimental acute and chronic ulcerative colitis in mice. *Gastroenterology* **98**, 694–702.

67. Murthy, S. N. S., Cooper, H. S., Shim, H., Shah, R. S., Ibrahim, S. A., and Sedergran, D. J. (1993) Treatment of dextran sulfate sodium-induced murine colitis by intracolonic cyclosporin. *Dig. Dis. Sci.* **38**, 1722–1734.

68. Wooley, P. H., Luthra, H. S., Stuart, J. M., and David, C. S. (1981) Type II collagen-induced arthritis in mice. Major histocompatability complex linkage and antibody correlates. *J. Exp. Med.* **154**, 688–700.

69. Bliven, M. L., Wooley, P. H., Pepys, M. B., and Otterness, I. G. (1986) Murine type II collagen arthritis: association of an acute-phase response with clinical course. *Arthritis Rheum.* **29**, 1131–1138.

70. Simons, M., Edelman, E. R., DeKeyser, J.-L., Langer, R., and Rosenberg, R. D. (1992) Antisense c-myb oligonucleotides inhibit arterial smooth muscle cell accumulation in vivo. *Nature* **359**, 67–70.

71. Morishita, R., Gibbons, G. H., Ellison, K. E., Nakajima, M., Zhang, L., Kaneda, Y., Ogihara, T., and Dzau, V. J. (1993) Single intraluminal delivery of antisense cdc2 kinase and proliferating-cell nuclear antigen oligonucleotides results in chronic inhibition of neointimal hyperplasia. *Proc. Natl. Acad. Sci.USA* **90**, 8474–8478.

72. Wagner, R. W. (1994) Gene inhibition using antisense oligodeoxynucleotides. *Nature* **372**, 333–335.

73. Chiasson, B. J., Hooper, M. L., Murphy, P. R., and Robertson, H. A. (1992) Antisense oligonucleotide eliminates in vivo expression of c-fos in mammalian brain. *Eur. J. Pharmacol.* **227**, 451–453.

74. Gillardon, F., Beck, H., Uhlmann, E., Herdegen, T., Sandkühler, J., Peyman, A., and Zimmermann, M. (1994) Inhibition of c-fos protein expression in rat spinal cord by antisense oligodeoxynucleotide superfusion. *Eur. J. Neurosci.* **6,** 880–884.
75. Zhou, L.-W., Zhang, S.-P., Qin, Z.-H., and Weiss, B. (1994) *In vivo* administration of an oligodeoxynucleotide antisense to the D_2 dopamine receptor messenger RNA inhibits D_2 dopamine receptor-mediated behavior and the expression of D_2 dopamine receptors in mouse striatum. *J. Pharmacol. Exp. Ther.* **268,** 1015–1023.
76. Akabayashi, A., Wahlestedt, C., Alexander, J. T., and Leibowitz, S. F. (1994) Specific inhibition of endogenous neuropeptide Y synthesis in arcuate nucleus by antisense oligonucleotides suppresses feeding behavior and insulin secretion. *Mol. Brain Res.* **21,** 55–61.
77. Dean, N. M. and McKay, R. (1994) Inhibition of protein kinase C-alpha expression in mice after systemic administration of phosphorothioate antisense oligodeoxynucleotides. *Proc. Natl. Acad. Sci. USA* **91,** 11,762–11,766.
78. Burns, A. R., Takei, F., and Doerschuk, C. M. (1994) Quantitation of ICAM-1 expression in mouse lung during pneumonia. *J. Immunol.* **153,** 3189–3198.
79. Quinlan, W. M., Doyle, N. A., Kumasaka, T., Wancewicz, E., Bennett, C. F., and Doerschuk, C. M. (1994) The effect of ICAM-1 anti-sense oligonucleotides on the expression of ICAM-1 mRNA induced by E. coli endotoxin. *Am. Rev. Respir. Dis.* **150,** A335.
80. Kumasaka, T., Quinlan, W. M., Doyle, N. A., Condon, T., Bennett, C. F., Sligh, J., Beaudet, A. L., and Doerschuk, C. M. (1995) The role of ICAM-1 in E. coli endotoxin-induced pneumonia evaluated using ICAM-1 mutant mice or ICAM-1 anti-sense oligonucleotides. *Am. Rev. Respir. Dis.* **151,** A456.
81. Stein, C. A. and Krieg, A. M. (1994) Problems in interpretation of data derived from in vitro and in vivo use of antisense oligodeoxynucleotides. *Antisense Res. Dev.* **4,** 67–69.
82. Cook, P. D. (1991) Medicinal chemistry of antisense oligonucleotides—future opportunities. *Anti-Cancer Drug Design* **6,** 585–607.
83. Goodchild, J. (1992) Enhancement of ribozyme catalytic activity by a contiguous oligodeoxynucleotide (facilitator) and by 2'-*O*-methylation. *Nucleic Acids Res.* **20,** 4607–4612.

3

Antisense Efficacy

Site-Restricted In Vivo and Ex Vivo Models

Leonard M. Neckers, Daniel Geselowitz, Christine Chavany, Luke Whitesell, and Raymond Bergan

1. Introduction

In the laboratory, antisense oligodeoxynucleotides (ODN) have repeatedly demonstrated efficacy in modulating the expression of various genes, thus providing important insights into their roles in tumorigenesis or normal growth and development *(1–3)*. Although attention has been focused recently on the development of antisense ODN as therapeutics for a variety of diseases, including cancer *(4)*, systemic application of ODN to treat tumors other than those of the hematopoietic system presents several problems, not least of which is the ability of systemically administered antisense to reach distant tumor sites. This is particularly true when considering tumors of the central nervous system (CNS), such as glioblastomas and HIV-associated B-cell lymphomas. In this case, the blood–brain barrier poses an additional obstacle to successful delivery of anionic ODN. On the other hand, the intractability of CNS tumors to standard chemotherapy makes them interesting candidates for antisense intervention. The first model system we will discuss involves direct infusion of ODN into the CNS for the purpose of continuous perfusion of tumor cells.

It is now clear that ODN are generally internalized by a process of endocytosis. This is true for both negatively charged phosphodiester and phosphorothioate ODN, as well as for neutral methylphosphonates and conjugated ODN *(5)*. Although in some cells, this form of uptake is sufficient to produce antisense effects, antisense efficacy would certainly be increased if endosomal internalization could be bypassed altogether. The second model we will discuss demonstrates how endocytosis can be bypassed in the *ex vivo* treatment of tumor cells.

From: *Methods in Molecular Medicine: Antisense Therapeutics*
Edited by: S. Agrawal Humana Press Inc., Totowa, NJ

2. Development of a CNS Model for In Vivo Analysis of Antisense Efficacy

To avoid obstacles associated with systemic ODN administration, we and others have focused on regional therapeutic strategies to evaluate ODN actions in vivo. We have previously reported that continuous local perfusion with an unmodified ODN can specifically modulate gene expression in a subcutaneous tumor model system *(6)*. Others have recently demonstrated the successful inhibition of intimal smooth muscle cell accumulation following vascular injury using local application of a c-*myb*-targeted ODN *(7)*. Recently, we have characterized the stability, disposition, and clearance of ODN within a clinically important, physiologically well-defined biologic compartment, namely the cerebrospinal fluid (CSF) space of the rat. We found that potentially therapeutic concentrations of intact phosphorothioate ODN can be maintained within the CSF without gross toxicity to the animal. Extensive penetration into brain parenchyma could also be demonstrated *(8)*.

When we began these studies, it was understood that both phosphodiester and phosphorothioate ODN are quite stable in CSF with little degradation noted after up to 24 h of incubation at 37°C. Using in vitro culture conditions with pure CSF, we were able to confirm these data (in order to detect oligonucleotides by gel electrophoresis, we labeled them internally with fluorescein). However, although phosphodiester ODN was quite stable in CSF alone, addition of brain tissue resulted in nearly complete loss of intact ODN within 24 h. Phosphorothioate ODN stability, on the other hand, was insensitive to the presence of tissue and ODN remained intact for at least 24 h. Consistent with these degradation findings, very little fluorescein label could be recovered from the brain slice when phosphodiester ODN was used, and no intact phosphodiester ODN could be seen on gel electrophoresis. However, when phosphorothioate ODN was used, a significant accumulation of label by the slice (almost 10% of the total incubated material) could be detected within 24 h. Moreover, gel electrophoresis of tissue extracts revealed that the label recovered represented intact phosphorothioate ODN.

We next wished to determine in vivo stability and clearance parameters of ODN administered as a bolus injection into the cerebral ventricles of rats. In order to standardize experiments and allow for meaningful comparison among ODN modifications, we coinjected (^{3}H)-inulin, a well-recognized marker of extracellular space, sharing the same approximate molecular weight (5 kDa) as ODN *(9)*. The level of (^{3}H)-inulin marker measured in CSF following bolus injection behaves quite consistently in this rat model. Generally, about 5 min after completion of a 50-µL injection, inulin concentration in a CSF sample is about 20–30% that of injectate. During the next 2 h, concentration falls with nonfirst-order kinetics. Inulin is thought to be cleared mainly by bulk flow;

we believe the initial apparent half-life observed reflects clearance of inulin from the ventricle, whereas the secondary half-life represents dispersion and bulk flow from the total CSF. However, for several reasons, determination of clearance $t_{1/2}$ values from such bolus data is subject to experimental artifact. First, distortion of CSF volume can occur owing to an initial administration of 10% of total volume, followed by repeated sampling of up to 4% of that volume. Second, the level of radioactivity sampled at later time-points is quite low and subject to error. Third, normal circulation of CSF in the rat depends in part on animal movement, and the animals are anesthetized during the course of the bolus experiments. Therefore, we feel that estimation of clearance $t_{1/2}$ values from bolus kinetic data is not reliable. Instead, these values can be estimated from the steady-state concentration of ODN and inulin obtained during Alzet pump infusion *(see below)*.

When phosphodiester ODN end-labeled with (^{32}P) at the 5' or 3' end was intraventricularly coinjected with inulin, the ratio of (^{32}P) in the CSF relative to injectate was reduced compared to inulin by about 20% at 5 min, and this relative depletion of (^{32}P) typically increased with time. This apparent discrepancy in clearance rates is most readily explained by degradation of phosphodiester ODN within the CSF space following injection. Alternatively, ODN might be cleared from the CSF by a mechanism(s) other than, or in addition to, bulk flow. The anionic nature of the ODN used raised the possibility that they were being cleared by the small anion efflux pump located in the choroid plexus. However, administration of a dose of probenecid sufficient to block this pump completely had no effect on the apparent rate of ODN clearance. However, gel electrophoresis of CSF samples following injection of fluorescein-labeled material demonstrated extensive degradation of ODN recovered 100 min after injection. Such a finding echoed the behavior of phosphodiester ODN cultured in vitro in the presence of brain tissue. When 3'-labeled (^{32}P) ODN was used, no size ladder of degradation products could be seen, although intact ODN rapidly disappeared. When a 5' (^{32}P) label was used, a ladder similar to that seen with internal fluorescein-labeled material became apparent. These results suggest that degradation of phosphodiester ODN readily occurs in brain and is initiated primarily at the 3' end of an ODN. If this were the case, phosphorothioate ODN should be cleared with kinetics similar to inulin following a bolus injection.

When phosphorothioate ODN with internal fluorescein label was injected as an intraventricular bolus, the level of fluorescence in CSF closely matched the (^{3}H) level over a 100-min time-course. Thus, <5% of the material sampled 5 min after injection could be detected by 100 min. In contrast to the findings with phosphodiester material, gel electrophoretic analysis of CSF samples following phosphorothioate bolus injection revealed only intact ODN even after

100 min. Again, these findings are consistent with the pattern of stability observed in vitro in the presence of fresh brain slices. Preliminary results have suggested that resistance to degradation does not require an all-phosphoro-thioate molecule. Protection of the 3' portion of the molecule by several phosphorothioate linkages, or other "capping" modifications, may be sufficient to confer protection.

Given the relatively rapid clearance demonstrated for phosphorothioate ODN after bolus administration, we explored the potential of a continuous infusion technique to maintain steady-state levels of ODN within the CSF of free-roaming rats. In this procedure, an indwelling cannula is stereotaxically implanted in the third ventricle and cemented in place. Polyethylene tubing is connected to the cannula, run under the skin, and connected to an Alzet microinfusion pump implanted subcutaneously between the rat's shoulder blades. Infusate concentrations >3 mM, administered at 1 μL/h, are toxic. Global CNS depression followed by death routinely occurred within 24 h of infusion initiation. Concentrations up to 1.5 mM, however, were well tolerated during a 2-wk period. ODN levels in the CSF of 0.1% of the infusate concentration can be readily maintained during infusion periods. Thus, a CSF concentration of up to 1.5 μM ODN can be maintained for at least 14 d without gross toxicity. As might be expected from bolus observations, ODN levels correlated almost identically with the fractional level of (^{3}H)-inulin observed in a particular animal. As expected, the material recovered in CSF appears completely intact by gel electrophoresis. Using the data obtained from continuous infusion experiments, we have calculated a clearance $t_{1/2}$ for phosphorothioate ODN based on steady-state values ($n = 3$) to be 17.2 ± 4.7 min. The corresponding clearance $t_{1/2}$ calculated for inulin ($n = 9$) is 23 ± 7.5 min. This value is consistent with previously published data on inulin clearance out of CSF (9).

Given that phosphorothioate ODN appears stable in the presence of brain tissue, we examined phosphorothioate ODN penetration into brain by direct fluorescence microscopy. Paraformaldehyde-fixed frozen sections were prepared from rat brains 30 min after bolus intraventricular injection of 40 nmol of a fluorescein-labeled phosphorothioate ODN. Analysis of these frozen sections demonstrated a readily detectable tissue signal. A gradient distribution was apparent with the highest signal noted at the ependymal surfaces of the ventricle, consistent with diffusion in from the surrounding CSF. The endothelium of small vessels also appeared highly fluorescent, suggesting uptake of material that had been cleared into the vasculature, but precise definition of clearance pathways has not been undertaken at this time. Following continuous infusion of phosphorothioate ODN for 1 wk (1.5 nmol/h), a gradient distribution of ODN was less apparent, but marked uptake of fluorescent material could now be observed in many cell bodies scattered throughout the tissue section. GFAP (a glial cell marker protein) immunostaining of sections prepared from such a phosphorothioate-perfused brain demonstrated colocalization of ODN uptake and GFAP positivity within a subset of astrocytic appearing cells. Indeed, we have found that the most strongly positive cells in brain following ODN administration appear to be glial in nature. Glial cells induced to proliferate by brain injury (i.e., cannula implantation) are the most avid in concentrating extracellular ODN.

Because many drugs, including ODN, show very poor penetration into the CNS after systemic administration *(10)*, regional therapeutic strategies are particularly attractive in this organ system *(11)*. The CSF space of the rat provides a practical, physiologically well-defined compartment in which to examine systematically ODN actions in vivo. Moreover, it is clearly clinically relevant, given that the intrathecal administration of classical chemotherapeutics is already extensively employed as CNS prophylaxis and treatment in the management of acute leukemias where it has markedly improved treatment outcome *(12)*. Technical advances, such as the Ommaya reservoir, have made repeated and/or prolonged access to the ventricular system a practical reality *(13,14)*. Novel intrathecal reagents, such as immunotoxins *(15)*, radioimmunoconjugates *(16)*, and interleukins *(17)*, are in the early stages of clinical investigation in the therapy of both primary CNS cancers and the leptomeningeal metastases of systemic cancers, such as breast and lung. The very poor outcome of current therapies for high-grade glioma as well as the rapidly increasing incidence of poor prognosis for AIDS-associated primary CNS lymphoma *(18)* emphasize the need for new therapeutic approaches.

The penetration into brain we observed in these studies suggested the possibility of targeting both leptomeningeal and intraparenchymal disease processes. Therefore, we turned next to the direct injection of ODN into brain parenchyma itself (the caudate nucleus). These studies again demonstrated the presence of significant nuclease activity, and even some degradation of phosphorothioate ODN was observed. However, when phosphorothioate ODN were used, we were struck by the high and circumscribed level of material achieved in the immediate vicinity of the infusion site, even after several days of continuous delivery. Intact ODN could be recovered from this site, as well as from more distant brain regions (Geselowitz et al., unpublished observations). Quantitation of ODN (by fluorescence) after 4 d of continuous infusion indicated that more than half of the total ODN infused remained within 2 mm of the infusion site. When equal amounts of fluorescent brain extract from phosphodiester and phosphorothioate infusates were examined by gel electrophoresis, little intact phosphodiester material was found, whereas the phosphorothioate ODN extracted was only partially degraded. McCarthy et al. *(19)* also reported that phosphorothioate ODN, injected into brain parenchyma, remained largely intact for at least 5 h. Continuous infusion of phosphorothioate ODN directly into brain tissue has the potential to deliver focally high concentrations of ODN, with a strong concentration gradient to the rest of the brain.

We have recently begun to examine administration of ODN to growing glioblastoma tumors by continuous infusion. We have been able to achieve a similar sort of focal distribution in the growing tumor as we saw in the healthy caudate. We believe that it should be possible to attain a very high ratio of ODN concentration in tumor relative to the remainder of the brain, especially since proliferating glial cells avidly internalize ODN. This should, in turn, be a very positive contribution to the therapeutic index.

Recent studies by many groups have implicated the over-expression of specific oncogene products in the transformed phenotype of malignant gliomas *(20,21)*. We are now evaluating antisense sequences targeted against EGF receptor, TGF-α, bFGF, VEGF, and c-*myc* for antitumor activity both in syngeneic rat and human xenograft brain tumor models. These in vivo functional studies should clarify the currently confusing role of these oncogenes in brain tumor initiation and/or malignant progression. For example, in initial studies targeting glioma bFGF, we observed a marked growth inhibition in vitro, whereas in vivo ODN administration into the CNS had no effect on either tumor growth or animal survival (Chavany et al., unpublished observations). Thus, although a useful target in vitro, bFGF autocrine production by glioma cells may not play a major role in their in vivo growth. Perhaps the tumor can obtain sufficient bFGF (and other cytokines) from surrounding normal stromal tissue. One of the advantages of these CNS tumor models is their ability to point out such discrepancies between in vitro and in vivo biological phenomena. Coupled to the power of antisense technology, we can now determine which genes are required for tumorigenesis in an in vivo setting.

3. An *Ex Vivo* Model to Study Efficacy of Electroporated Antisense ODN

Because endogenous uptake of ODN by cells, at least in vitro, is primarily endocytotic in nature, a means of getting ODN into cells that bypasses the normal uptake machinery should be beneficial. Since it appears that the nucleus is the final destination of ODN released from cytoplasmic endosomal entrapment, a technique allowing for direct administration of ODN to cell nuclei should increase antisense efficacy. This is in fact the case. As we recently reported *(22)*, electroporation can be utilized as a simple and rapid technique to introduce a high concentration of ODN directly into cell nuclei, resulting in almost immediate antisense effects. The results are similar to microinjection studies with the major difference that one can readily load up to five million cells at a time with uniformly high amounts of ODN. Although phosphodiester ODN so introduced into cells is rapidly degraded in most cases, phosphorothioate ODN remains at a consistently elevated level in cell nuclei for up to 3 d after a single electroporation. With c-*myc* as a target, one can observe a significant, sequence-specific reduction in protein level within 90 min of electroporation. Because the half-life of the c-*myc* protein is 20–30 min, these data imply that the electroporated ODN begin to exert an antisense effect almost at once.

In certain B-cell lymphomas, c-*myc* is overexpressed, and its expression appears critical for cell growth and survival *(23)*. These tumor cells often are present in bone marrow, and an effective *ex vivo* purging regimen would be of great clinical significance. Other *ex vivo* models have been described, but these rely on the endogenous cellular uptake of ODN and require prolonged in vitro culture for this purpose. We have designed an *ex vivo* model that incorporates electroporation as a means of ODN delivery. In order to validate this approach, we have electroporated B-lymphoma cells with phosphorothioate c-*myc* antisense ODN (1–10 µ*M*) under conditions in which loss of viability owing to electroporation remains minimal (20–25%). We have confirmed that such treatment results in a sequence-specific loss of c-*myc* protein within 90 min to 2 h *(22)*. After allowing the cells to recover overnight, we have inoculated athymic mice with equal numbers of viable tumor cells treated with antisense or control ODN and monitored for animal survival (Bergan et al., unpublished observations). These tumors normally prove fatal to 100% of mice within 3–4 wk. In three separate experiments, 80% of mice (*n* = 20) receiving antisense-electroporated tumor cells survived longer than 3 mo (at which point the experiment was terminated), whereas mice receiving control ODN-electroporated or untreated tumor cells (*n* = 20) all died within 1 mo. ODN were only administered by electroporation and were not present at any other time.

In order to validate this model, we had to determine whether the conditions used for tumor cell electroporation were harmful to normal bone marrow. Initially, we determined that normal murine bone marrow viability was not compromised under the electroporation conditions used (<10% nonviable cells). In fact, these cells were more resistant to electroporation than the B-lymphoma cells. We then determined that electroporation of normal human marrow with c-*myc* antisense did not compromise the marrow's colony-forming ability. Thus, colony formation was normal in sham electroporated and antisense electroporated normal marrow. Taken together, these experiments validate the use of electroporation to deliver antisense ODN in an *ex vivo* setting. Additionally, c-*myc* is suggested as a useful target for the *ex vivo* purging of B-lymphoma cells from bone marrow.

4. Summary

These studies demonstrate the utility of restricted site models for evaluation of antisense antitumor therapy. Not only can clinically relevant tumors be targeted for which there is currently no effective long-term therapy, but other advantages accrue as well. Sequestration and clearance by liver and kidney *(10,24,25)* are not an issue, as they are in the systemic administration of ODN. The large quantities of material necessary to achieve potentially therapeutic concentrations by systemic delivery need not be considered. Comparably minimal amounts of material are required for even continuous perfusion in the site-restricted models described here. Finally, potential systemic toxicity owing to widely distributed, and perhaps chronically administered, ODN need not be considered.

Other site-restricted models can be utilized for ODN studies. These include intraperitoneal ODN delivery to treat intraperitoneal tumor, intraocular ODN to treat retinoblastoma, and ODN applied topically to skin to treat melanoma, as well as benign keratinocyte proliferations, such as psoriasis and dermatitis. Antisense efficacy is currently being tested in several of these model systems, and such studies will no doubt lead to clinically useful antisense strategies in years to come.

References

1. Cohen, J. S. (1991) Chemically modified oligodeoxynucleotide analogs as regulators of viral and cellular gene expression, in *Gene Regulation. Biology of Antisense RNA and DNA* (Erickson, R. P. and Izant, J. G., eds.), Raven, New York, pp. 247–260.
2. Goodchild, J. (1989) Inhibition of gene expression by oligonucleotides, in *Oligodeoxynucleotides. Antisense Inhibitors of Gene Expression* (Cohen, J. S., ed.), Macmillan, London, pp. 53–77.
3. Neckers, L., Whitesell, L., Rosolen, A., and Geselowitz, D. A. (1992) Antisense inhibition of oncogene expression. *CRC Crit. Rev. Oncogenesis* **3**, 175–231.
4. Zamecnik, P. C. (1991) Oligonucleotide based hybridization as a modulator of genetic message readout, in *Prospects for Antisense Nucleic Acid Therapy of Cancer and AIDS* (Wickstrom, E., ed.), Wiley-Liss, New York, pp. 1–6.
5. Neckers, L. M. (1993) Cellular internalization of oligodeoxynucleotides, in *Antisense Research and Applications* (Crooke, S. T. and Lebleu, B., eds.), CRC, Boca Raton, FL, pp. 451–460.
6. Whitesell, L., Rosolen, A., and Neckers, L. M. (1991) *In vivo* modulation of N-myc expression by continuous perfusion with an antisense oligonucleotide. *Antisense Res. Dev.* **1**, 343–350.
7. Simons, M., Edelman, E. R., DeKeyser, J.-L., Langer, R., and Rosenberg, R. D. (1992) Antisense c-myb oligonucleotides inhibit intimal arterial smooth muscle cell accumulation in vivo. *Nature* **359**, 67–70.

8. Whitesell, L., Geselowitz, D., Chavany, C., Fahmy, B., Walbridge, S., Alger, J. R., and Neckers, L. M. (1993) Stability, clearance, and disposition of intraventricularly administered oligodeoxynucleotides: implications for therapeutic application within the central nervous system. *Proc. Natl. Acad. Sci. USA* **90**, 4665–4669.
9. Bass, N. and Lundborg, P. (1973) Postnatal development of bulk flow in the cerebrospinal fluid system of the albino rat: clearance of carboxyl-[^{14}C]inulin after intrathecal infusion. *Brain Res.* **52**, 323–332.
10. Agrawal, S., Temsamani, J., and Tang, Y. T. (1991) Pharmacokinetics, biodistribution, and stability of oligodeoxynucleotide phosphorothioates in mice. *Proc. Natl. Acad. Sci. USA* **88**, 7595–7599.
11. Poplack, D. G. and Riccardi, R. (1987) The role of pharmacology in pediatric oncology, in *The Role of Pharmacology in Pediatric Oncology* (Poplack, D. G., Massimo, L., and Cornaglia-Ferraris, P., eds.), Martinus Nijhoff, Boston, pp. 137–156.
12. Blaney, S. M., Balis, F. M., and Poplack, D. G. (1991) Pharmacologic approaches to the treatment of meningeal malignancy. *Oncology* **5**, 107–116.
13. Machado, M., Salcman, M., Kaplan, R. S., and Montgomery, E. (1985) Expanded role of the cerebrospinal fluid reservoir in neurooncology: indications, causes of revision, and complications. *Neurosurgery* **17**, 600–603.
14. Sundaresan, N. and Suite, N. D. A. (1989) Optimal use of the omaya reservoir in clinical oncology. *Oncology* **3**, 15–22.
15. Urch, C., George, A., Stevenson, G., Bolognesi, A., Stirpe, F., Weller, R., and Glennie, M. (1991) Intra-thecal treatment of leptomeningeal lymphoma with immunotoxin. *Int. J. Cancer* **47**, 909–915.
16. Moseley, R. P., Papanastassiou, V., Zalutsky, M. R., Ashpole, R. D., Evans, S., Bigner, D. D., and Kemshead, J. T. (1992) Immunoreactivity, pharmacokinetics and bone marrow dosimetry of intrathecal radioimmunoconjugates. *Int. J. Cancer* **52**, 38–43.
17. List, J., Moser, R. P., Steuer, M., Loudon, W. G., Blacklock, J. B., and Grimm, E. A. (1992) Cytokine responses to intraventricular injection of interleukin 2 into patients with leptomeningeal carcinomatosis: rapid induction of tumor necrosis factor alpha, interleukin 1 beta, interleukin 6, gamma-interferon, and soluble interleukin 2 receptor (M$_r$ 55,000 protein). *Cancer Res.* **52**, 1123–1128.
18. Pluda, J. M., Yarchoan, R., Jaffe, E. S., Feurstein, I. M., Solomon, D., Steinberg, S. M., Wyvill, K. M., Raubitschek, A., Katz, D., and Broder, S. (1990) Development of non-Hodgkin lymphoma in a cohort of patients with severe human immunodeficiency virus (HIV) infection on long-term antiretroviral therapy. *Ann. Intern. Med.* **113**, 276–282.
19. McCarthy, M. M., Brooks, P. J., Pfaus, J. G., Brown, H. E., Flanagan, L. M., Schwartz-Giblin, S., and Pfaff, D. W. (1993) Antisense oligodeoxynucleotides in behavioral neuroscience. *Neuroprotocols* **2**, 64–66.
20. La Rocca, R. V., Rosenblum, M., Westermark, B., and Israel, M. A. (1989) Patterns of proto-oncogene expression in human glioma cell lines. *J. Neurosci Res.* **24**, 97–106.

21. Morrison, R. S. (1991) Suppression of basic fibroblast growth factor expression by antisense oligodeoxynucleotides inhibits the growth of transformed human astrocytes. *J. Biol. Chem.* **266,** 728–734.

22. Bergan, R., Connell, Y., Fahmy, B., and Neckers, L. M. (1993) Electroporation enhances c-myc antisense oligodeoxynucleotide efficacy. *Nucleic Acids Res.* **21,** 3567–3573.

23. McManaway, M. E., Neckers, L. M., Loke, S. L., Al-Nasser, A. A., Redner, R. L., Shiramizu, B. T., Goldschmidts, W. L., Huber, B. E., Bhatia, K., and Magrath, I. T. (1990) Tumor-specific inhibition of lymphoma growth by an antisense oligodeoxynucleotide. *Lancet* **335,** 808–811.

24. Goodarzi, G., Watabe, M., and Watabe, K. (1992) Organ distribution and stability of phosphorothioated oligodeoxynucleotides in mice. *Biopharm. Drug Dispos.* **13,** 221–227.

25. Inagaki, M., Togawa, K., Carr, B. I., Ghosh, K., and Cohen, J. S. (1992) Antisense oligonucleotides: inhibition of liver cell proliferation and *in vivo* disposition. *Transplant. Proc.* **24,** 2971–2972.

4

Antisense Blockade of Expression

SNAP-25 In Vitro and In Vivo

Marina Catsicas, Astrid Osen-Sand, Julie K. Staple, Kenneth A. Jones, Guidon Ayala, Jonathan Knowles, Gabriele Grenningloh, Emilio Merlo Pich, and Stefan Catsicas

1. Introducing the Challenge

With the advent of modern molecular genetics and molecular biology, we will face more and more situations where novel gene products with unknown functions are identified. Genetic linkage analysis will allow the association of novel or known genes to important diseases *(1)*. Similarly, sensitive differential cloning procedures will identify rare genes expressed in specific physiological or pathological situations *(2,3)*. In both cases, establishing the precise function of the identified gene is an essential step for the understanding of the cellular mechanisms that either lead to the disease or are pivotal in important physiological processes.

In this chapter, we describe experiments designed to identify the function of genes expressed in the developing nervous system. More specifically, one of our goals is to identify genes involved in axonal growth and synapse formation. We chose the chick visual system for the cloning of target genes for the following reasons:

1. The size of the embryonic chick retina allows the extraction of large quantities of mRNA at early stages of development;
2. The calendar of differentiation of the visual pathway is well known; and
3. This system provides unique opportunities for experimental approaches *(4)*, as well as for the correlation between the distribution of molecular markers and the organization of the underlying neuronal networks *(5,6)*.

In the chick retina, synaptogenesis begins at embryonic day (E) 11 for most classes of neurons *(7,8)*, and the density of synapses reaches adult levels in the

From: *Methods in Molecular Medicine: Antisense Therapeutics*
Edited by: S. Agrawal Humana Press Inc., Totowa, NJ

central retina by E15 *(8)*. To identify candidate genes, we have developed a sensitive differential screening procedure, based on DNA–DNA competitive hybridization *(9)*, and we have used it to clone cDNAs of transcripts more abundant at E15 than at E9. We have cloned 65 independent partial cDNAs, of which 15 encode known proteins and 50 are novel. All 65 cDNAs were screened in selection assays, and the cDNA encoding Synaptosomal Associated Protein of 25 KD (SNAP-25) was one of a series corresponding to genes potentially involved in synapse formation and selected for further functional analysis *(10)*. We describe here experiments based on the use of antisense oligonucleotides that have allowed us to identify SNAP-25 as a key gene in the process of axonal elongation and synapse formation *(11)*.

2. Cellular and Molecular Biology of SNAP-25

SNAP-25 was first cloned by M. C. Wilson and his colleagues, based on its differential expression in several brain areas of the adult mouse *(12,13)*. The authors generated a polyclonal antibody to the carboxy terminus of the protein and were able to localize the protein at nerve terminals *(13,14)*. SNAP-25 was then found to be translocated to terminals by fast axonal transport *(15)* and to show a striking degree of evolutionary conservation. The chick and mouse proteins were found to be identical *(16)*, and the *Drosophila* SNAP-25 shows 61% identity with the mammalian protein *(17)*. Two regions of SNAP-25 have been suggested to play a role in protein–protein interactions: a potential amphipatic α-helix in the amino terminus *(13)* and a stretch of four clustered cysteines in the middle portion of the protein that may be the site for fatty acylation *(13,15)*. No significant level of identity with other known proteins was found. Thus, SNAP-25 was considered as a novel gene product in search of a function.

2.1. Patterns of Expression

The first steps in the functional analysis of a gene must include a detailed study of the spatial and temporal patterns of expression of the gene of interest. These correlative studies will not identify any precise function, but should give valuable information on the cellular processes in which the gene is involved. Three experimental systems are used in our laboratory to study neurite outgrowth and synapse formation: rat primary neuronal cultures, rat pheochromocytoma 12 (PC12) cells, and chick retinas in vivo. We have generated an antibody against SNAP-25 *(16)*, and we have analyzed SNAP-25 expression at the protein and mRNA levels in these three systems.

2.1.1. Expression by Primary Cortical Cells

We studied SNAP-25 expression in newborn rat cortical cells in culture and compared it to that of a series of cytoskeletal and synaptic proteins *(11*, and Staple

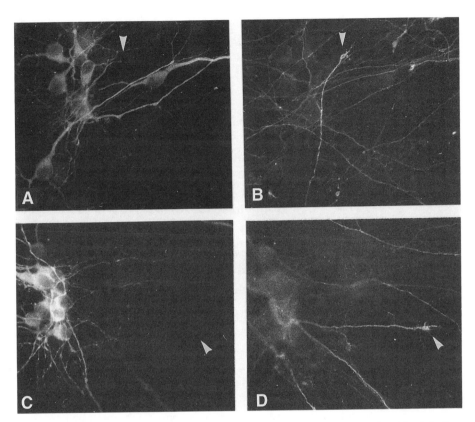

Fig. 1. SNAP-25 expression by cortical neurons in culture. Double-labeling after 5 d in culture with MAP2 **(A, B)** and SNAP-25 **(C, D)** antibodies. MAP2 is localized in cell bodies and thick processes, whereas SNAP-25 is found selectively in thin axon-like neurites. The arrowheads indicate the position of SNAP-25-positive growth cones. Scale bar: 10 μm.

et al., submitted). We found that SNAP-25 is expressed in the growth cone and distal segment of developing axon-like neurites, after their initial outgrowth, but before the formation of synaptic boutons (Fig. 1). These observations were consistent with a potential role in axonal growth and synapse formation.

2.1.2. Expression by Cell Lines

A second useful system to study SNAP-25 is the PC12 cell line, since:

1. All of the cells express SNAP-25 *(18)*;
2. All the NGF-induced neurites (*see* Section 3.2.) are SNAP-25 positive *(18)*; and
3. The cells can be incubated with oligonucleotides before the induction of neurite growth with NGF (*see* Section 3.2.).

We found that the levels of SNAP-25 protein were already high in proliferating PC12 cells not treated with NGF. Following induction with NGF, SNAP-25 was detectable in the perikarya and along the entire length of the processes of most PC12 cells.

2.1.3. Expression In Vivo

In the retina of adult chick, SNAP-25 protein and mRNA are mainly expressed by ganglion cells and amacrine cells (Fig. 2A). During development, the protein was first detectable at about E9, when the inner plexiform layer (where amacrine cells make synapses; *see* Section 4.1.) and the retino-tectal connection form *(16)*. These observations were confirmed with *in situ* hybridization (Fig. 2B,C) as well as with immunoblots and Northern blots. Again, these data suggested a role in axonal growth and identified the most appropriate stages to inhibit the expression of the protein.

3. Use of Antisense Oligonucleotides in Tissue Culture

3.1. Use of Antisense Oligonucleotide in Primary Tissue Culture

Our goal was to determine the relevance of SNAP-25 to axonal growth and synaptogenesis in the central nervous system (CNS) in vivo. Nonetheless, in vitro systems are the systems of choice to ascertain the effects of antisense oligonucleotides on protein expression, as well as their possible toxic effects. Cell cultures allow strict control over the concentration of the oligonucleotides, and protein expression can be rapidly characterized with immunochemical techniques. A second useful piece of information can be obtained by assessing morphological changes produced by antisense oligonucleotide treatment on the developing neurons in cultures. Using appropriate imaging methodology, semi-quantitative evaluation of neurite growth can give useful readouts regarding the effects of the treatment on axonal growth and synaptogenesis.

As a first approach to the study of the role of SNAP-25 in neuronal development, we chose to assess the ability of antisense oligonucleotides to inhibit its expression by cortical neurons in vitro.

3.1.1. Oligonucleotide Design

We considered several parameters in our design of oligonucleotides, including proximity of the sequence chosen to the translation start site, GC content, length, melting temperature, and secondary structure. Potential oligonucleotides were checked using the program Primer (Scientific and Educational Software), and those sequences likely to form homodimers or that had internal complementary regions were eliminated. Specific sequences, known to be toxic owing to interactions with proteins, such as guanine tetramers, were avoided *(19; see also 20)*. Since efficient protein downregulation has been obtained

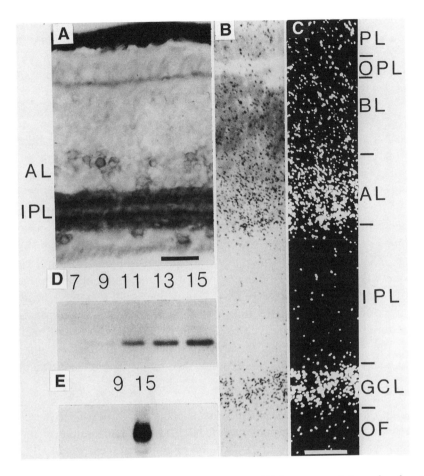

Fig. 2. SNAP-25 expression in the chick retina. **(A)** Immunostaining of embryonic retina at E17 by using avidin–biotin–peroxidase complex. SNAP-25 staining outlines some cell bodies in the amacrine and ganglion cell layers, and is heavily expressed in the fibers in the inner plexiform layer. **(B, C)** Localization of SNAP-25 mRNA showing bright-field (B) and dark-field (C) photomicrographs of *in situ* hybridization using an antisense complementary RNA probe. AL, amacrine cell layer; BL, bipolar cell layer; GCL, ganglion cell layer; IPL, inner plexiform layer; OF, optic fiber layer; OPL, outer plexiform layer. Scale bar: 20 μ*M*. **(D)** Immunoblots of retina protein extracts prepared from E7, 9, 11, 13, 15 staged embryos. **(E)** Northern blots of RNA extracts of E9 and E15 retina. Reproduced from ref. *(16)* with permission.

with oligonucleotides targeted to mRNA sites both close to the translation start site *(21,22)* and further downstream *(23,24)*, we chose two different oligonucleotides: PSNAP1 (5'-ATG TCT GCG TCC TCG GCC AT-3') includes the translation start site of the murine SNAP25 protein (nucleotides 1–20) *(12,13)* whereas SNAP2AS (5'-CTTCTC CAT GAT CCT GTC AA-3') is complementary to a site toward the 3' end of the protein (nucleotides 533–552), which is identical in the chick and mouse cDNA sequences *(16)*. Several nonsense oligonucleotides were used, including RANDOM1.SNAP (5'-ATC CCT CCG TGT AGC GCG TT) and PKC8GMN (5'-ACT GCT ACA CCT CAC GTG TT-3').

3.1.2. Testing Antisense Oligonucleotides in Primary Tissue Culture

We chose to study the activity of antisense oligonucleotides in dissociated rat cortical primary cultures because the morphological stages of neuronal differentiation in these cultures have been well characterized in relation to the patterns of expression of endogenous SNAP-25 (Staple et al., submitted). Cultures were prepared from the neocortices of 2–3 d-old rats as described by Baughman et al. *(25)*. Dissociated cells were plated on poly-L-lysine (Sigma, St. Louis, MO) treated coverslips fixed to the undersides of 35-mm culture dishes having a precut 10–12 mm central hole. This combination of coverslip and hole created a well of approx 250 μL in which small quantities of growth medium containing oligonucleotides could be fed. In later experiments, we used commercially available culture dishes that contained a central well and grid (Milian Instruments, S.A., Geneva, Switzerland). Neurons were allowed to adhere and grow overnight prior to the addition of oligonucleotides.

3.1.3. Choice of the Oligonucleotide Type, Purification, and Toxicity

In addition to nonmodified phosphodiester oligonucleotides, it is possible to use different types of oligonucleotides, such as phosphorothioates, methylphosphonates, 2'-*O*-alkyl-, and 2'-*O*-fluoro-oligonucleotides, where either the phosphodiester backbone or the sugar moiety has been modified to confer better nuclease resistance. Each has advantages and disadvantages, as shown by results obtained in different experimental systems. Unmodified oligonucleotides are degraded by nucleases, and therefore, high concentrations are usually required *(26–28)*, whereas oligonucleotides composed of phosphorothioate-substituted nucleotides (PONs) are nuclease-resistant *(29–31)*, but may have nonspecific toxic effects, such as the nonsequence-specific inhibition of RNaseH *(32,33; see also 20)*. In order to determine which would be most effective and least toxic in our system, we tried unmodified oligonucleotides, PONs, and oligonucleotides that had only three phosphorothioate-modified residues at each end of the 20-mer oligonucleotide. All oligonucleotides were synthesized on an automated DNA synthesizer (Applied Biosystems 394-8, Perkin Elmer Int., Rothkreuz, Switzerland) using standard phosphoramidate chemistry. The oxidation step in the synthesis of PONs was achieved using tetraethylthiuram disulfide (TETD; *34*). We found that the method of purification of the final product was critical in reducing nonspecific oligonucleotide toxicity. Some batches of oligonucleotides purified by *n*-butanol precipitation *(35)* caused degeneration and death of neurons in culture within 24–48 h. Further purification over G50 Sephadex columns (Pharmacia, Brussels, Belgium) considerably reduced these problems, although they were not entirely eliminated. Finally, coevaporation of oligonucleotides, after elution from Sephadex columns, with 3 × 0.2 mL of deionized water or dialysis against deionized water for 48 h consistently resolved problems of toxicity.

Each type of oligonucleotide was tested by adding varying concentrations of the antisense PSNAP1 to cultures of rat cortical neurons immediately after plating and readding oligos every day for 1 wk. The cells were then lysed for use in Western blot quantitation of SNAP-25 and neuron-specific enolase (NSE). Antisense PONs downregulated SNAP-25 by 95% within 7 d at a final concentration of 2 μ*M*, whereas other types of antisense oligonucleotides had no effect at 2 μ*M* or at higher concentrations (*see* Section 3.1.7.).

3.1.4. Determination of Oligonucleotide Uptake by Cells

We used PONs that were labeled at the 5'-end with fluorescein isothiocyanate (FITC) to visualize the accumulation of oligonucleotides into primary neurons. FITC was added to the 5' end of oligonucleotides via an amino group that was introduced using Aminolink II (Applied Biosystems). Fluorescence was visible in all cells 24 h after addition of PONs-FITC (not shown) at a final concentration of 2 μ*M* and increased in intensity after 48 h of incubation. When FITC alone was added to the cultures, no fluorescence was observed in the cells. Solubilized protein extracts of neuronal cultures showed that the PONs-FITC were as efficient as PONs in inhibiting SNAP-25 expression (not shown).

3.1.5. Evaluation of Dose–Response and Time-Course

In order to optimize the dose of PONs to be used in our assays, various concentrations of oligonucleotides were added to cultured neurons, and the effects on SNAP-25 levels were evaluated by immunoblotting. PONs were added directly to culture medium once a day at final concentrations that varied between 0.25 and 10 μ*M*. Doses of PONs of 5 or 10 μ*M* were toxic to both neurons and glial cells within 3–4 d. The glial cells appeared more sensitive to the toxic effects than the neurons (not shown). Small decreases in SNAP-25 were detectable after 3 d of treatment with 2 μ*M* PONs. After 5 and 7 d of treatment with 2 μ*M* PONs, SNAP-25 was decreased to 10% and undetectable levels, respectively, compared with untreated or nonsense PON-treated cultures. Cells looked healthy and expressed normal levels of NSE for at least 7 d of treatment. (Fig. 3A).

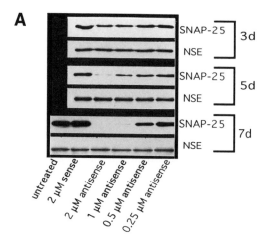

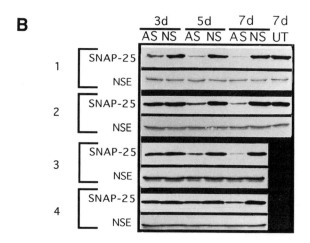

Fig. 3. Design of PON treatment in rat primary cortical neurons. **(A)** Time-course and effective concentrations of PONs: Cortical neurons were plated in 24-well dishes at 4×10^4 cells/well. PSNAP1 PON was added at a final concentration of 0.25, 0.5, 1, or 2 μM immediately after plating and the same dose of PON was readded each day for up to 7 d. After 3, 5, or 7 d of treatment, cells were solubilized and protein extracts were used for immunoblots probed with SNAP-25 and NSE (Seralab) antibodies. **(B)** Effective schedule of PON addition: Cortical neurons were plated as above, and PSNAP1 PONs were added at a final concentration of either 2 (#3 and #4) or 10 μ*M* (#1 and #2) immediately after plating. Further addition of 2 μM PONs occurred every day (#1 and #3) or every other day (#2 and #4). After 7 d of treatment, cells were solubilized and protein extracts were used for immunoblots probed with SNAP-25 and NSE antibodies.

3.1.6. Schedule of Oligonucleotide Addition

Since SNAP-25 is expressed at high levels by many cortical neurons when they are dissociated from newborn rats, it was essential to downregulate the protein as quickly as possible to evaluate changes in neuronal development. Since addition of 2 μM PONs to cultures each day efficiently downregulated SNAP, and without apparent toxicity, we chose to work with this concentration further to determine the most efficient schedule of PONs addition. On the first day after dissociation and plating of neurons, PONs were added at a final concentration of either 2 or 10 μM. PONs were then added at 2 μM every day, every other day, or every 2 d. Immunoblots of proteins extracted from treated cultures after 7 d showed that initial addition of 10 μM final concentration of PONs and subsequent addition of 2 μM every other day resulted in optimal SNAP-25 downregulation with the least amount of total PONs added (Fig. 3B).

3.1.7. Analysis of Rat Neurons Incubated with Oligonucleotides In Vitro

In our experiments, inhibition of SNAP-25 expression reached 94% and was dose-dependent (Fig. 4). To assess further the specificity of the inhibition, we checked the level of expression of a series of control proteins that included neuron-specific proteins as well as general cytoskeletal markers. The only protein tested that showed a reduction of expression (more that 60% in this case) was synaptophysin (Fig. 4). Interestingly, the effects on the level of expression of synaptophysin were less pronounced if the density of neurons in culture was increased (not shown). Synaptophysin is a marker of differentiated synapses *(36)*, so the effects on its level of expression, which cannot be attributed to a general effect on protein synthesis, suggested that inhibition of SNAP-25 expression reduced synapse formation. In addition, qualitative analysis of the cultures showed a strong reduction in the network of thin, axon-like neurites following antisense treatment (not shown). However, a precise quantification of axonal growth with image analysis (*see* Section 3.2.1.) was difficult, because axons and dendrites intermingled. Also, natural heterogeneity of the levels of SNAP-25 expression by different types of cortical cells may cause problems of interpretation, since some cells express less (and therefore presumably need less) SNAP-25. Finally, the role of SNAP-25 in axonal elongation may be underestimated in these cultures, because several days of PONs treatment are necessary to inhibit SNAP-25 expression and by that time neurons have already sent out axons and started to form synaptic contacts.

3.2. Use of Antisense Oligonucleotides with the PC12 Cell Line

For a more quantitative assay of SNAP-25's role in neurite elongation, it was necessary to use a culture system with a more homogeneous population of

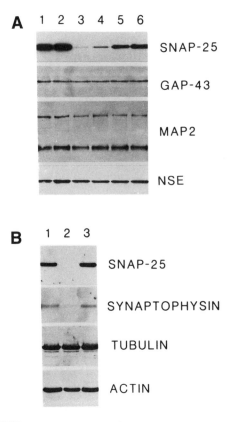

Fig. 4. Effect of PON treatment on protein expression in primary cortical neurons. **(A)** Dose–response and specificity of the antisense PON treatment. Western blots of protein extracts from cultures incubated for 7 d with 2 μ*M* control (sense) PONs (lane 1), without PONs (lane 2) or with 2, 1, 0.5, and 0.25 μ*M* antisense PONs, lanes 3, 4, 5, and 6, respectively. **(B)** Effects of SNAP-25 expression inhibition on the synaptophysin expression. Western blots of protein extracts incubated for 7 d without PONs or with 2 μ*M* antisense and control PONs, lanes 1, 2, and 3, respectively. The same blots were reprobed with antibodies against control proteins GAP-43, MAP2, NSE (A) actin, and tubulin (B). Reproduced from ref. *11* with permission.

cells and, ideally, where the levels of SNAP-25 could be downregulated before the onset of neurite outgrowth. We have used PC12 rat pheochromocytoma cells because they fulfilled all these criteria. PC12 cells derive from a transplantable tumor of rat adrenal medulla *(37)*. They can exist in two states, either undifferentiated/proliferating or differentiated. Following exposure to NGF, PC12 cells stop proliferating, differentiate, and assume a neuron-like morphology *(37,38)*. All PC12 cells express SNAP-25 *(18)*, and all the NGF-induced neurites are SNAP-25-positive *(18)*. In addition, and most importantly, the cells can be incubated with PONs before the induction of neurite growth with NGF. Therefore, it was possible to study a situation where the levels of SNAP-25 were already low before the onset of neurite outgrowth.

3.2.1. Inhibition of Neurite Extension by SNAP-25 Antisense Oligonucleotides in PC12 Cells

We studied the effect of SNAP-25 depletion on the capacity of PC12 cells to generate neurites. Rat pheochromocytoma cells have been an especially useful model system in analyzing molecular mechanisms implicated in neurite outgrowth *(39,40)*. PC12 cells express SNAP-25 before induction with NGF, and the level of SNAP-25 protein is unaffected by NGF treatment. The first step was to establish a dose–response curve in this culture system. The cells were plated in medium containing NGF (25 ng/mL) and 10 μM PON. Then, PONs (0.2–2 μM final concentration each time) were added every day. Higher concentrations were also tested, but were found to have nonspecific effects on protein synthesis (not shown). Total proteins were extracted, and antibody binding visualized and quantified on Western blots. Based on these data, for the neurite growth assays, cells were plated on collagen-coated 24-well Costar (Cambridge, MA) plates with medium containing 10 μM PONs, and 2 μM PONs were added every 2 d. The same PONs used for primary neurons were used for the PC12 cells. An unrelated PON was used of sequence 5'-ATGCTGTGCTGTATG AGAAG-3'. NGF (25 ng/mL) was added to the cultures for the first time on d 2 and then every 2 d. Each plate contained wells with different culture conditions in duplicate (antisense, sense, and unrelated PONs); controls comprised NGF-stimulated as well as non-NGF-stimulated wells that were not treated with PONs. In all cultures, the cells started to extend neurites 3 d after NGF induction. After four additional days, the cultures were fixed with 4% paraformaldehyde and immunostained to check for the inhibition of SNAP-25 expression. The plates were then dried, and the cultures analyzed under phase-contrast using a semiautomatic computer-assisted image analyzer (Vidas, Kontron, AG, Zürich, Switzerland). All the images were digitized under constant conditions, and the boundaries of the cell bodies were determined on the digitized image. The index of neurite growth, defined as the ratio between the surface area of neurites and the surface area of cell bodies, was then calculated automatically. The number of measurements in each well ranged from 43 to 74.

Both SNAP-25 antisense PONs, but not control PONs, specifically reduced SNAP-25 expression by more than 75%, and this effect was dose-dependent (Figs. 5 and 6). In the presence of SNAP-25 antisense PONs, neurite extension was nearly prevented (Figs. 6 and 7). No cells with long neurites could be observed, but the number of neurites and the surface area of the cell bodies were not affected

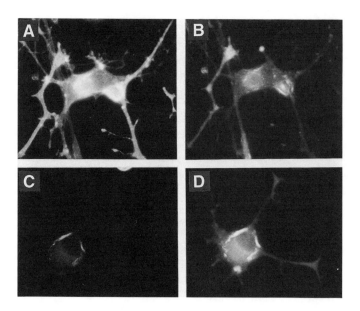

Fig. 5. Selective inhibition of SNAP-25 expression in PC12 cells. PC12 cells were cultured in the presence of control (sense) PONs **(A, B)** or antisense PONs **(C, D)**. After 7 d, the cultures were fixed and double-labeled with antibodies against SNAP-25 (A, C) and synaptophysin (B, D). Treatment with antisense PONs resulted in the inhibition of SNAP-25 expression and the lack of long neurites in these cultures.

by PON treatment. The inhibition of neurite extension was also dose-dependent. Using antisense concentrations of 1 μM or less had no effect on the level of SNAP-25 or on neurite outgrowth. Control cultures treated with sense or unrelated oligos were indistinguishable from control NGF-induced PC12 cells (Fig. 6).

These results show that SNAP-25 has a major role in neurite elongation and is apparently not involved in initial outgrowth. In addition, inhibition of SNAP-25 expression had no effect on the elaboration of dibutyryl cAMP-induced spikes (not shown). The formation of these short processes occurs through a different mechanism than the NGF-induced neurite elongation *(41)*; therefore, this result does not contradict our interpretation.

Fig. 6. *(opposite page)* Effect of PON treatment on neurite extension by PC12 cells. **(A)** Doseresponse curve of SNAP-25 expression inhibition in PC12 cells with antisense and control sense and unrelated PONs. a.u., Normalized mean gray value of the signal in Western blots in arbitrary units. **(B)** Inhibition of neurite extension by PC12 cells. SNAP-25 antisense PONs prevent NGF-induced neurite elongation, whereas control PONs do not. To display the data on the same figure, the index of neurite growth *(see*

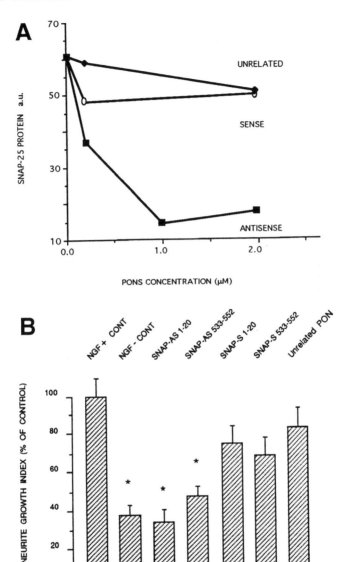

text) is shown as percentage of the NGF-stimulated control for all culture conditions tested. The asterisks indicate statistically significant differences from the NGF-stimulated controls with ($P < 0.05$) and without ($P < 0.01$) PONs, as calculated with the original data (*t*-test). Error bars indicate standard error of the mean. NGF + CONT, NGF-stimulated control; NGF − CONT, non-NGF-stimulated control. Reproduced from ref. *11* with permission.

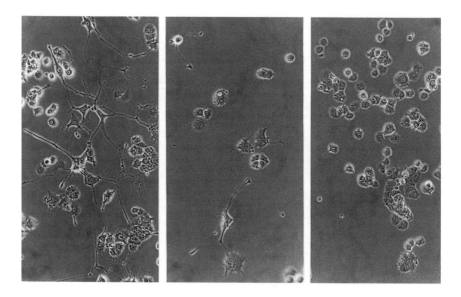

Fig. 7. Morphology of PC12 cells following antisense PON treatment. Representative fields of control PC12 cultures **(left)**, and cultures treated with 1 μM **(middle)** or 4 μM **(right)** antisense PONs. *See text* for details.

4. Use of Antisense Oligonucleotides for In Vivo Studies
4.1. The Chick Retina as an Experimental Model

The chicken offers several advantages as an experimental model to study the development of the CNS. Although it is a higher vertebrate, its development, which takes 21 d, occurs completely *in ovo*, and allows direct and easy access to the embryo at any time. We have taken advantage of the fact that the prominent developing eye provides a closed chamber that can be easily injected and thus allows the pharmacological manipulation of the neural retina, which is a part of the CNS.

The vertebrate retina is a highly organized, stratified structure (Fig. 8) comprising a relatively well-characterized number of neuron types. Retinal neurons are arranged into three layers of neuronal cell bodies alternating with two so-called plexiform layers, which contain most of the neuronal processes and synapses in the retina (*see* ref. 42). One of these two plexiform layers, the inner plexiform layer (IPL), is comprised of the processes of three types of retinal neurons, namely bipolar, ganglion, and amacrine cells. Therefore, the thickness of the IPL is a direct measure of process outgrowth in these cells. Finally, the retina of birds offers the additional advantage of being poorly vascularized, which prevents any rapid metabolization of intraocularly injected drugs or, in our case, oligonucleotides.

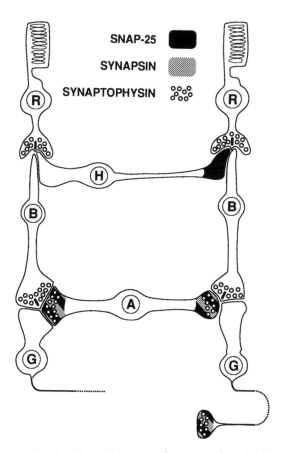

Fig. 8. Diagram showing the major synaptic connections in the retina and the differential expression of SNAP-25, synapsin, and synaptophysin. Amacrine and ganglion cells display conventional synapses, whereas photoreceptors and bipolar cells have ribbon synapses. Horizontal cells display nonvesicular synapses. Open circles represent synaptic vesicles. A, amacrine cells; B. bipolar cells; G, ganglion cells; H, horizontal cells; R, photoreceptor cells; FL, fiber layer; GCL, ganglion cell layer; INL, inner nuclear layer; IPL, inner plexiform layer; ONL, outer nuclear layer; OPL, outer plexiform layer. Reproduced from ref. *51* with permission.

The bulk of synapse formation in the chick embryo retina takes place between E12 and E16. Earlier work had shown that the level of both the mRNA and the protein SNAP-25 showed a dramatic upregulation in chick retinal tissue between E9 and E15 *(16,43)*. We performed a detailed immunocytochemical analysis of the normal expression of SNAP-25 protein in the chick embryo retina in order to determine which areas—and at what stages—were most likely to be affected by the antisense oligonucleotide treatment. We used cryostat

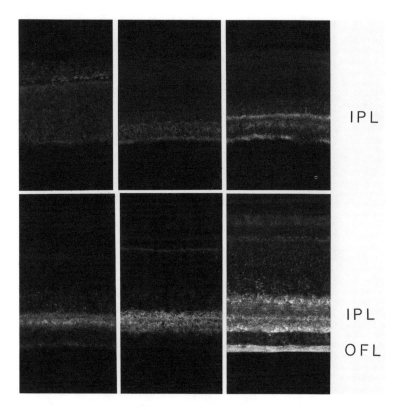

Fig. 9. Pattern of SNAP-25 expression in the developing chick retina. Twelve-micrometer cryostat sections were cut from E6, 8, 10, 12, 14, and hatchling retinas (**A–F**, respectively) and immunolabeled with SNAP-25 antibody. Note that the protein is localized in the nascent and mature IPL and OFL. IPL, inner plexiform layer; OFL, optic fibers layer. Scale bar: 20 μM.

sections from E6, 8, 10, 12, 14 and 19 chick retinas (Fig. 9). At E6, no SNAP-25 immunoreactivity was detected (not shown). At E9, the staining outlined all ganglion cell bodies and most amacrine cells, and it was already detectable in the processes of amacrine and ganglion cells. By E19, the labeling was very strong in both layers of fibers and was similar in intensity to that seen in adults (Fig. 9).

4.2. Injections of Antisense Oligonucleotides and Uptake In Vivo

All the chick embryos used for this study were of the strain White-Leghorn. Fertile eggs were kept in an incubator at 38°C and 60% relative humidity. On sacrifice, the stage of embryonic development was determined according to anatomic criteria described by Hamburger and Hamilton *(44)*.

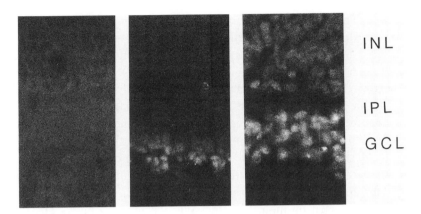

Fig. 10. FITC-PON uptake by retinal neurons in vivo. Twelve-micrometer cryostat sections were cut from E11 retinas and viewed with fluorescence optics, 8 h **(middle)** or 24 h **(right)** after injection of FITC-PONS **(left)**. Control retina, 48 h following injection of FITC alone.

In order to determine the effectiveness of PONs in vivo and the ensuing differences in retinal development, we administered a unilateral intraocular injection of either sense (533–552) or antisense (533–552, SNAP2AS) PONs at E9 and E11. The volume of the vitreous body was estimated for these stages, and the final PON concentration in the eye after each injection ranged from 1 to 10 μM, doses that proved efficient in blocking SNAP-25 expression in vitro (*see* Sections 3.1.5. and 3.2.1.). The PONs were resuspended in physiological saline at a concentration allowing injections of approx 5 μL.

To monitor the fate of the injected PONs after various survival times, fluorescently labeled (FITC-conjugated) PONs were used, at the same concentration as above. FITC-labeled SNAP2AS PON was injected into one eye of chicks at E9 in the same manner as described below. Controls included injection of unconjugated FITC and uninjected eyes. Animals were returned to the incubator, *in ovo*, until sacrifice after 8, 24, and 48 h of treatment. The eyes were removed after decapitation of the animal, rinsed briefly in phosphate-buffered saline, and then frozen in Tissue Tek OCT Compound (Miles, Inc., Elkhart, IN) on dry ice. Twelve-micron sections were collected on poly-L-lysine-coated slides (Sigma, St. Louis, MO) and viewed on a Zeiss Axiophot with fluorescein optics. After 8 h of exposure to FITC-oligonucleotides, ganglion cell bodies were clearly labeled, although their processes were not. After 24 h, cells of the INL had also accumulated fluorescent PONs (Fig. 10). The fluorescence was present in cell bodies throughout the retina 48 h after injection, although neurites in plexiform layers did not appear to contain much PONs at any time. This is consistent with the finding that, in vitro, oligonucleotides rapidly translocate to the nucleus once they have entered a cell *(45,46)*. Retinas that had been injected with unconjugated FITC showed no fluorescence above background levels.

To perform the injections, eggs were taken out of the incubator and candled in order to locate the air chamber. A hole was drilled in the eggshell on top of the air chamber with a fine forceps, and a small window was opened in the shell. The shell membrane was removed, and a small slit was cut in the underlying vascularized chorioallantoid membrane (taking care to avoid any major blood vessel) to allow access to the embryo. The embryo's head was gently raised with an L-shaped plastic tool, and the needle of a 10-µL Hamilton syringe was inserted through the sclera into the vitreous body. The solution containing the PONs was slowly delivered, and the needle was delicately pulled out. The opening was then resealed with adhesive tape, and the eggs returned to the incubator. At E12 or E13 the embryos were sacrificed by decapitation, and the eyes fixed in 4% paraformaldehyde and processed for immunocytochemistry as described below. Controls comprised sense PON-injected eyes as well as the contralateral noninjected eye.

4.3. Effects of Blockade of SNAP-25 Expression In Vivo

4.3.1. Processing the Retinas for Morphological Analysis

The eyes were dissected out and postfixed for 3 h with 4% paraformaldehyde in PBS, at 4°C. They were then cryoprotected in graded series of sucrose solutions. Twenty-micrometer cryostat sections were mounted on poly-L-Lysine-coated slides (Sigma), with sections from both eyes of each embryo mounted on the same slide. Serial sections were stained with the appropriate antibody, and immunoreactivity was viewed with the avidin–biotin peroxidase method (Vector, Peterborough, UK) using DAB (Sigma) as a substrate for HRP. Controls were performed where the primary antibody was omitted.

SNAP-25 immunodetection was performed with a rabbit polyclonal antibody directed against a synthetic peptide representing the 12 carboxy terminal residues of mouse SNAP-25 *(13)*; the antibody was affinity-purified to the immunizing peptide. All other antibodies, directed against synaptophysin, tubulin (Boehringer Mannheim Biochemica, Mannheim, Germany), and calbindin (Swiss Antibodies, Swant, Bellinzona, Switzerland), were used at the recommended dilutions.

The effects of the injections on the thickness of the retinal layers and on the intensity of immunoreactivity with the antibodies used were measured with computer-assisted image analysis (Vidas, Kontron). The thickness of the retinal layers was measured on tubulin-reacted sections viewed with Nomarski optics. The intensity of immunoreactivity in the IPL was assessed as the average of the mean gray value in each of three strips 60-μM wide for each retina.

Table 1
Chick Retina Analysis After SNAP-25 Antisense PON Treatment[a]

Embryo	PON	Densitometric analysis			Thickness of IPL, %	Length of processes (calbindin), %
		SNAP-25, %	Tubulin, %	Synaptophysin, %		
I. 1	Sense 533–552	+9	−8	+12	−4	−5
II. 5	Sense 533–552	−15	−2	−8	−10	+2
II. 9	Sense 533–552	−10	−3	+6	+13	+6
I. 2	Antisense 533–552	−67	+2	−37	−87	−73
II. 3	Antisense 533–552	−47	−2	−35	−70	−61
II. 7	Antisense 533–552	−77	−3	−33	−69	−63
III. 1	Antisense 533–552	−71	−2	−29	−52	−48
III. 2	Antisense 533–552	−72	+10	−56	−57	−42

[a]Chick embryos were given an intraocular injection of sense or antisense PON, complementary to positions 533–552 of SNAP-25 sequence, on E9 and E11 and then sacrificed at E12 or E13. The sectioned retinas were stained with the appropriate antibody, and then used either for densitometric analysis of the IPL (where SNAP-25 immunoreactivity is predominant) after SNAP-25, tubulin, or synaptophysin staining, or measured for the thickness of the IPL or length of calbindin-positive processes *(see text)*. The results are expressed as percentage inhibition (−) or increase (+) relative to similar measurements in the contralateral, control retina. The retinas contralateral to sense or antisense PON injected retinas were shown not to be different from control, uninjected retinas (not shown). IPL, inner plexiform layer; SNAP-25, synaptosomal-associated protein 25.

All measurements were made in corresponding areas of the central retina in the experimental and control eyes.

4.3.2. Effects of Antisense Oligonucleotides to SNAP-25 on Neurite Outgrowth In Vivo

Mounting the sections from both retinas of the same animal on the same slide allowed us direct comparison between treated and nontreated retinas. Serial cryostat sections were stained with antibodies against SNAP-25, tubulin, calbindin, or synaptophysin, and the following five parameters were assessed. Table 1 summarizes these results.

1. In the antisense PON-injected retinas ($n = 5$), the intensity of SNAP-25 immunoreactivity in the IPL (where SNAP-25 is predominantly expressed; *see* refs. *16*

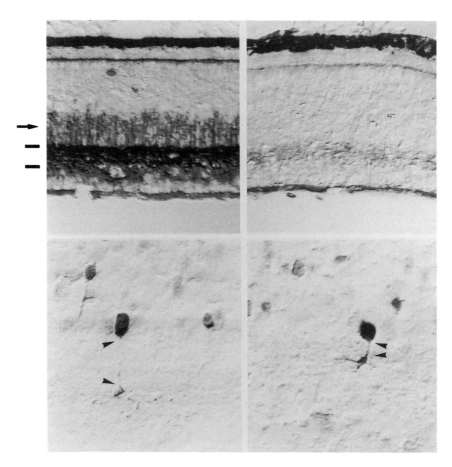

Fig. 11. Downregulation of SNAP-25 expression after antisense PON injection in vivo, **(A, B)** and effects on the morphology of identified neurons **(C, D)**. (A, B) Transverse sections of the retinas of one embryo that received two unilateral intraocular injections of 10 μ*M* antisense PON (oligo 533–552) at E9 and E11 and was fixed at E13. SNAP-25 immunostaining of the control (A) and antisense-injected (B) retina. Note the decrease in immunostaining in the IPL (bars) and in the cell bodies of the amacrine cells (arrow); the thickness of the IPL is also clearly reduced after antisense PON treatment. (C, D) The cell bodies of a subtype of calbindin-positive amacrine cells are located at the inner border of the INL. Note the difference in the length of the main process (arrowheads) that crosses the IPL. Scale bar: 50 μ*M*. Reproduced from ref. *11* with permission.

and *47*), as measured by densitometric analysis, was reduced by 49–77% when compared to the corresponding contralateral retina mounted on the same slide (Fig. 11 and Table 1). This suggests that these PONs were efficient in blocking SNAP-25 expression in vivo.

2. Inhibition of SNAP-25 expression did not correlate with any significant modification in the intensity of tubulin immunoreactivity in the IPL, since the latter ranged from a 2% decrease to a 10% increase when compared to the noninjected retinas (Table 1). However, it resulted in a decrease in the intensity of synaptophysin immunoreactivity (29–56%), suggesting that a lower number of differentiated synapses are present in these retinas.
3. In these same retinas, the thickness of the IPL was reduced after antisense PON treatment in each case analyzed, the decrease reaching 52–87% (Fig. 11, Table 1).
4. Labeling of chick retinas with an antibody directed against the calcium-binding protein calbindin resulted in the staining of a discrete number of neurons. A specific population was particularly heavily stained and easy to identify. Their cell body is located in the innermost part of the INL, and they send a single process that crosses the entire IPL, forms a bulge, and then branches (Fig. 11). This characteristic pattern of staining allowed us to assess the effects of inhibiting SNAP-25 expression on the morphology of single, identified neurons. Treatment with antisense PONs did not alter the staining, number, and distribution of these neurons, but their processes were significantly shorter (Fig. 11). The distance between the cell body and the branching point was measured in an average of 30 cells/retina, and there was a mean reduction of 42–73% (Table 1). The length of the terminal branches was not measured, but seemed also shorter.
5. We also measured the intensity of synaptophysin immunoreactivity as an indicator of the stage of differentiation of the terminals *(see above)*, and found a reduction ranging from 29–56% in the remaining IPL of the antisense PON-injected retinas (Table 1).

The retinas injected with the sense PON ($n = 3$) were similar to the control noninjected retinas for all the parameters tested *(see* Table 1). These results taken together suggest a specific positive correlation between SNAP-25 expression and axonal elongation.

5. Discussion
5.1. Antisense Oligonucleotides as a Laboratory Tool

The main result of the work presented here is that inhibition of SNAP-25 expression prevents neurite elongation in vitro and in vivo. However, we have also learned important features of the use of antisense oligonucleotides.

The experiments in vitro have allowed us to evaluate the effectiveness, specificity, and toxicity of various types of antisense oligonucleotides. Phosphorothioate-substituted oligos throughout the entire length of the molecule were preferred because they were the only ones that blocked efficiently SNAP-25 expression at concentrations that were not toxic *(see also* ref. *47)*. Presumably this is because of the combined resistance to exo- and endonucleases *(29–31)* and to their ability to activate RNaseH when they form the DNA–mRNA duplex *(32)*. Finally, and most importantly, our multisteps purification procedure *(see* Section 3.1.3.) solved all the problems of toxicity in vitro.

Our data are also consistent with the importance of aiming at relatively high melting temperatures (T_m) for the hybrid when designing oligonucleotides. It has been shown that the T_m for PONs is lower than that of the corresponding phosphodiester oligos *(30,47)*. The T_m being influenced by the length of the oligos, we chose 18–20-mer PONs; also, we kept the guanine-cytidine content between 45 and 60%, since more than 60% GC seems not to increase the T_m significantly but compromises specificity *(47)*, whereas <45% would decrease the T_m to levels not compatible with hybridization in these experimental conditions.

Targeting mRNA sites in proximity to the translation initiation site (ATG) has proven succesful in this and other studies *(21,22)* (*see also* Fig. 12). Also, mRNA regions characterized by a potential sharp change in orientation and a theoretical secondary structure associated with a loop have been used often (including in our study; *see* Fig. 12). Although this configuration implies that the primary sequence of the mRNA is already involved in base-pairing interactions, such oligos have already proven more successful than other sequences targeted at predicted single-stranded regions *(23)*.

However, these results should be considered with caution if one wants to draw conclusions about general rules for oligonucleotide antisense strategy. Successful knockouts (including in this study) have been reported with antisense oligonucleotides targeted to sequences quite far from the translation site *(23,24)*, and with other mRNAs we have experienced lack of effects even when selecting the sequence close to the translation site and in regions located in proximity to loop stems (Ayala and Merlo Pich, unpublished results). Thus, it is not possible to conclude that the criteria used in the present investigation to select the oligos are general rules for all the experiments of transient gene-product knockout with antisense oligonucleotides. Sharing the opinion of other investigators *(20)*, we believe that, so far, the information available is not sufficient to provide a reliable strategy for antisense oligonucleotide design that would work in any situation. However, we can now walk along the SNAP-25 mRNA in search of the golden rule that relates sequence to structure to success.

5.2. The Function of SNAP-25

During early stages of the work described here, we showed that SNAP-25 is strongly upregulated during synaptogenesis *(16)* and that presynaptic proteins, including SNAP-25, are differentially expressed in specific synaptic terminals *(48, see also* ref. *49)*. These observations raised the possibility that the synthesis of SNAP-25 may be triggered by specific interactions between an afferent neuron and its target cell. More recently, we have shown that the onset of SNAP-25 expression precedes synapse formation and that the levels of the protein are probably regulated by electrical activity *(43,50,* and this study). These data suggest that SNAP-25 may be associated with the electrical

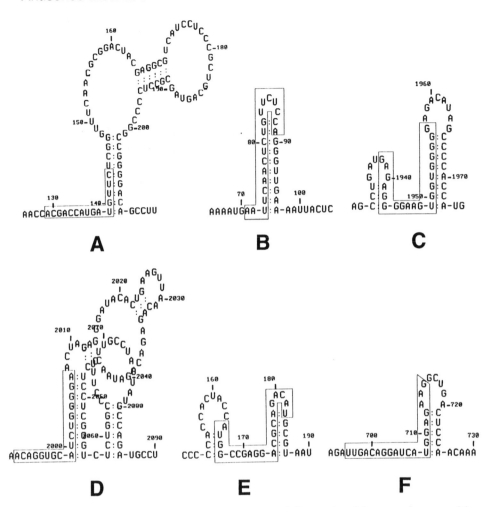

Fig. 12. Examples of mRNA regions successfully targeted in experiments with antisense oligonucleotides. The PON target-region is represented as a box along the mRNA sequence. **(A)** c-*fos* in rat brain *(58)*; **(B)** NPY-Y1 receptors in rat brain *(22)*; **(C)** ICAM-1 in human endothelial cells *(23)*; **(D)** E-selectin in human endothelial cells *(24)*; **(E, F)** SNAP-25 in PC12, hippocampal rat neurons, and chick retina *(11)*.

activity-dependent transition between the growth cone and the synapse, a prerequisite for the formation of functional and stable connections *(see* ref. *51)*. This hypothesis is also consistent with the predominant localization of SNAP-25 in the distal segment of growing axons and, in particular, in the growth cones.

The qualitative survey of the primary cultures described here suggested that the treatment with SNAP-25 antisense PON resulted in a strong reduction of the network of thin, axon-like neurites. This was quantitatively confirmed when SNAP-25 expression was inhibited by antisense PON treatment in PC12 cells induced to differentiate by NGF. These results show that SNAP-25 has a major role in neurite elongation, but is apparently not involved in initial outgrowth. A member of the family of growth-associated proteins, GAP-43, has been shown to be expressed since the earliest stages of neurite elongation in primary neuron cultures, and to be subsequently restricted to the growing axon-like process *(52)*. This observation led the authors to propose that GAP-43 could thus be implicated in early neurite outgrowth *(see* ref. *53)*. The balanced expression of proteins like SNAP-25 and GAP-43 may reflect, or dictate, the level of differentiation and the capacity to form stable synapses of the growing axons.

The results that we obtained in vitro prompted us to test the function of SNAP-25 in vivo. Following SNAP-25 antisense PON injections, the level of expression of the protein was specifically reduced in the IPL, indicating that these PONs were effective in vivo. This decrease in SNAP-25 immunoreactivity was concomitant with a reduction in the thickness of this layer of fibers, possibly owing to a reduction in the length of the individual fibers. This interpretation was further supported by the observation that the antisense PON treatment did reduce the length of the IPL-spanning process of a subpopulation of amacrine cells (calbindin-positive). Thus, our results in vivo are in agreement with those obtained in vitro, and demonstrate that SNAP-25 plays an essential role in axonal elongation during development and in final stages of differentiation of the synapses.

Although these results indicate that SNAP-25 is involved in axonal elongation, they do not address the question of its mechanisms of action. Recent experiments by Rothman and collaborators have shed light on this issue by identifying SNAP-25 as one member of a key protein complex involved in the fusion of synaptic vesicles to their target membrane in adult neurons *(54)*. In view of the relatively early expression of SNAP-25 in our model systems, the most straightforward interpretation of our results is that SNAP-25 is essential for the fusion of vesicles supplementing membrane components to the growing axon. Söllner et al. *(54)* suggested that, in the adult, SNAP-25 may be involved in regulated exocytosis *(see also* ref. *55)*. Functional evidence that SNAP-25 plays an essential role in transmitter release came from the demonstration that botulinum neurotoxin A blocks neurotransmitter release through specific proteolytic cleavage of the protein *(see* ref. *56* for review). We also showed that cleavage of SNAP-25 by the toxin in developing neurons inhibited axonal growth, a result clearly consistent with the data obtained using antisense oligonucleotides *(57)*.

In the adult, SNAP-25 could contribute to axonal terminal plasticity in much the same way as it contributes to axonal elongation during development. This hypothesis is consistent with the observation that SNAP-25 is highly expressed in regions of the adult brain related to cognitive processes *(13)* for which synaptic plasticity could be a morphological substrate *(see* ref. *51)*. In the current state of our knowledge, both roles (elongation and release) are possible for SNAP-25, and they are not mutually exclusive. Recent studies have shown that SNAP-25 has two isoforms that are generated by differential splicing using either of two homologous exons that encode the cystein-rich domain responsible for fatty-acylation *(58)*. These two isoforms are differentially regulated during development, SNAP-25a being expressed early on, whereas SNAP-25b expression begins with synaptogenesis and is the predominant form in the adult *(58,59)*. SNAP-25a and SNAP-25b might serve related, but slightly different roles in development and adulthood, and it would be interesting to investigate this issue by specifically blocking the expression of either one of them. Here, isoform-specific antisense PONs would be an ideal tool and probably one of the few possible experimental approaches.

References

1. Corder, E. H., Saunders, A. M., Strittmatter, W. J., Schmechel, D. E., Gaskell, P. C., Small, G. W., Roses, A. D., Haines, J. L., and Pericak-Vance, M. A. (1993) Gene dose of Apolipoprotein E type 4 allele and the risk of Alzheimer's disease in late onset families. *Science* **261,** 921–923.
2. Miller, F. D., Naus, C. C. G., Higgins, G. A., Bloom, F. E., and Milner, R. J. (1987) Developmentally regulated rat brain mRNAs: molecular and anatomical characterization *J. Neurosci.* **7,** 2433–2444.
3. Liang, P. and Pardee, A. B. (1992) Differential display of eukaryotic messenger RNA by means of the polymerase chain reaction. *Science* **257,** 967–971.
4. Catsicas, S., Thanos, S., and Clarke, P. G. H. (1987) Major role for neuronal death during development: refinement of topographical connections. *Proc. Natl. Acad. Sci. USA* **84,** 8165–8168.
5. Karten, H. J. and Brecha, N. (1983) Localization of neuroactive substances in the vertebrate retina: evidence for lamination of the inner plexiform layer. *Vision Res.* **23,** 1197–1205.
6. Brecha, N., Eldred, W., Kuljis, R. O., and Karten, H. J. (1984) Identification and localization of biologically active peptides in the vertebrate retina, in *Progress in Retinal Research,* vol. 3 (Osborne, N. N. and Chader, G. Y., eds), Pergamon, Oxford, pp. 185–226.
7. Rager, G. (1976) Morphogenesis and physiogenesis of the retino-tectal connection in the chicken. I. The retinal ganglion cells and their axons. *Proc. R. Soc. Lond. Ser. B* **192,** 331–352.
8. Daniels, M. P. and Vogel, Z. (1980) Localization of alpha-bungarotoxin binding sites in synapses of the developing chick retina. *Brain Res.* **201,** 45–56.
9. Lebeau, M.-C., Alvarez-Bolado, G., Wahli, W., and Catsicas, S. (1991) PCR-driven DNA-DNA competitive hybridization: a new method for sensitive differential cloning. *Nucleic Acids Res.* **19,** 4778.

10. Alvarez-Bolado, G., Lebeau, M.-C., Braissant, O., Wahli, W., and Catsicas, S. (1991) Electrical activity regulates the expression of synapse-specific genes during neuronal development. *Soc. Neurosci. Abstr.* **17,** 215.

11. Osen-Sand, A., Catsicas, M., Staple, J. K., Jones, K. A., Ayala, G., Knowles, J., Grenningloh, G., and Catsicas, S. (1993) Inhibition of axonal growth by SNAP-25 antisense oligonucleotides in vitro and in vivo. *Nature* **364,** 445–448.

12. Branks, P. L. and Wilson, M. C. (1986) Patterns of gene expression in the murine brain revealed by in situ hybridization of brain-specific mRNAs. *Mol. Brain Res.* **1,** 1–16.

13. Oyler, G. A., Higgins, G. A., Hart, R. A., Battenberg, E., Billingstey, M. L., Bloom, F. E., and Wilson, M. C. (1989) The identification of a novel synaptosomal associated protein, SNAP-25, differentially expressed by neuronal populations. *J. Cell. Biol.* **109,** 3039–3052.

14. Geddes, J. W., Hess, E. J., Hart, R. A., Kesslak, J. P., Cotman, C. W., and Wilson, M. C. (1990) Lesions of hippocampal circuitry define synaptosomal-associated protein 25 (SNAP-25) as a novel presynaptic marker. *Neuroscience* **38,** 515–528.

15. Hess, E. J., Slater, T. M., Wilson, M. C., and Pate-Skene, J. H. (1992) The 25 kDa synaptosomal-associated protein SNAP-25 is the major methionine-rich polypeptide in rapid axonal transport and a major substrate for palmitoylation in adult CNS. *J. Neurosci.* **12,** 4634–4641.

16. Catsicas, S., Larhammar, D., Blomqvist, A., Sanna, P.-P., Milner, R. J., and Wilson, M. C. (1991) Expression of a conserved cell-type specific protein in nerve terminals coincides with synaptogenesis. *Proc. Natl. Acad. Sci. USA* **88,** 785–789.

17. Risinger, C., Blomqvist, A. G., Lundell, I., Lambertsson, A., Nassel, D., Pieribone, V. A., Brodin, L., and Larhammar, D. (1993) Evolutionary conservation of synaptosome-associated protein 25 kDa (SNAP-25) shown by Drosophila and Torpedo cDNA clones. *J. Biol. Chem.* **268,** 24,408–24,414.

18. Sanna, P.-P., Bloom, F. E., and Wilson, M. C. (1991) Dibutyryl-cAMP induces SNAP-25 translocations into the neurites in PC12. *Dev. Brain Res.* **59,** 104–108.

19. Yaswen, P., Stampfer, M., Ghosh, K., and Cohen, J. (1993) Effects of sequence of thioated oligonucleotides on cultured human mammary epithelial cells. *Antisense Res. Dev.* **3,** 67–77.

20. Stein, C. A. and Cheng, Y.-C. (1993) Antisense oligonucleotides as therapeutic agents—Is the bullet really magical? *Science* **261,** 1004–1012.

21. Heilig, M., Engel, J. A., and Soderplam, B. (1993) c-fos antisense in the nucleus acumbens blocks the locomotor stimulant action of cocaine. *Eur. J. Pharmacol.* **236,** 339–340.

22. Wahlestadt, C., Merlo Pich, E., Koob, G. F., Yee, F., and Heilig, M. (1993) Modulation of anxiety and neuropeptide Y-Y$_1$ receptors by antisense oligodeoxynucleotides. *Science* **259,** 528–531.

23. Chiang, M. Y., Chan, H., Zounes, M. A., Freier, S. M., Lima, W. F., and Bennett, C. F. (1991) Antisense oligonucleotides inhibit intercellular-adhesion molecule 1 expression by two distinct mechanisms. *J. Biol. Chem.* **266,** 18,162–18,171.

24. Bennett, C. F., Condon, T. P., Grimm, S., Chan, H., and Chiang, M. Y. (1994) Inhibition of endothelial cell adhesion molecule expression with antisense oligonucleotides. *J. Immunol.* **11,152,** 3530–3540.

25. Baughman, R. W., Huettner, J. E., Jones, K. A., and Khan A. A. (1991) Cell culture of neocortex and basal forebrain from postnatal rats, in *Culturing Nerve Cells* (Banker, G. and Goslin, K., eds.), MIT Press, Cambridge, MA, pp. 227–249.

26. Caceres, A., Potrebic, S., and Kosik, K. S. (1991) The effect of tau antisense oligonucleotides on neurite formation of cultured cerebellar macroneurons. *J. Neurosci.* **11**, 1515–1523.

27. Caceres, A., Mautino, J., and Kosik, K. S. (1992) Suppression of MAP2 in cultured cerebellar macroneurons inhibits minor neurite formation. *Neuron* **9**, 607–618.

28. Ferreira, A., Kosik, K. S., Greengard, P., and Han, H. Q. (1994) Aberrant neurites and synaptic vesicle protein deficiency in synapsin II-depleted neurons. *Science* **264**, 977–979.

29. Bayever, E., Iversen, P., Smith, L., Spinolo, J., and Zon, G. (1992) Systemic human antisense therapy begins. *Antisense Res. Devel.* **2**, 109–110.

30. Stein, C. A., Subasinghe, C., Shinozuka, K., and Cohen, J. S. (1988) Physicochemical properties of phosphorothioate oligodeoxynucleotides. *Nucleic Acids Res.* **16**, 3209.

31. Crooke, S. T. (1992) Therapeutic applications of oligonucleotides. *Ann. Rev. Pharmacol. Toxicol.* **32**, 329–376.

32. Gao, W.-Y., Han, F. S., Storm, C., Egan, W., and Cheng, Y. C. (1992) Phosphorothioate oligonucleotides are inhibitors of human DNA polymerases and RNAse H—implications for antisense technology. *Mol. Pharmacol.* **41**, 223–229.

33. Cazenave, C., Stein, C. A., Loreau, N., Thuong, N. T., Neckers, L. M., Subashinge, C., Helene, C., Cohen, J. S., and Toulme, J. J. (1989) Comparative inhibition of rabbit globin mRNA translation by modified antisense oligodeoxynuclotides. *Nucleic Acids Res.* **17**, 4255–4273.

34. Vu, H. and Hischbein, B. (1991) Internucleotide phosphite sulfurization with tetraethylthiuram disulfide phosphorothioate oligonucleotide synthesis via phosphoramidite chemistry. *Tetrahedron Lett.* **32**, 3005–3008.

35. Sawadogo, M. and Van Dyke, M. W. (1991) A rapid method for the purification of deprotected oligodeoxynucleotides. *Nucleic Acids Res.* **19**, 674.

36. Sudhof, T. C., Lottspeich, F., Greengard, P., Mehl, E., and Jahn, R. (1987) A synaptic vesicle protein with a new cytoplasmic domain and four transmembrane regions. *Science* **238**, 1142–1144.

37. Greene, L. A. and Tischler, A. S. (1976) Establishment of a noradrenergic clonal line of rat adrenal pheochromocytoma cells which respond to nerve growth factor. *Proc. Natl. Acad. Sci. USA* **73**, 2424–2428.

38. Tischler, A. S. and Greene, L. A. (1978) Morphologic and cytochemical properties of a clonal line of rat adrenal pheochromocytoma cells which respond to nerve growth factor. *Lab. Invest.* **39**, 77–89.

39. Jap Tjoen San, E. R. A., ScmidtMichels, M. H., Spruijt, B. M., Oestreicher, A. B., Schotman, P., and Gispen, W. H. (1991) Quantitation of the growth-associated protein B-50/GAP-43 and neurite outgrowth in PC12 cells. *J. Neurosci. Res.* **29**, 149–154.

40. Burry, R. W. and PerroneBizzozero, N. I. (1993) Nerve growth factor stimulates GAP-43 expression in PC12 cell clone independently of neurite outgrowth. *J. Neurosci. Res.* **36**, 241–251.

41. Gunning, P. W., Landreth, G. E., Bothwell, M. A., and Shooter, E. M. (1981) Differential and synergistic actions of nerve growth factor and cyclic AMP in PC12 cells. *J. Cell Biol.* **89,** 240–245.

42. Rodieck, R. W. (1973) *The Vertebrate Retina: Principles of Structure and Function.* W. H. Freeman, San Francisco.

43. Alvarez-Bolado, G., Lebeau, M.-C., Braissant, O., Wahli, W., and Catsicas, S. (1991) Electrical activity regulates the expression of synapse-specific genes during neuronal development. *Soc. Neurosci. Abstr.* **17,** 215.

44. Hamburger, V. and Hamilton, H. L. (1951) A series of normal stages in the development of the chick embryo. *J. Morphol.* **88,** 49–92.

45. Leonetti, J. P., Mechti, N., Degols, G., Gagnor, C., and Lebleu, B. (1991) Intracellular distribution of microinjected antisense oligonucleotides. *Proc. Natl. Acad. Sci. USA* **88,** 2702–2706.

46. Zamecnik, P., Aghajanian, J., Zamecnik, M., Goodchild, J., and Witman, G. (1994) Electron micrographic studies of transport of oligodeoxynucleotides across eukaryotic cell membranes. *Proc. Natl. Acad. Sci. USA* **91,** 3156–3160.

47. Reed, J. C., Stein, C., Subashinge, S., Haldar, S., Croce, C. M., Yum, S., and Cohen, J. (1990) Antisense-mediated inhibition of bcl-2 protooncogene expression and leukemic cell growth and survival—comparisons of phosphodiester and phosphorothioate oligodeoxynucleotides. *Cancer Res.* **50,** 6565–6570.

48. Catsicas, S., Catsicas, M., Keyser, K. T., Karten, H. J., Wilson, M. C., and Milner, R. J. (1992) Differential expression of the presynaptic protein SNAP-25 in mammalian retina. *J. Neurosci. Res.* **33,** 1–9.

49. Mandell, J. W., Townes-Anderson, E., Czernik, A. J., Cameron, R., Greengard, P., and DeCamilli, P. (1990) Synapsins in the vertebrate retina: absence from ribbon synapses and heterogeneous distribution among conventional synapses. *Neuron* **5,** 19–33.

50. Braissant, O., Wilson, M. C., Wahli, W., and Catsicas, S. (1990) Expression of presynaptic and cytoskeletal proteins following early target removal in the chick embryo. *Soc. Neurosci. Abstr.* **16,** 42.

51. Catsicas, S., Grenningloh, G., and Merlo Pich, E. (1994) Nerve-terminal proteins: to fuse to learn. *Trends Neurosci.* **17,** 368–373.

52. Goslin, K., Schreyer, D. J., Pate Skene, J. H., and Banker, G. (1990) Changes in the distribution of GAP-43 during the development of neuronal polarity. *J. Neurosci.* **10,** 588–602.

53. Jap Tjoen San, E. R. A., ScmidyMichels, M., Oestreicher, A. B., Gispen, W. H., and Schotman, P. (1992) Inhibition of nerve growth factor-induced B-50/GAP-43 expression by antisense oligomers interferes with neurite outgrowth of PC12 cells. *Biochem. Biophys. Res. Commun.* **187,** 839–846.

54. Söllner, T., Whiteheart, S. W., Brunner, M., Erdjument-Bromage, H., Geromanos, S., Tempst, P., and Rothman, J. E. (1993) SNAP receptors implicated in vesicle targeting and fusion. *Nature* **362,** 318–324.

55. De Camilli, P. (1993) Exocytosis goes with a SNAP. *Nature* **364,** 387–388.

56. Montecucco, C. and Schiavo, G. (1994) Mechanism of action of tetanus and botulinum neurotoxins. *Mol. Microbiol.* **13**, 1–8.

57. Osen-Sand, A., Staple, J. K., Naldi, E., Schiavo, G., Grenningloh, G., Malgaroli, A., Montecucco, C., and Catsicas, S. (1995) Common and distinct SNARES for axonal growth and transmitter release. *Soc. Neurosci. Abst.* **21**, 327.

58. Bark, I. C. (1993) Structure of the chicken gene for SNAP-25 reveals duplicated exons encoding distinct isoforms of the protein. *J. Mol. Biol.* **233**, 67–76.

59. Bark, I. C. and Wilson, M. C. (1994) Regulated vesicular fusion in neurons: snapping together the details. *Proc. Natl. Acad. Sci. USA* **91**, 4621–4624.

60. Chiasson, B. J., Hooper, M. L., Murphy, P. R., and Robertson, H. A. (1992) Antisense oligonucleotides eliminates in vivo expression of c-fos in mammalian brain. *Eur. J. Pharmacol.* **227**, 451–453.

5

Antisense Tumor Therapy

Activated C-Ha-ras Oncogene in the Mouse

Eric Wickstrom

1. Introduction

1.1. The ras Oncogene Family

Cancerous cells display overexpression or mutant expression of one or more of the genes normally used in cell proliferation. Such genes are called proto-oncogenes *(1)*. The implication is that the targets that must be attacked in neoplastic cells are normal cellular genes that have sustained some activating lesion. The *ras* family of mammalian proto-oncogenes includes three members, termed Ha-*ras*, Ki-*ras*, and N-*ras*, that are found to be activated very often in human solid tumors and leukemias *(2)*.

The human c-Ha-*ras* proto-oncogene codes for an evolutionarily conserved GTP binding protein, p21, which is associated with the inner surface of the plasma membrane owing to farnesyl and palmitoyl adducts, and is involved in a cascade transducing signals from growth factors binding to cell-surface receptors, which ultimately turns on expression of proliferative genes *(3)*.

A broad range of eukaryotes carry the *ras* gene family, whose gene products are all immunologically related and code for proteins with 188–189 amino acid residues, with molecular weights of about 21 kDa. Many varieties of *ras* have been found, differing the most at their carboxy termini. Point mutations in *ras* oncogenes, which alter the enzymatic properties and cause an inappropriate response of the mutant c-Ha-*ras* protein in the pathway for activation of cell proliferation, may be causative or closely linked to the onset of malignant transformation *(2)*.

The first such gene to be characterized was a c-Ha-*ras* oncogene from a human bladder cancer, mutated in the 12th codon from GGC to GTC, coding for valine at that position instead of glycine *(4)*. This oncogene provided a logical candidate for antisense inhibition.

From: *Methods in Molecular Medicine: Antisense Therapeutics*
Edited by: S. Agrawal Humana Press Inc., Totowa, NJ

1.2. Antisense DNA Inhibitors of Gene Expression

All oncogenes and most of their protein products operate inside the cell, and they are therefore not accessible to protein-based drugs whose specificity derives from antibody–ligand-related recognition. The ability of antisense DNA to turn off individual genes at will in growing cells provides a powerful tool for therapeutic intervention. In principle, one needs to identify a unique target sequence in the gene of interest and prepare a complementary oligo-nucleotide against the target sequence *(5)*. Antisense DNAs were first success-fully utilized against Rous sarcoma virus *(6)*. The efficacy of antisense DNA inhibition has been demonstrated in cell culture against a wide variety of genes *(7–9)*, and a broad array of derivatives have been prepared in an effort to increase potency *(10)*.

In the case of activated human c-Ha-*ras* oncogene *(11)*, translation of c-Ha-*ras* p21 mRNA in rabbit reticulocyte lysate was specifically inhibited by antisense DNA methylphosphonates at 50–200 µ*M (12)*. The oligomers were comple-mentary to eight nucleotides spanning the frequently mutated 12th codon of c-Ha-*ras* mRNA. Similarly, translation of murine Balb-*ras* p21 mRNA in a rabbit reticulocyte lysate and expression of normal c-Ha-*ras* p21 in RS485 cells were both specifically inhibited by DNA methylphosphonates comple-mentary to the first 11 nucleotides of the c-Ha-*ras* mRNA coding region *(13)*. In a direct comparison of the antisense efficacy of normal DNAs with methylphosphonates, alternating methylphosphonate-phosphodiester, and phosphorothioates, it was observed that a phosphodiester antisense DNA directed against 11 nucleotides of the initiation codon region inhibited p21 mRNA translation in rabbit reticulocyte lysate one-third as well as a methyl-phosphonate of the same sequence and 1/12 as well as a phosphorothioate of the same sequence *(14)*.

Comparable results were observed with the 61st codon mutants in cell cul-ture *(15)*. In all four papers, inhibition of p21 expression was dose-dependent and sequence-specific (except for phosphorothioates at higher concentrations); single and double mismatches were significantly less effective. Antisense DNA inhibition of c-Ha-*ras* RNA in cell-culture studies correlated with a decrease in cell proliferation rate. Similarly, an acridine-conjugated nonamer was observed to inhibit proliferation of 12th codon mutant human T24 cells, compared with nontransformed cells *(16)*.

1.3. Message Walking, Focus Formation, and Tumorigenesis

In applying antisense DNA inhibition to an oncogene, it is of interest to identify the most sensitive targets in the messenger, and to discern whether there is any correlation between antisense effectiveness and mRNA secondary structure. In the first experiment, the effect of three antisense DNAs targeted against the human c-Ha-*ras* mRNA, and two control sequences, on expression of c-Ha-*ras* p21 and focus formation was studied in NIH3T3 cells transformed by the T24-activated c-Ha-*ras* (Val12) oncogene *(17)*. Inhibition of p21 expression was sequence-specific and dose-dependent. The sequence complementary to a site within an intron of the 5' flanking region was the most effective, followed by the initiation AUG codon region, whereas a target 60 nucleotides upstream of the initiation codon was least effective. Inhibition of focus formation was studied with the antisense oligomer specific for the initiation codon, and one control; the specific oligomer was much more inhibitory than the control.

It remained to be determined, however, whether antisense DNA treatment can have any sustained effect on tumor growth in vivo. In an attempt to address this question, the second experiment examined tumor growth in athymic nude mice following implantation of transformed cells pretreated in vitro with the most potent anti-c-Ha-*ras* DNA from the first experiment. Tumor growth of cells treated with anti-c-Ha-*ras* DNA was significantly reduced for up to 14 d following the end of treatment and implantation into the mice, whereas the nonspecific control DNA had no significant effect.

2. Materials and Methods

2.1. Cell Lines

NIH3T3 mouse fibroblasts transformed by the T24 human bladder cancer c-Ha-*ras* oncogene with a G12→V12 point mutation *(18)* were employed as the experimental cell line. The original NIH3T3 mouse fibroblast line (American Type Culture Collection [Rockville, MD], #CLR 1658) was used as a nontransformed control cell line. Cells were grown in Dulbecco's Modified Eagle's Medium (DMEM) supplemented with 10% heat-inactivated calf serum, penicillin (100 U/mL), streptomycin (100 µg/mL), and 2 mM glutamine and maintained at 37°C in an incubator with 5% CO_2. The Y13-259 rat monoclonal hybridoma cell line *(19)* was maintained as described for the preceding cell lines, except that the medium was supplemented with 10% heat-inactivated fetal bovine serum rather than calf serum. Filtered media from the cells served as the source for the anti-*RAS* p21 Y13-259 antibody.

2.2. Oligodeoxynucleotide Synthesis

Five sequences were utilized in these studies: 5'-d(CAGCTGCAACCCAGC)3' (*RAS*2), 5'-d(GCCCCACCTGCCAAG)3' (*RAS*3), 5'-d(TTATATTCCGTCATC)3' (*RAS*4), 5'-d(TTGGGATAACACTTA)3' (VSVM), and 5'-d(CATTTCTTGC-TCTCC)3' (*TAT*9). The antisense DNA oligomers were synthesized on a Millipore 8750 DNA synthesizer using standard phosphoramidite chemistry *(20)*. Small-scale (1-µmol) preparations for cell-culture studies were purified by reversed-phase liquid chromatography *(21)*. Large-scale (15-μmol) preparations for animal studies were purified by *n*-butanol precipitation *(22)*, and purity was verified by high-performance liquid chromatography using a reversed-phase C_{18} column *(21)*. Following purification, the DNAs were dissolved in deionized water, sterilized by filtration through 0.2-μm filters, and frozen at −80°C for storage. DNAs were 5'-labeled with 5'-[γ-^{35}S]thioATP (New England Nuclear [Boston, MA], #NEG 27H; >1100 Ci/mmol) using bacteriophage T4 polynucleotide kinase.

2.3. Oligodeoxynucleotide Stability, Uptake, and Compartmentalization

For each time-point, 5×10^5 cpm of 5'-^{35}S-labeled DNAs were added to 2×10^6 T24-transformed NIH3T3 cells *(18)*, the kind gift of Stuart Aaronson, in 0.5 mL of DMEM (Sigma [St. Louis, MO], # D5523) with 10% heat-inactivated fetal bovine serum (FBS) (Sigma, St. Louis, MO, #F4135). Each sample was incubated at 37°C for 1, 2, 4, 8, or 24 h in 5% CO_2 and 95% air. Cells were sedimented for 3 min at 15,000*g*, and the supernatant was removed and saved (SN1). Cellular pellets were washed once in 0.5 mL of PBS (10 m*M* Na_2HPO_4, pH 7.4, 150 m*M* NaCl) and sedimented again. The supernatant was removed and saved (SN2) and cells were lysed in 0.1 mL of Tris-buffered saline (10 m*M* Tris-HCl, pH 7.4, 150 m*M* NaCl) containing 1% $NaDodSO_4$, and then extracted with 0.1 mL of phenol. The aqueous phase (AQ1) was removed and organic phase was extracted again with 0.1 mL of H_2O. The aqueous phase was removed (AQ2) from the organic phase (ORG1), and aliquots were analyzed by liquid scintillation counting. Cellular uptake (CU) was calculated using the following formula:

$$CU = cpm\ (AQ1 + AQ2)/cpm\ (AQ1 + AQ2 + SN1 + SN2 + ORG1) \times 100\% \quad (1)$$

To determine oligodeoxynucleotide stability, aliquots of the combined aqueous phases and of the culture-medium supernatant fraction were lyophilized, redissolved in 30 μL of 80% deionized formamide containing 0.01% xylene cyanole FF and 0.01% bromophenol blue, then electrophoresed in a denaturing 20% polyacrylamide gel, and autoradiographed. Stability of DNAs was also tested at different temperatures: 4, 15, 25, and 37°C. Additional experiments were conducted to test the exposure of DNAs associated with cellular membranes. Experiments were performed essentially as described above with an additional step. At the end of each incubation period, 1 U of mung bean nuclease (New England Biolabs [Beverly, MA], #250) was directly added to the cells, and mixtures were further incubated at 37°C for 30 min. To compensate for the pH difference, a 500-fold increase of the recommended enzyme concentration was used (P. Muralikrishna, unpublished results).

Experiments were performed using differential centrifugation methods *(23)*. 2.0 × 10⁷ T24-transformed NIH3T3 cells were suspended in 2.0 mL of DMEM with 10% heat-inactivated FBS and 1.0×10^6 cpm of $5'$-^{35}S-labeled d(TTATATTCCGTCATC)3' (*RAS*4). Cells were incubated at 4°C for 3 h, and then sedimented at 15,000*g* for 3 min. The supernatant was removed, and pellets were washed twice with ice-cold PBS. Pellets were then resuspended in lysis buffer (20 m*M* HEPES-KOH, pH 6.8, 5 m*M* KCl, 5 m*M* MgCl₂, 0.1 m*M* phenylmethylsulfonyl fluoride) and incubated at ambient temperature for 10 min. Cells were then disrupted with 0.5% NP-40 and 0.1% sodium deoxycholate for 20 min at 4°C. Nuclear (N), mitochondrial (MT), and cytoplasmic membrane (CM) fractions were obtained by sedimentation at 300, 4000, and 20,000*g*, respectively. The final supernatant, the cytoplasmic fraction (CT), was saved. Each pellet was resuspended in 100 µL PBS and analyzed by liquid scintillation counting.

2.4. Inhibition of p21 Translation

Aliquots of 1.0×10^5 T24-transformed NIH3T3 cells were seeded into 33-mm diameter wells (Costar [Cambridge, MA], #760270) for 24 h at 37°C. The various DNAs were added to final concentrations of 25 and 50 µ*M* each, and cells were incubated for additional 24 h at 37°C. Next, cells were metabolically labeled and immunoprecipitated *(18)*. Briefly, cells were preincubated with DMEM lacking methionine and cysteine, supplemented with 10% heat-inactivated FBS, for 30 min at 37°C. Cell-culture medium was replaced with -Met, -Cys DMEM supplemented with 150 µCi/mL ^{35}S-labeled methionine and cysteine (ICN [Costa Mesa, CA], #51006; >1100 Ci/mmol) containing 10% heat-inactivated FBS, and cells were incubated for 4 h at 37°C. Cells were then lysed, and samples containing $1–10 \times 10^6$ cpm trichloroacetic acid-precipitable cpm were immunoprecipitated with Y13-259 MAb from the culture medium of Y13-259 rat hybridoma cells *(19)*, the kind gift of Stuart Aaronson. Because many rat IgG molecules do not bind well to *Staphylococcus aureus* protein A, the fixed *S. aureus* cells were first coated with 0.2 mg/mL rabbit antirat IgG (Sigma, #R-3756) *(19)*. The final pellets were resuspended in 15–25 µL Laemmli *(24)* sample buffer, boiled for 5 min, and then sedimented at 15,000*g* for 10 min. The supernatants were electrophoresed on a denaturing 18% polyacrylamide gel *(24)* followed by fluorography. Bands were quantitated by scanning of the film using a Zeineh SLR-2D-1 D soft laser scanning densitometer.

2.5. Focus Formation Assay

Semisolid medium was prepared by adding 10 g of 4000 mPa methylcellulose (Fluka [Ronkonkoma, NY], #CH-9470) to 250 mL of sterile boiling water, mixing with 250 mL of 2X DMEM with 20% newborn calf serum (NCS)

(Sigma, #N4637), and stirring overnight at 0–4°C *(25)*. 1×10^4 T24-transformed NIH3T3 cells were resuspended in <10 μL of DMEM, and added to 3.0 mL of semisolid DMEM supplemented with 10% NCS, with no oligomer addition, or final concentrations of 50 μM *RAS*4, or VSVM. Cells were incubated at 37°C in 95% air and 5% CO_2 saturated with water, and inspected for focus formation 3–5 d later. Cells were examined microscopically, and colonies with a cluster of 10 or more cells were scored as foci.

2.6. Antisense DNA Treatment of Cells Before Implantation

Two DNA pentadecamer sequences were employed in the animal study. One sequence, termed *RAS*2, was the most efficacious anti-c-Ha-*ras* DNA in the cell-culture experiments. The second sequence, VSVM, served again as a nonspecific control. Four experimental groups were included in the in vitro DNA treatment of cells:

1. T24-T24/NIH3T3 cells, which received no antisense DNA treatment;
2. *RAS*-T24/NIH3T3 cells, which received *RAS*2;
3. VSV-T24/NIH3T3 cells, which received VSVM; and
4. 3T3-NIH3T3 control cells, which received no DNA treatment.

All cells were first plated in 25-mm^2 flasks at a concentration of 1×10^6 cells/flask. On the next three successive days, the old medium was removed, and fresh medium containing either no DNA or the appropriate antisense DNA at a concentration of 50 μM was added. On the fourth day, the cells were harvested using 0.1% trypsin in PBS, washed twice with PBS, and either placed in Hank's Balanced Salt Solution for immediate injection into nude mice or lysed for subsequent measurement of *RAS* p21 protein. A portion of cells from each flask was counted in a hemocytometer using trypan blue exclusion to identify viable cells. Cells of all four treatment groups evidenced >90% viability based on trypan blue exclusion.

2.7. Tumor Growth in Nude Mice

Tumor growth was investigated in two trials. In the first trial, 3 female Balb/c athymic nude mice (6 wk old)/treatment group were injected sc in the flank region with 5×10^5 cells. In the second trial, 7 or 8 C57BL/Balb/c male athymic nude mice (6 wk old)/treatment group were similarly injected. Tumor growth was monitored daily beginning several days after injection in a single blind protocol. Two perpendicular measurements of the diameter of any palpable nodule were obtained, and an estimated volume was calculated by the formula lw2/2. In the case of multiple nodules, the total volume was recorded. The animals were sacrificed at the end of the trial, and examined for any intrusion of tumor through the body wall or evidence of metastases to various body organs. The tumors were removed, fixed in buffered formalin for 24 h, and imbedded in paraffin. For experiment 2, three tumors were selected from each treatment group, sectioned at 5 μm, stained with hematoxylin/eosin, and examined for morphological characteristics.

2.8. Chemiluminescent Measurement of RAS p21 Protein

Evaluation of the *RAS* 21 kDa protein products of both the mutated and wild-type c-Ha-*ras* genes in cells used for animal experiments was conducted by Western blot analysis of immunoprecipitated cell lysates. Approximately 5×10^6 cells were lysed in ice cold lysis buffer (28 mM Tris-HCl, pH 7.1, 100 mM NaCl, 1 mM MgCl$_2$, 1% NP-40, 0.5% sodium deoxycholate, 1 mM dithiothreitol, 1 mM phenylmethanesulfonyl fluoride, and 6 mg% aprotinin) for 10 min, and the lysate was collected after centrifugation of particulate matter. Total protein in the lysate was measured on a Cobas Fara analyzer based on the method of Bradford *(26)*. Cell lysate containing 500 μg total protein was immunoprecipitated using rat anti-*RAS* antibody Y13-259 followed by antirat IgG agarose bead conjugate. Controls included a negative control that contained no cell lysate, but was immunoprecipitated in the same manner as the cell samples and an IgG control which contained untreated T24/NIH3T3 cell lysate but was immunoprecipitated with 1 μg of rat IgG rather than Y13-259 antibody. Following washing, the immune complex containing *RAS* p21 and anti-*RAS* antibody was eluted from the agarose conjugate, denatured using Laemmli sample buffer, and heating at 90° C for 5 min *(24)*. The resulting sample was subjected to polyacrylamide gel electrophoresis and electroblotted onto a nitrocellulose membrane *(27)*. Immunodetection of blotted *RAS* p21 protein was accomplished using anti-*RAS* Y13-259 antibody, antirat IgG horseradish peroxidase conjugate, and horseradish peroxidase amplification (Blast amplification kit, NEN Research Products, Boston, MA). Visualization was achieved by chemiluminescence (Amersham, Arlington Heights, IL) followed by X-ray film exposure, and the resulting bands were quantitated by optical scanning and analytical imaging using a Bio-Image Scanner. The results yielded a total measure of mutated and wild-type *RAS* p21.

3. Results

3.1. Prediction of Antisense Targets and Thermodynamics

From the 1177 nucleotide sequence of human c-Ha-*ras* mRNA *(28)*, a secondary structure was predicted. Three targets were selected from this structure: the first 15 nucleotides from the 5' cap, nucleotide 1455–1469 of the Reddy sequence, which were predicted to be involved in a hairpin loop; nucleotide 1609–1623, a region of weak base pairing 60 nucleotides upstream from the initiation codon, which an alternative calculation placed in a hairpin loop

(D. Konings, personal communication), the initiation codon region, nucleotide 1669–1683, predicted to occur in a large bulge. The complementary pentadecadeoxynucleotides of these targets, 5'-d(CAGCTGCAACCCAGC)3' (*RAS*2), 5'-d(GCCCCACCTGCCAAG)3' (*RAS*3), and 5'-d(TTATATTCCGTCATC)3' (*RAS*4), were synthesized. Subsequently, it became clear that the actual 5' end of the transcript occurs much farther upstream *(29)*, and that the *RAS*2 target exists within the first intron. Thus, the original hairpin calculated for that target is even less likely. Two control pentadecamers were prepared: 5'-d(TTGGGATAACACTTA)3' (VSVM), complementary to nucleotides 17–31 of VSV M-protein mRNA *(30)*, which nonspecifically inhibited translation of VSV mRNAs *(31)* but differed from the anti-c-Ha-*ras* sequences in at least 10 out of 15 residues; 5'-d(CATTTCTTGCTCTCC)3' (*TAT*9), complementary to nucleotide 5399–5413 of HIV-1 RNA *(32)*, the initiation codon region of *tat* mRNA, and differing from the anti-c-Ha-*ras* sequences in at least 9 out of 15 residues.

3.2. Oligodeoxynucleotide Uptake and Stability

To examine oligodeoxynucleotide uptake by T24-transformed NIH3T3 cells, aliquots of 5'-^{35}S-labeled *RAS*4 oligomer were added to cells in DMEM with 10% heat-inactivated FBS and incubated for up to 24 h. Radioactivity retained by the washed cell pellets was compared with that remaining in the culture medium. In DMEM with 10% heat-inactivated FBS, 1–2% of the labeled oligomers were found to be associated with the cell pellet after 3 h and this amount increased very slowly up to 24 h (Fig. 1). The 5'-^{35}S-labeled *RAS*4 was equally stable in DMEM with 10% heat-inactivated FBS at 4, 15, 25, and 37°C for up to 3 h (not shown). Denaturing gel electrophoresis of labeled oligomers remaining in the culture medium supernatant revealed significant loss of oligodeoxynucleotides within 1 h and disappearance was virtually complete by 4 h (not shown). Mung bean nuclease addition to oligodeoxynucleotide uptake experiments had no effect on radioactivity retained by the cell pellets (not shown).

3.3. Compartmentalization Experiments

T24-transformed NIH3T3 cells were treated with 5'-^{35}S-labeled *RAS*4 and fractionated using differential centrifugation methods *(23)*. The distribution results are shown in Fig. 2. Percentages of oligodeoxynucleotide compartmentalization in each fraction were as follows: nucleus, 28.8 ± 8.8%; mitochondria, 4.4 ± 1.0%; cytoplasmic membrane, 0.85 ± 0.05%; cytoplasm, 70.4 ± 4.5%. Although most oligodeoxynucleotides were detected in the cytoplasm, it is significant that a substantial fraction of the oligodeoxynucleotides migrated into the nucleus.

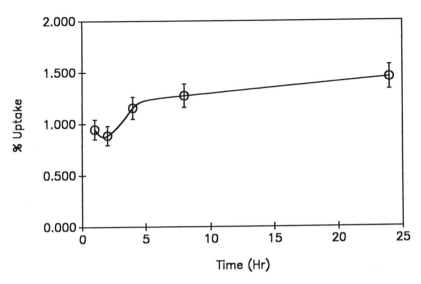

Fig. 1. Uptake of oligodeoxynucleotides by T24-transformed NIH3T3 cells. Samples of 5'-^{35}S-labeled *RAS*4 oligomer were incubated with cells for 0, 1, 4, 8, and 24 h at 37°C. Cells were separated from culture medium by sedimentation, lysed, and extracted with phenol. Oligodeoxynucleotide uptake was determined by liquid scintillation counting of the aqueous phases. Points shown are the average of three repetitions, and the error bars represent 1 SD.

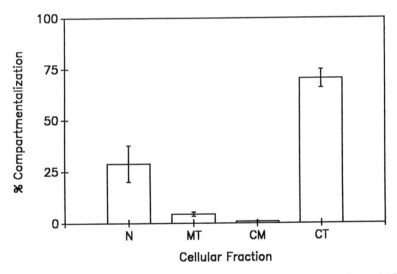

Fig. 2. Compartmentalization of oligodeoxynucleotides in T24-tranformed NIH3T3 cells. Cells were incubated with 5'-^{35}S-labeled *RAS*4 oligomer for 3 h at 37°C, followed by two washes with ice-cold PBS. Cellular pellets were lysed and sedimented at 300, 4000, and 20,000g to obtain the nuclear (N), mitochondrial (MT), and cytoplasmic membrane (CM) pellets, respectively. The final supernatant represents the cytoplasmic fraction (CT). The histogram displays averages of three repetitions, and the error bars represent 1 SD.

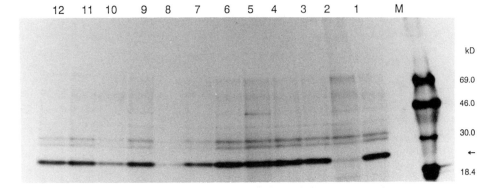

Fig. 3. Expression of *RAS* p21 protein in treated and untreated T24-transformed NIH3T3 cells, measured by radioimmunoprecipitation and fluorography. Antisense oligodeoxynucleotides were added directly to the suspensions at final concentrations of 25 and 50 μM for 24 h. Lane M, ^{14}C-labeled mol-wt markers; lane 1, no oligomer, anti-p21 antibody; lane 2, no oligomer, normal rat serum; lanes 3 and 4, 25 and 50 μM VSVM oligomer, anti-p21 antibody; lanes 5 and 6, 25 and 50 μM TAT9 oligomer, anti-p21 antibody; lanes 7 and 8, 25 and 50 μM *RAS*2 oligomer, anti-p21 antibody; lanes 9 and 10, 25 and 50 pM *RAS*4 oligomer, anti-p21 antibody; lanes 11 and 12, 25 and 50 μM *RAS*3 oligomer, anti-p21 antibody.

3.4. Antisense Inhibition of RAS p21 Expression

Immediately after addition of the antisense DNAs, T24-transformed NIH3T3 cells were incubated for 24 h at 37°C, and then processed for radio-immunoprecipitation of p21 antigen (Fig. 3). Cells treated to a final concentration of 25 and 50 μM *RAS*2, *RAS*3, and *RAS*4 oligomers showed dose-dependent reduction in the expression of *RAS* p21. Laser densitometry of p21 bands from three repetitions of this experiment allowed quantitation of the inhibitory effects (Fig. 4). No significant effect on p21 levels was observed when cells were treated with the nonspecific control oligomers, VSVM or *TAT*9. In all cases, TCA-precipitable counts/min in cellular lysates remained constant within 20% of each other, varying with no correlation to the added oligomer. This implies that cellular protein synthesis was not generally inhibited by any of the oligomers.

3.5. Antisense Inhibition of Focus Formation

Growth potential of a cell line can be assessed by its ability to form foci *(25)*. T24-transformed NIH3T3 cells were treated with 50 μM *RAS*4, and VSVM oligomers and foci were counted. Cells treated with *RAS*4 oligomer exhibited a sequence-specific decrease in focus formation after 3–5 d of growth (Fig. 5). Colony size also decreased significantly, to almost half the size of

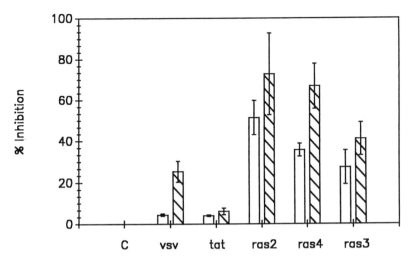

Fig. 4. Relative intensities of *RAS* p21 protein expression from laser densitometry of three repetitions of the experiment in Fig. 4, with error bars representing 1 SD.

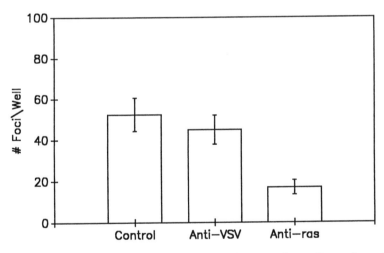

Fig. 5. Effect of antisense oligodeoxynucleotides on focus formation of T24-transformed NIH3T3 cells. Transformed cells were resuspended in <10 µL of DMEM, with no addition, *RAS*4, or VSVM, and added to 3.0 mL of semisolid DMEM supplemented with 10% NCS. Oligomers were added to a final concentration of 50 µ*M*. Cells were incubated at 37°C and inspected for focus formation 3–5 d later. Cells were examined microscopically, and colonies with a cluster of 10 or more cells were scored as foci. Duplicate-well experiments were carried out three times for the control, and twice for each oligomer; the error bars represent 1 SD.

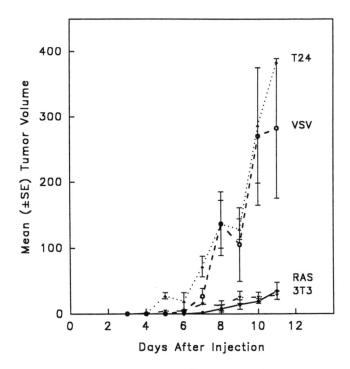

Fig. 6. Mean (± SE) tumor volume (mm³) in nude Balb/c female mice following sc injection of pretreated cells from the four treatment groups: **(A)** T24-untreated T24/NIH3T3 cells (*n* = 2 mice); **(B)** VSV-T24/NIH3T3 cells treated with the nonspecific VSVM DNA (*n* = 3 mice); **(C)** *RAS*-T24/NIH3T3 cells treated with *RAS2* DNA (*n* = 3 mice); **(D)** 3T3-untreated, nontransformed NIH3T3 cells (*n* = 3 mice). Day 0 marks the end of DNA pretreatment and the day of injection.

untreated cells. This effect was uniform throughout the culture. In contrast, no significant decrease in focus number or size was observed in cells treated with 50 μ*M* anti-VSVM oligomer.

3.6. Antisense Inhibition of Tumor Growth in Nude Mice

In the first trial of the second experiment, tumor growth in the Balb/c females injected with cells from the four treatment groups is shown in Fig. 6. All females except one evidenced some tumor development in the 11 d following injection, and this one animal (from the untreated T24 group) was excluded from the results. Statistical analyses were not performed because of the small number of animals in the experiment. It is apparent, however, that pretreatment of T24/NIH3T3 cells with *RAS2* anti-c-Ha-*ras* DNA substantially reduced subsequent tumor growth in the mice. The *RAS* treatment group had much

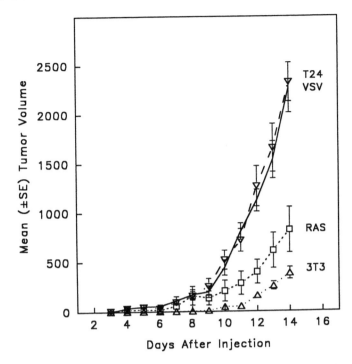

Fig. 7. Mean (± SE) tumor volume (mm³) in nude C57BL/Balb/c male mice following sc injection of pretreated cells from the four treatment groups: **(A)** T24-untreated
T24/NIH3T3 cells (*n* = 8 mice); **(B)** VSV-T24/NIH3T3 cells treated with the nonspecific VSVM DNA (*n* = 7 mice); **(C)** *RAS*-T24/NIH3T3 cells treated with *RAS*2 DNA
(*n* = 7 mice); **(D)** 3T3-untreated, nontransformed NIH3T3 cells (*n* = 7 mice). Day 0
marks the end of DNA pretreatment and the day of injection.

smaller tumors than either the untreated T24 or VSV-treated groups throughout the period of observation. Tumor size in the *RAS* treatment group was still
less than one-eighth that of the VSV or untreated T24 groups on the last day of
observation, 11 d after the end of DNA pretreatment and injection of the cells
into the mice. In contrast, the tumor growth of T24/NIH3T3 cells pretreated
with the nonspecific VSVM DNA was as pronounced as that of untreated T24/
NIH3T3 cells.

Because of the small number of mice in the first trial, the second trial was
carried out to confirm the reliability and validity of the above results. This
second trial involved essentially the same protocol, but employed a larger number and a different strain and sex of nude mice. The results are shown in Fig. 7.
All mice evidenced some tumor development in the course of the trial. Analysis of variance was conducted on the data, and in light of a significant treat-

ment effect ($p < .01$), multiple comparisons among the four treatment groups were performed using the Tukey-Kramer procedure in the SAS data analysis software (SAS Institute).

Pretreatment with *RAS*2 DNA significantly reduced tumor growth as compared to either the untreated T24 group ($p < .01$) or the VSV group ($p < .01$). The effect extended throughout the trial so that even on d 14, tumor volume in the *RAS* group was substantially less than that of the other two groups. However, tumor growth in the *RAS* group did significantly exceed that of the 3T3 group ($p < .01$). The nonspecific VSVM DNA had no significant effect on subsequent tumor growth, since this group did not differ significantly from the untreated T24 group.

Histological examination of randomly selected tumors of the four treatment groups revealed a morphology typical of vigorous fibrosarcomas, with hypercellularity and a relatively high mitotic index (15–$40/mm^2$). There were no obvious differences among the groups. Even the 3T3 group had a relatively high mitotic index, although the overall tumor growth rate for this group was by far the lowest among the four groups.

3.7. Measurement of RAS p21 Protein in Cells Treated for Implantation

A portion of the cells from the second trial were lysed for subsequent measurement of *RAS* p21 protein on the day DNA pretreatment ended and cells were injected into the mice (d 0). The resulting Western blot revealed a pronounced suppression of *RAS* p21 levels in the group treated with *RAS*2 DNA as compared to either the untreated T24 or VSVM-treated group (Fig. 8). Scanning results indicated that the suppression amounted to a 95% reduction in total *RAS* p21 relative to the level in untreated T24/NIH3T3 cells (Fig. 9). In fact, *RAS*2-treated cells had p21 levels below that of NIH3T3 cells. It should be noted, however, that in the case of the *RAS*2-treated T24/NIH3T3 cells, the levels reflect mutated as well as normal p21, whereas NIH3T3 cells contain only the normal p21. The T24/NIH3T3 cells treated with VSVM DNA had *RAS* p21 levels somewhat lower than the untreated T24/NIH3T3 cells, but it is not clear whether the 20% reduction reflects a real effect of the DNA or simply a variability in the Western blot procedure. In either case, the reduction was not associated with any significant change in tumor growth (Fig. 7).

4. Discussion

4.1. Antisense Efficacy

The present studies were directed toward exploring further the approach of gene control using antisense oligodeoxynucleotides against the T24-activated c-Ha-*ras* oncogene in transformed NIH3T3 cells. In the first experiment

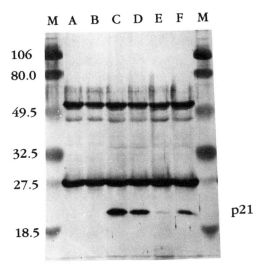

Fig. 8. Western blot results for total cellular *RAS* p21 protein levels following 3 d of in vitro DNA treatment. The lanes include: (M) protein mol-wt standards; **(A)** immunoprecipitation control—immunoprecipitation was conducted in the absence of any cell lysate; **(B)** IgG control—immunoprecipitation was conducted with rat IgG rather than anti-RAS Y13-259 antibody; **(C)** T24—lysate of untreated T24/NIH3T3 cells; **(D)** VSV—lysate of T24/NIH3T3 cells treated with the nonspecific VSVM DNA; **(E)** *RAS*—lysate of T24/NIH3T3 cells treated with *RAS2* DNA; **(F)** 3T3—lysate of untreated, nontransformed NIH3T3 cells.

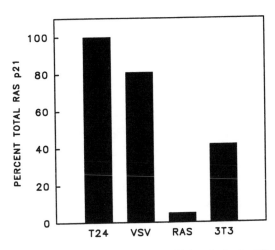

Fig. 9. Densitometric measurement of *RAS* p21 bands in the Western blot of Fig. 8. For each lane, optical scanning of the p21 band was performed, and the percentage of the band's integrated optical density relative to that of the untreated T24/NIH3T3 cells is presented.

described above, it was observed that normal oligodeoxynucleotides are stable in cell-culture medium containing 10% serum for up to 3 h, are distributed among the various compartments of the cell, and inhibit both *RAS* p21 expression and focus formation in c-Ha-*ras* transformed mouse fibroblasts. In the second experiment, it was observed that pretreatment with anti-c-Ha-*ras* DNA effectively inhibited tumor growth of c-Ha-*ras*-activated cells.

4.2. Nuclear Localization of Internalized DNA

In studies utilizing oligodeoxynucleotide probes, the first questions to arise concern the cellular uptake and survival of the oligodeoxynucleotides. The low levels of oligodeoxynucleotide binding to the NIH3T3-attached fibroblasts used in this study are comparable to those observed previously in HL-60-suspended hematopoietic cells *(21)*, and in L929 fibroblasts and Krebs ascites cells *(33)*. It is noteworthy that the bound oligodeoxynucleotides were resistant to added mung bean nuclease, as well as the endogenous serum nucleases.

It has been demonstrated that antisense DNA inhibition is at least in part dependent on RNase H hydrolysis of DNA–mRNA hybrids *(34)*. Different compartments in the cell have various concentrations of RNase H. As a result, it is necessary to determine the distribution of oligomers inside the cell in order to assess their effectiveness. The fact that about a quarter of the labeled oligomer studied above migrated into nuclei indicates the importance of using antisense sequences targeted against regions of the premessenger located in the nucleus. Additionally, the fact that RNase H exists in its highest concentration in the nucleus *(35)* makes this organelle of prime interest.

4.3. Limited Usefulness of RNA Structure Predictions

In the absence of data from nuclease mapping, chemical probing, base-pair replacement, or phylogenetic comparisons, secondary-structure predictions are often misleading. The free energies used for base pairing are still approximate, and we lack algorithms to predict tertiary-structure interactions such as knots or pseudoknots *(36)*. Nevertheless, calculation of the c-Ha-*ras* mRNA secondary structure provided a starting point for selecting initial targets for antisense inhibition. In the cell culture experiments described above, the ability of sequence-specific antisense DNAs to inhibit expression of *RAS* p21 was explored as a function of target location.

The effectiveness of the anti-c-Ha-*ras* oligomers at inhibiting *RAS* protein expression showed little correlation with predicted secondary structure, except for the case of the initiation codon region. The *RAS*2 and *RAS*4 antisense oligomers targeted against the first intron target and the initiation codon region, respectively, effectively inhibited p21 expression. However, the *RAS*2 sequence was significantly more effective. There is no obvious explanation for the efficacy of the first intron target. The initiation codon region, on the other hand, is also likely to be a sensitive target for antisense hybridization arrest, because this region is involved in the assembly of the 80S initiation complex *(37)* and may be specifically recruited for ribosome binding by eukaryotic initiation factor 4B *(38)*. A third sequence, *RAS*3, targeted against a possibly weakly base paired or loop region upstream from the initiation codon was less effective in inhibiting p21 expression than the other two anti-c-Ha-*ras* oligomers. This leads to the suggestion that the real mRNA secondary structure may actually lack an accessible site for hybridization at that position or that tertiary structure may limit access by the oligomer. The dependence of antisense inhibition on the nucleotide sequence of the oligomer was verified by testing two oligomers that had little complementarity to c-Ha-*ras* mRNA. Anti-VSV and anti-*tat* oligomers had no significant effect on either p21 expression or focus formation in T24-transformed NIH3T3 cells.

4.4. Greater Potency of Methylphosphonates

In a similar cell culture system, sequence-specific, dose-dependent inhibition of the expression of RAS p21 was observed on using antisense oligodeoxynucleoside methylphosphonates complementary to the first 11 coding nucleotides of Balb-*ras* mRNA, i.e., a shorter version of *RAS*4 *(13)*. The cell-culture studies described above using normal antisense DNAs targeted against 15 residues of the initiation codon region did not show inhibitory effects as large as the methylphosphonate undecamer at identical concentrations. This difference of inhibitory potency could be explained by the fact that oligodeoxynucleoside methylphosphonates are almost completely resistant to nuclease digestion and, hence, survive much longer in cell culture medium containing serum *(39)*. Moreover, it is observed that the uncharged oligodeoxynucleoside methylphosphonates enter living cells at a higher percentage than the normal polyanionic oligodeoxynucleotides *(40)*.

4.5. Inhibition of Focus Formation

Transformed cells are distinguished from immortalized cells by several criteria. One such criterion is the ability to form foci in semisolid cell-culture medium, which correlates closely with tumorigenesis *(25)*. Results from the focus formation assays were consistent with the radioimmunoprecipitation experiments. Normal antisense oligodeoxynucleotides at a 50-μM concentration targeted against the initiation codon had a specific inhibitory effect on the ability of T24-transformed NIH3T3 cells to form foci. The inhibition was seen not only on the number of foci formed, but also on the size of the foci. *RAS*4-treated cells had fewer cells per focus when compared to untreated or to VSVM-treated cells. The specific inhibition of focus formation by *RAS*4 supports the model that expression of activated *RAS* p21 plays a pivotal role in cellular transformation and that antisense inhibition of p21 expression may have therapeutic value in limiting tumor growth.

4.6. Inhibition of Tumorigenesis

The results clearly demonstrate the pronounced effectiveness of anti-c-Ha-*ras*-DNA in inhibiting tumor growth of c-Ha-*ras*-activated cells. The effect was evident in two different strains of nude mice, and in both males and females. It was specific to the anti-c-Ha-*ras* DNA, since the nonspecific anti-VSV DNA had no effect. What is apparent in the results is not just the degree of inhibition, but also its duration. Tumor growth was suppressed in both experiments throughout the period of observation. In the second trial, this was 14 d after DNA treatment was concluded and the cells were injected into the mice. Even after 14 d, the *RAS*-treated group still had substantially smaller tumors and a slower rate of tumor growth than either the untreated T24 or VSV-treated groups.

The effects of the anti-c-Ha-*ras* DNA on tumor growth are associated with its ability to suppress *RAS* p21 levels. Total *RAS* p21 was reduced by more than 90% with the 3-d treatment regimen. This is not surprising, since in the first experiment above, comparable concentrations of the same *RAS*2 antisense DNA produced a substantial inhibition of radiolabeled *RAS* p21 synthesis *de novo*, and the 3-d treatment regimen is several times the reported 24 h cellular half-life of the *RAS* protein *(41)*. The DNA presumably exerts its effect by binding specifically to the c-Ha-*ras* RNA transcript, and inhibition of synthesis is maintained as long as antisense DNA is present to bind the RNA transcript. Inhibition of *RAS* p21 expression should therefore have relaxed within a few days following the end of antisense DNA treatment, since the intracellular half-life of DNA oligomers is on the order of 12 h *(21)*. Yet, the effect on tumor growth extended for up to 14 d, suggesting more than a simple relationship between inhibition of *RAS* p21 and suppression of tumor growth. *RAS* p21 has been shown to affect the expression of a number of genes, including *MDR1*, actin, *fos*, and transforming growth factor β, and the sequences of at least two transcriptional elements have been identified that mediate some of the *ras* effects on transcription *(2)*. A sustained reduction of *RAS* p21 levels may therefore result in an extensive reorganization of cellular genetic expression and behavior, a reversal of the transformation phenotype of the cell. Such a reversal may interfere not just with cellular replication, but also with such tumor-related processes as colonization, vascularization, and invasion. The time involved in reversing the transformation process as a result of *RAS* p21 inhibition and the subsequent retransformation following restoration of the protein levels may account for the prolonged duration of anti-c-Ha-*ras* DNA effects on tumor growth.

4.7. Prospects for Therapeutic Applications

The possibility of reversing the transformation of cancer cells may be a critical advantage of drugs targeted specifically against oncogenes. Most current drugs affect only replication nonspecifically. An antisense DNA able to reverse the transformation process may produce a reorganization of cellular genetic expression that affects a number of neoplastic and malignant attributes. Even a partial re-establishment of normal cellular controls on growth could markedly alter the behavior of cancer cells. Many, if not most, neoplasms have identifiable oncogenes that may serve as targets for specific antisense DNAs. Although the oncogenes presumably serve to trigger neoplastic behavior, what remains to be determined is the degree to which neoplastic and malignant characteristics of transformed cells remain dependent on particular oncogenes. For example, the neoplastic behavior of some NIH3T3 cells transformed by the T24 c-Ha-*ras* oncogene eventually became independent of c-Ha-*ras* expression itself *(42)*, suggesting that the c-Ha-*ras* oncogene may act only to trigger cellular changes initially, and it is these changes that subsequently served to maintain neoplastic behavior independent of the oncogene. On the other hand, microinjection of anti-*RAS* p21 antibodies directly into c-Ha-*ras* transformed cells was observed to reverse transformation as measured by morphology and growth rate *(43)*, and in the present study, anti-c-Ha-*ras* DNA was capable of suppressing tumor growth in c-Ha-*ras*-transformed cells. These results imply that neoplastic growth is dependent on continued expression of the c-Ha-*ras* oncogene.

Positive results from this animal study, as well as in trials in Eμ-*myc* transgenic mice *(44,45)*, and those in other laboratories *(46–48)*, suggest that application to human therapy should be pursued. A wide variety of possible modifications, formulations, and routes of administration must be explored in order to maximize potency and minimize toxicity *(49)*.

Acknowledgments

The author wishes to thank his associates who have worked on this investigation over the years: Yehia Daaka, Audrey Gonzalez, Gary Gray, Denise Hebel, Olga Hernandez, Christine O'Connell, Julio Pow-Sang, and Kara Sessums. I would also like to thank Stuart Aaronson for cultures of T24-transformed NIH3T3 cells and Y13-259 cells. This work was supported by grants from the American Cancer Society.

References

1. Bishop, J. M. (1991) Molecular themes in oncogenesis. *Cell* **64,** 235–248.
2. Lowy, D. R. and Willumsen, B. M. (1993) Function and regulation of *ras*, in *Annual Review of Biochemistry*, vol. 62 (Richardson, C. C., Abelson, J. N., Meister, A., and Walsh, C. T., eds.), Annual Reviews, Palo Alto, pp. 851–891.

3. McCormick, F. (1993) How receptors turn Ras on. *Nature* **363**, 15,16.
4. Tabin, C. J., Bradley, S. M., Bargmann, C. I., Weinberg, R. A., Papageorge, A. G., Scolnick, E. M., Dhar, R., Lowy, D. R., and Chang, E. H. (1982) Mechanism of activation of a human oncogene. *Nature* **300**, 143–149.
5. Belikova, A. M., Zarytova, V. F., and Grineva, N. I. (1967) Synthesis of ribonucleosides and diribonucleoside phosphates containing 2-chloroethylamine and nitrogen mustard residues. *Tet. Lett.* **37**, 3557–3562.
6. Zamecnik, P. C. and Stephenson, M. L. (1978) Inhibition of Rous sarcoma virus replication and cell transformation by a specific oligodeoxynucleotide. *Proc. Natl. Acad. Sci. USA* **75**, 280–284.
7. Wickstrom, E. (ed.) (1991) *Prospects for Antisense Nucleic Acid Therapy of Cancer and AIDS.* Wiley-Liss, New York.
8. Murray, J. A. H. (ed.) (1992) *Antisense RNA and DNA.* Wiley-Liss, New York.
9. Crooke, S. T. and Lebleu, B. (eds.) (1993) *Antisense Research and Applications.* CRC, Boca Raton, FL.
10. Agrawal. S., ed. (1993) *Methods in Molecular Biology, vol. 20: Protocols for Oligonucleotides and Analogs,* Humana, Totowa, NJ.
11. Chang, E. H. and Miller, P. S. (1991) *Ras*, an inner membrane transducer of growth stimuli, in *Prospects for Antisense Nucleic Acid Therapeutics for Cancer and AIDS* (Wickstrom, E., ed.), Wiley-Liss, New York, pp. 115–124.
12. Yu, Z., Chen, D., Black, R. J., Blake, K., Ts'o, P. O. P., Miller, P., and Chang, E. H. (1989) Sequence specific inhibition of in vitro translation of mutated or normal ras p21. *J. Exp. Pathol.* **4**, 97–108.
13. Brown, D., Yu, Z. P., Miller, P., Blake, K. R., Wei, C., Kung, H. F., Black, R., Ts'o, P. O. P., and Chang, E. H. (1989) Modulation of ras expression by antisense, nonionic deoxyoligonucleotide analogs. *Oncogene Res.* **4**, 243–252.
14. Chang, E. H., Yu, Z., Shinozuka, K., Zon, G., Wilson, W. D., and Strekowska, A. (1989) Comparative inhibition of ras p21 protein synthesis with phosphorus-modified antisense oligonucleotides. *Anti-Cancer Drug Design* **4**, 221–232.
15. Chang, E. H., Miller, P. S., Cushman, C., Devadas, K., Pirollo, K. F., Ts'o, P. O. P., and Yu, Z. P. (1991) Antisense inhibition of RAS p21 expression that is sensitive to a point mutation. *Biochemistry* **30**, 8283–8286.
16. Saison-Behmoaras, T., Tocque, B., Rey, I., Chassignol, M., Thuong, N. T., and Hélène, C. (1991) Short modified antisense oligonucleotide directed against Ha-*ras* point mutation induces selective cleavage of the mRNA and inhibits T24 cell proliferation. *EMBO J.* **10**, 1111–1116.
17. Reddy, E. P., Reynolds, R. K., Santos, E., and Barbacid, M. (1982) A point mutation is responsible for the acquisition of transforming properties by the T24 human bladder carcinoma oncogene. *Nature* **300**, 149–152.
18. Srivastava, S. K., Yuasa, Y., Reynolds, S. H., and Aaronson, S. A. (1985) Effects of two major activating lesions on the structure and conformation of human ras oncogene products. *Proc. Natl. Acad. Sci. USA* **82**, 38–42.
19. Furth, M. E., Davis, L. J., Fleurdelys, B., and Scolnick, E. M. (1982) Monoclonal antibodies to the p21 products of the transforming gene of Harvey murine sarcoma virus and of the cellular ras gene family. *J. Virol.* **43**, 294–304.

20. Sinha, N. D., Biernat, J., McManus, J., and Koster, H. (1984) Polymer support oligonucleotide synthesis XVIII: use of G-cyanoethyl-N,N-dialkylamino-/N-morpholino phosphoramidite of deoxynucleosides for the synthesis of DNA fragments simplifying deprotection and isolation of the final product. *Nucleic Acids Res.* **11,** 4539–4557.

21. Wickstrom, E. L., Bacon, T. A., Gonzalez, A., Freeman, D. L., Lyman, G. H., and Wickstrom, E. (1988) Human promyelocytic leukemia HL-60 cell proliferation and c-myc protein expression are inhibited by an antisense pentadecadeoxynucleotide targeted against c-myc mRNA. *Proc. Natl. Acad. Sci. USA* **85,** 1028–1032.

22. Sawadogo, M. and Van Dyke, M. W. (1991) A rapid method for the purification of deprotected oligodeoxynucleotides. *Nucleic Acids Res.* **19,** 674.

23. de Duve, C., Pressman, B. C., Gianetto, R., Wattiaux, R., and Appelmans, F. (1955) Tissue fractionation studies: intracellular distribution patterns of enzymes in rat-liver tissue. *Biochem. J.* **60,** 604–617.

24. Laemmli, U. K. (1970) Cleavage of structural proteins during the assembly of the head of bacteriophage T4. *Nature* **227,** 680–685.

25. Graf, T. (1973) Two types of target cells for transformation with avian myelocytomatosis virus. *Virology* **54,** 398–413.

26. Bradford, M. M. (1976) Simple colorimetric assay for measuring total protein concentration. *Anal. Biochem.* **72,** 248–253.

27. Towbin, J., Staehelin, T., and Gordon, J. (1979) Electrophoretic transfer of proteins from polyacrylamide gels to nitrocellulose sheets: procedures and some applications. *Proc. Natl. Acad. Sci.* **76,** 4350–4354.

28. Reddy, E. P. (1983) Nucleotide sequence analysis of the T24 human bladder carcinoma oncogene. *Science* **220,** 1061–1063.

29. Ishii, S., Merlino, G. T., and Pastan, I. (1985) Promoter region of the human Harvey ras proto-oncogene: similarity to the EGF receptor proto-oncogene promoter. *Science* **230,** 1378–1382.

30. Rose, J. K. and Gallione, C. J. (1981) Nucleotide sequences of the mRNAs encoding the vesicular stomatitis G and M proteins determined from cDNA clones containing the complete coding regions. *J. Virol.* **39,** 519–528.

31. Wickstrom, E., Simonet, W. S., Medlock, K., and Ruiz-Robles, I. (1986) Complementary oligonucleotide probe of vesicular stomatitis virus matrix protein mRNA translation. *Biophys. J.* **49,** 15–17.

32. Ratner, L., Haseltine, W., Patarca, R., Livak, K. J., Stacich, B., Josephs, S. F., Doran, E. R., Rafalski, J. A., Whitehorn, E. A., Baumeister, K., Ivanoff, L., Petteway, S. R., Pearson, M. L., Lautenberger, J. A., Papas, T. S., Ghrayeb, J., Chang, N. T., Gallo, R. C., and Wong-Staal, F. (1985) Complete nucleotide sequence of the AIDS virus, HTLV-III. *Nature* **313,** 277–284.

33. Vlassov, V. V. and Yakubov, L. A. (1991) Oligonucleotides in cells and in organisms: pharmacological considerations, in *Prospects for Antisense Nucleic Acid Therapeutics for Cancer and AIDS* (Wickstrom, E., ed.), Wiley-Liss, New York, pp. 243–266.

34. Walder, R. Y. and Walder, J. A. (1988) Role of RNase H in hybrid-arrested translation by antisense oligonucleotides. *Proc. Natl. Acad. Sci. USA* **85,** 5011–5015.

35. Haberken, R. C. and Cantoni, G. L. (1973) Studies on a calf thymus RNase specific for ribonucleic acid-deoxyribonuclease and hybrids. *Biochemistry* **12,** 2389–2395.
36. Puglisi, J. D., Wyatt, J. R., and Tinoco, I., Jr. (1988) A pseudoknotted RNA oligonucleotide. *Nature* **331,** 283–286.
37. Kozak, M. (1983) Comparison of initiation of protein synthesis in prokaryotes, eukaryotes, and organelles. *Microbiol. Rev.* **47,** 145.
38. Goss, D. J., Woodley, C. L., and Wahba, A. J. (1987) A fluorescence study of the binding of eucaryotic initiation factors to messenger RNA and messenger RNA analogues. *Biochemistry* **26,** 1551–1556.
39. Miller, P. S., McParland, K. B., Jayaraman, K., and Ts'o, P. O. P. (1981). Biochemical and biological effects of nonionic nucleic acid methylphosphonates. *Biochemistry* **20,** 1874–1880.
40. Miller, P. S. (1989) Non-ionic antisense oligonucleotides, in *Oligodeoxynucleotides: Antisense Inhibitors of Gene Expression* (Cohen, J. S., ed.), CRC, Boca Raton, pp. 79–95.
41. Magee, A. I., Gutierrez, L., McKay, I. A., Marshall, C. J., and Hall, A. (1987) Dynamic fatty acylation of p21 N-ras. *EMBO J.* **6,** 3353–3357.
42. Gilbert, P. X. and Harris, H. (1988) The role of the ras oncogene in the formation of tumours. *J. Cell Sci.* **90,** 433–446.
43. Feramisco, J. R., Clark, R., Wong, G., Arnheim, N., Milley, R., and McCormick, F. (1986) Transient reversion of ras oncogene-induced cell transformation by antibodies specific for amino acid 12 of RAS protein. *Nature* **314,** 639–641.
44. Wickstrom, E., Bacon, T. A., and Wickstrom, E. L. (1992) Down-regulation of c-*myc* antigen expression in lymphocytes of Eµ-c-*myc* transgenic mice treated with anti-c-*myc* DNA methylphosphonate. *Cancer Res.* **52,** 6741–6745.
45. Huang, Y., Snyder, R., Kligshteyn, M., and Wickstrom, E. (1995) Prevention of tumor formation in a mouse model of Burkitt's lymphoma by six weeks of treatment with anti-c-*myc* DNA phosphorothioate. *Mol. Med.* **1,** 647–658.
46. Whitesell, L., Rosolen, A., and Neckers, L. M. (1991) *In vivo* modulation of N-*myc* expression by continuous perfusion with an antisense oligonucleotide. *Antisense Res. Dev.* **1,** 343–350.
47. Ratajczak, M. Z., Kant, J. A., Luger, S. M., Hijiya, N., Zhang, J., Zon, G., and Gewirtz, A. M. (1992) *In vivo* treatment of human leukemia in a scid mouse model with c-*myb* antisense oligodeoxynucleotides. *Proc. Natl. Acad. Sci. USA* **89,** 11,823–11,827.
48. Skorski, T., Nieborowska-Skorska, M., Nicolaides, N. C., Szczylik, C., Iversen, P., Iozo, R. V., Zon, G., and Calabretta, B. (1994) Suppression of Philadelphia[1] leukemia cell growth in mice by *BCR-ABL* antisense oligodeoxynucleotide. *Proc. Natl. Acad. Sci. USA* **91,** 4504–4508.
49. Wickstrom, E. (1995) Nuclease resistant derivatives of antisense DNA, in *Delivery Systems for Antisense Oligonucleotide Therapeutics* (Akhtar, S., ed.), CRC, Boca Raton, FL, pp. 85–104.

6

Antisense Inhibition
of Protein Synthesis and Function

Rabbit Retinal Protein

Anil P. Amaratunga, Kenneth S. Kosik, Peter A. Rittenhouse, Susan E. Leeman, and Richard E. Fine

1. Introduction
1.1. Advantages of Vitreal Injection of Antisense Oligonucleotides

When considering the use of antisense technology for in vivo application, whether for therapeutic development or for the creation of animal models for human diseases, a major problem is that most extracellular compartments are constantly mixing with the blood, lymph, cerebrospinal fluid (CSF), and so forth, and thus constantly diluting the oligonucleotide. A possibly unique exception to this problem is the vitreous, a gelatinous fluid overlying the inner retinal surface. The vitreous is formed early in development and is in essence a relatively closed compartment with no active transport of fluids or ions. Forensic pathologists make use of this fact in determining the composition of electrolytes at the time of death by sampling the vitreal compartment.

The closed nature of the vitreal compartment and the fact that it overlies the retinal ganglion cell (RGC) layer of the retina has made it a very useful injection site for radioisotopes and more recently for drugs used to treat retinal diseases. The anatomy of the retina and of the RGCs in particular has again provided an ideal system to investigate the transport of membranes from the cell body where they are synthesized and assembled to the axon and subsequently to the axon terminals in the brain *(1)*. The retinal ganglion cell axons comprise the optic nerve, which carries all the visual information collected by the retina to the brain. It has thus provided us with an excellent system to probe

From: *Methods in Molecular Medicine: Antisense Therapeutics*
Edited by: S. Agrawal Humana Press Inc., Totowa, NJ

the function of proteins involved in membrane sorting and rapid axonal transport in vivo, using antisense oligonucleotides.

1.2. Animal Model

We use the young adult New Zealand White Rabbit as the animal model for several reasons. The rabbit has very large eyes for its size, comparable to those of humans. This makes injection of oligonucleotides, radioactive isotopes, and so on, extremely easy. Relatively large quantities (up to 100 µL of fluid) can be injected with very little leakage from the vitreous, whose volume is approx 1–2 mL. All of the axons of two branches of the optic nerve of this albino breed of rabbit cross at the optic chiasm, in contrast to those of other species, making the use of one retina and optic nerve of an individual an ideal control for the other. This is essential in our experiments since there are large variations in the amount of radioactivity incorporated into proteins between retinas of different individuals. Finally, the length of the rabbit optic nerves, 3 cm, and their relatively large diameter allow us the ability to obtain sufficient material for biochemical analysis from a single retina/optic nerve, again avoiding the pooling of data from groups of animals.

Labeled amino acids are administered to the vitreal chamber of the eye of an animal, where they are taken up predominantly by cell bodies of retinal ganglion cells, which directly contact the vitreous, and are rapidly incorporated into proteins. Newly synthesized proteins in the retinal ganglion cell bodies that are destined to be rapidly transported are sorted into small vesicles and are carried along the axons in the optic nerve and optic tract (ON/OT) toward the axon terminals in the lateral geniculate and superior colliculus at the rate of 1 cm/h. Proteins in the rapid phase of transport, i.e., mainly membrane-associated proteins, show a peak of label in the 3-cm long rabbit ON/OT about 2–3 h following radiolabel injection *(2)*. Examination of the labeled proteins in the ON/OT at this time allows one to study specifically proteins rapidly transported from the cell body to axons of retinal ganglion cells.

1.3. Choice of a Protein to Target

Kinesin is a microtubule-stimulated ATPase, consisting of two heavy and two light chains *(3)*. The heavy chains contain the microtubule binding site as well as the ATPase consensus sequences *(4)*. Kinesin was originally isolated from squid-extruded axoplasm as a fraction capable of moving vesicles or fluorescent beads along microtubules at speeds that approximate that of anterograde rapid transport (about 1 cm/h in mammals) *(5)*. In vitro evidence suggests that kinesin in neurons, which have the highest expression level of any cell type *(6)*, serves as a motor for anterograde rapid transport of synaptic vesicle precursors, plasma membrane components, and possibly mitochondria from the cell body to the axon and the presynaptic nerve terminals *(2,7,8)*. If this were the case, then the specific block-age of kinesin synthesis should inhibit rapid axonal transport in vivo.

In this regard, Ferreira and colleagues have employed an antisense oligo-nucleotide directed against the region surrounding the start codon of the kinesin heavy chain (KHC) mRNA in cultured hippocampal neurons *(8)*. This antisense oligonucleotide inhibits the synthesis of kinesin and appears to block the trans-port of synapsin I and GAP43, two axon terminal membrane-associated pro-teins, to the neurite tip. This finding suggests that kinesin serves as a motor for vesicles moving from the cell body to the neurite ending. However, it is impos-sible to determine the speed of vesicle transport in this system, which is essen-tial to demonstrate the proposed function of kinesin as the motor for rapid anterograde transport in axons.

1.4. Choice of a Functional Assay

For an antisense oligonucleotide to have a measurable functional effect, it must significantly decrease the synthesis of a targeted protein. However, if the life-time of a molecule is very long and/or if there is a large excess of a given protein with regard to its function, it may be possible to achieve a large inhibi-tion of the synthesis of a particular molecule with no apparent decrease in func-tion. This consideration is of great importance in the nervous system. Most proteins that supply the needs of the dendrites and all of the proteins supplied to the axon are synthesized in the cell body. In addition to the rapidly trans-ported proteins, which are all membrane-associated or secreted molecules, and move at the speed of 1 cm/h, the bulk of the proteins supplied to the axon and nerve terminals thereof move at about 1 mm/d *(9)*. This means that even if an antisense was completely effective at blocking the synthesis of a molecule that functions in the axon, it would take many days to achieve a significant decrease in the concentration of the protein. Therefore, a key consideration in our choice of a protein to target was its speed of axonal transport. This can be easily deter-mined if an antibody against the protein is available by vitreal injection of ^{35}S-methionine and immunoprecipitation from the axon 3 h later. If a labeled band with the correct molecular weight is seen after SDS-PAGE, it provides strong evidence that the protein is rapidly transported *(10)*.

Using the retinal ganglion/optic nerve preparation, we have recently pro-vided evidence that kinesin itself is rapidly transported through the optic nerve in tight association with vesicles following its synthesis in retinal ganglion cells *(11)*. This is critical with respect to our subsequent experiments.

1.5. Choice of an Antisense Nucleotide

We used a sequence that was stabilized by sulfur-modified phosphate link-ages. This sequence was employed by one of us (K. S. K.) and shown to inhibit effectively both rat and human KHC synthesis even though there were two differences between the species *(8)*. We did not determine the rabbit KHC sequence. Because of the likely possibility of mismatches, we chose to use this 25-nucleotide sequence. Another reason for choosing a longer sequence is that even if a few nucleotides on either end were removed by exonucleases, a sufficiently long oligonucleotide would remain potentially to inhibit KHC synthesis.

2. Experimental Procedures

Oligonucleotides were synthesized as described previously *(8)* and stabilized by sulfur-modified phosphate linkages. Antisense oligonucleotide, GCCGGGTCCGCCATC TTTCTGGCAG, is the inverse complement of nucleotides −11 to +14 in the rat KHC sequence *(8)*. The sense oligonucleotide is the exact inverse complement of the antisense oligonucleotide.

Use of rabbit retinal ganglion cell for in vivo radioactive labeling of proteins has been described previously *(10)*. Briefly, albino rabbits are anesthetized by nembutal (iv) and proparacaine-HCl (topical eye anesthetizer). Labeled amino acids are introduced into the vitreous chamber of the eye (close to the retina) by U-100 insulin syringes with 28-gage needles. In experiments described here, this method was used for a single administration of 0.10 µmol of either antisense ($n = 4$ animals) or sense ($n = 3$ animals) oligonucleotide in 50 µL of PBS into one eye and 50 µL PBS into the contralateral eye of the animal. Sixteen hours later, 0.5 mCi (50 µL) ^{35}S-methionine-cysteine (New England Nuclear, Boston, MA) was injected into the vitreous chamber of each eye. Animals were sacrificed 3 h later by lethal injection, and retinas, ONs, and OTs were dissected out.

Two retinas were separately homogenized in ice-cold homogenization buffer (PBS, containing 30 µg/mL paramethylsulfonylfluoride, 0.5 µg/mL leupeptin, and 0.5 µg/mL aprotinin) (5 mL/retina) in a motor-driven Teflon™ glass homogenizer. Homogenates were diluted 1:1 by adding 2X Immunomix (1.0% [v/v] Triton X-100, 0.5% [v/v] Deoxycholate, and 0.1% [w/v] [SDS] in PBS), incubated on ice for 1 h, and clarified by centrifuging at 25,000g in a Ti 70 rotor for 10 min. The resulting solubilized protein preparations from the two retinas were used for immunoprecipitations and total labeled protein detections, as described below.

For detecting kinesin in immunoprecipitations, SUK4, an MAb raised against the sea-urchin KHC *(12)* (a generous gift by J. M. Scholey) was used. SUK4 was added (1:250) to 1-mL aliquots of retinal protein preparations and incubated overnight at 4°C. Sepharose-linked antimouse IgG beads (or antirabbit IgG for polyclonal antibodies [Abs] (Organon Teknika) were then added to samples (15 µL beads/1 µL primary antibody) and incubated for 3 h at

4°C with agitation. Beads were separated in an Eppendorf centrifuge and washed 5 times using Immunomix. Washed beads were suspended in sample buffer and analyzed by 10% SDS-PAGE *(13)*. Gels were incubated in an enhancing solution for 1 h, dried under vacuum for 2 h, and exposed to films for appropriate times. Similar immunoprecipitations were carried out by using an MAb against tubulin (1: 1000) (Sigma Chemical Co., St. Louis, MO) and a polyclonal antibody against the 20 c-terminal amino acids of the amyloid precursor protein (1:200). Immunoprecipitations of labeled amyloid precursor protein and synaptophysin from ON/OT membranes were performed as previously described *(10,14)*. Determination of the total amount of substance P in the axons was carried out by radioimmune assay *(10)*. Control experiments to determine nonspecific immunoprecipitations were carried out by using an MAb against the insulin-regulatable glucose transporter, which is not found in the central nervous system *(15)* (a generous gift by P. Pilch) (1:250) and pre-immune normal rabbit serum (1:200) (Rockland Company).

The ON from one eye and the contralateral OT were combined and homogenized in homogenization buffer (3 mL/ON/OT). The ON from the other eye was combined with its contralateral OT and homogenized in the same way. Homogenates were centrifuged at 100,000*g* in a Ti 70 rotor for 1 h at 4°C. The pellets were resuspended in Immunomix (3 mL/ON/OT) and then clarified by centrifugation at 25,000*g* for 10 min. The resultant ON/OT membrane preparations and retinal homogenates (obtained as described above) were processed for total labeled protein detections. To equal volumes of samples from the two retinas or ON/OTs of an animal, acetone (kept at −20°C) was added (1:7 by volume) and incubated at −20°C for 2 h. Proteins were precipitated by centrifugation at 20,000*g* for 10 min. Gel electrophoresis and autoradiography were carried out as described above.

Scanning of autoradiographs were performed by volume integration of protein bands using a densitometer (Molecular Dynamics). For immunoprecipitations, individual protein bands were scanned, and for total protein determinations, the entire profile of protein bands in a lane was scanned. Background values were subtracted and the resulting OD values were standardized to compensate for differences in the radioactivity associated with each experiment.

Results of densitometry were statistically analyzed. Mean differences between oligonucleotide-treated and control experiments were verified by using the paired *t*-test. The following differences were found to be statistically significant at the 99% confidence level: (1) Labeled KHC levels immunoprecipitated in the antisense injected retina and in the contralateral control retina; and (2) total labeled protein levels in the antisense ON/OT and in the contralateral control ON/OT. All other differences between antisense and control or sense and control experiments were not significant.

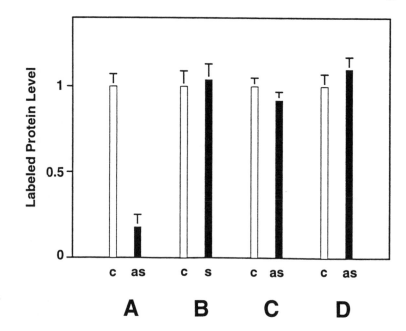

Fig. 1. Densitometric analysis of labeled protein levels immunoprecipitated in the retinas. Autoradiographs of antisense and control experiments ($n = 4$ animals) and sense and control experiments ($n = 3$ animals) were used for scanning of protein bands, KHC **(A,B)**, tubulin **(C)**, and amyloid precursor protein **(D)**. Results were standardized and mean values ± SEM were graphed as: antisense-treated retinas, s: sense-treated retinas, and c: control retinas. This figure was reproduced from ref. *13* with permission.

3. Results

As demonstrated in our recent publication *(14)*, an injection of 0.1 μmol of the synthetic KHC antisense oligonucleotide very effectively inhibited the synthesis of kinesin in the retina, whereas vehicle or sense injection had no effect (Fig. 1). There was no effect on total protein synthesis as shown by trichloroacetic acid precipitation. The synthesis of several other proteins, including tubulin and the amyloid precursor protein, was similarly unaffected *(14)*.

Since we now had an effective method to inhibit kinesin synthesis specifically in the retinal ganglion cells, we then determined the effect of antisense injection on rapid transport. As demonstrated in Fig. 2, there is a very significant decrease in the amount of rapidly transported protein in the optic nerve following antisense injection, whereas vehicle or sense injection had no effect.

After collecting data from seven animals, we found that the antisense oligonucleotide produced an 82 ± 7% reduction in kinesin synthesis and a concomi-

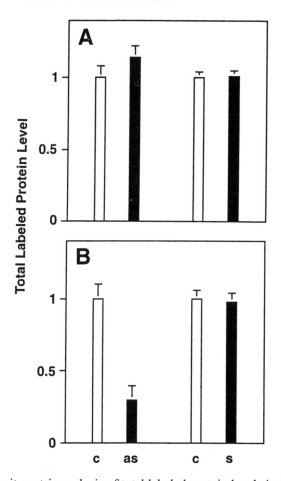

Fig. 2. Densitometric analysis of total labeled protein levels in the retinas **(A)** and ON/OTs **(B)**. Autoradiographs of antisense and control experiments (*n* = 4 animals) and sense and control experiments (*n* = 3 animals) showing retinal and ON/OT total labeled protein detections were used for scanning of total labeled protein profiles in each lane. Results were standardized, and mean values ± SEM were graphed. as: antisense-treated retinas or ON/OTs, s: sense-treated retinas or ON/OTs, and c: control retinas or ON/OTs. This figure was reproduced from ref. *13* with permission.

tant 70 ± 10% reduction in the level of rapidly transported protein. These results were statistically significant at the 99% confidence level. Several other results from this study should be noted here. We injected the antisense oligonucleotide and, 16 h later, injected the radioactivity into the vitreous. This interval was based on our experience with in vitro experiments with cultured neurons (e.g., ref. *8*). When we employed a much shorter interval, 3 h, no decrease was found in the amount of kinesin synthesized. We have not tried longer intervals to determine how long the antisense effect persists. This is an important consideration for therapeutic applications.

Reduced amounts of antisense oligonucleotide were employed in similar experiments to obtain a dose-response effect. When 0.05 μmol of oligonucleotide was injected, the level of labeled KHC was decreased by 42% compared to the control retina (n = 1 animal) and 0.03 μmol (n = 1 animal) or 0.02 μmol (n = 1 animal) of oligonucleotide failed to elict a response (data not shown). In these experiments, there was no decrease in the total retinal labeled protein level in the antisense ON/OT compared to the control ON/OT (data not shown). Of particular interest was the finding that although a 42% decrease in kinesin synthesis was found using 0.5 μmol of antisense, no change in the amount of rapid transport was found. Another interesting observation was the finding that injection of an 18-mer missing 4 nucleotides at the 5' end and 3 at the 3' end produced no decrease in kinesin synthesis.

Although the results summarized above suggested that kinesin is the motor protein for axonal rapid transport in general, we have recently performed an experiment that indicates that kinesin is the motor for at least three discrete classes of axonal transport vesicles, neurotransmitter-containing synaptic vesicles, peptide-containing synaptic vesicles, and plasma membrane precursors *(10,15)*. We have employed immunoprecipitations to quantitate inhibition of the rapid axonal transport of three proteins characteristic of these vesicle classes, synaptophysin, substance P, and amyloid precursor protein *(10)*. These results are summarized in Table 1. There is now evidence that there are at least four other kinesin-like heavy chain genes, with significant differences in amino acid sequence from KHC expressed in mammalian cells *(16)*. At least three of these are expressed in a single neuron *(16)*. Our results suggest that the other kinesin-like proteins may function in the cell body and dendrites, but probably not in the axon.

4. Discussion

We will limit our discussion to general considerations with respect to the use of antisense nucleotides in in vivo situations, especially in targeting retinal proteins. The finding that the 25-mer KHC antisense works effectively in the in vivo retina, whereas the 18-mer does not work at all may be generalized to other situations. In addition to the greater specificity of the 25-mer, it is also possible that in the in vivo situation there is a significant degradation of the oligonucleotide owing to exonucleases. Therefore, it is possible that a 25-mer, even though it is degraded to some degree, still maintains its ability to bind to the targeted sequence. The 18-mer, however, once degraded, loses its binding affinity. This advantage may outweigh any decreased uptake owing to

Table 1
Proteins Characteristic
of Three Different Transport Vesicles
Are Inhibited from Rapid Axonal Transport
by a Kinesin Antisense Oligonucleotide[a]

	Antisense	Sense
Amyloid precursor protein	0.24	0.95
Synaptophysin	0.07	0.85
Substance P	0.57	1.16

[a]Quantitation of representative rapid axonally transported proteins in the optic nerve/tract following oligonucleotide injection into the vitreous. For each experiment, antisense or sense kinesin oligonucleotide-treated optic nerve/tract was compared with the contralateral control optic nerve/tract of the same animal. For quantitative analysis, autoradiographs of immune precipitated ^{35}S-labeled protein bands were scanned by a densitometer. For substance P, total amounts of protein were determined by a radioimmunoassay. Result with each oligonucleotide-treated optic nerve/tract is expressed as a fraction of its control.

increased length. We have not tried the unmodified oligonucleotide in these experiments. It is possible that it may work in the vitreous as well.

The concentration dependency of the antisense effects on both inhibition of kinesin synthesis and inhibition of rapid axonal transport may also be of general interest. The maximum dose we injected, 0.1 µm, probably produces a vitreal antisense concentration of between 50 and 100 µM, since the vitreal volume is between 1 and 2 mL. This is much higher than the concentration necessary to produce in vitro effects on cells in culture *(8)*. At present, we do not understand the reason for this discrepancy; 0.05 µmol produced almost exactly 50% of the kinesin inhibition seen at 0.1 µm, but had no effect on rapid transport. This finding provides suggestive evidence for a threshold effect. The retinal ganglion cell may contain about two times as much kinesin as needed to carry out its function as the motor for rapid transport. If one only inhibits 40% of kinesin synthesis, no effect on transport is seen, but above 50%, one sees a measurable decrease in rapid transport. This type of result may be seen in other in vivo situations with other targeted proteins.

Our results shed some light on the duration of the antisense effect in an in vivo situation. A single antisense injection produces a very significant, 80%, inhibition of the synthesis of a targeted protein for at least 19 h after injection. We have no evidence as yet on how long the antisense effect persists. This is obviously very important information to obtain in considering therapeutic applications.

Even at the relatively high concentration we employed in our protocol, 50–100 μM, the phosphorothioate oligonucleotide appeared to show no detectable short-term toxicity. The rabbits' eyes showed no redness, and the rabbits did not demonstrate any discomfort, such as eye rubbing. The levels of total protein synthesis seen in the oligonucleotide-injected and noninjected retinas were identical, which again indicates that there is no short-term toxicity.

Although our results suggest that the vitreous may be an excellent compartment for antisense oligonucleotide injection to treat retinal diseases, such as virus infections, many questions still remain unanswered. For example, can we target proteins in the nuclear and photoreceptor layers of the retina that are not in direct contact with the vitreous? Even though these and many other questions need to be answered, we are optimistic that vitreal antisense delivery will prove to be useful in treatment of acute and chronic retinal disease.

Acknowledgments

This work was supported by National Institute of Health Grants R01EY 08535 (to R. E. F.) and R0INS 29031 (to K. S. K.). We thank Simona Zacarian for expert editorial assistance.

References

1. Sjostrand, J. and Karlsson, J.-O. (1969) Axoplasmic transport in the optic nerve and tract of the rabbit: a biochemical and radioautographic study. *J. Neurochem.* **16**, 833–844.
2. Lorenz, T. and Willard, M. (1978) Subcellular fractionation of intra-axonally transported polypeptides in the rabbit visual system. *Proc. Natl. Acad. Sci. USA* **75**, 505–509.
3. Bloom, G. S., Wagner, M. C., Pfister, K. K., and Brady, S. T. (1988) Native structure and physical properties of bovine brain kinesin and identification of the ATP-binding subunit polypeptide. *Biochemistry* **27**, 3409–3416.
4. Kuznetsov, S., Vaisberg, Y., Rothwell, S., Murphy, D., and Gel, V. (1989) Isolation of a 45 kDa fragment from the kinesin heavy chain with enhanced ATPase and microtubule binding activities. *J. Biol. Chem.* **264**, 589–595.
5. Vale, R. D., Reese, T. S., and Sheetz, M. P. (1985) Identification of a novel force-generating protein, kinesin, involved in microtubule-based motility. *Cell* **42**, 39–50.
6. Hollenbeck, P. (1989) The distribution, abundance, and subcellular localization of kinesin. *J. Cell Biol.* **108**, 2335–2342.
7. Hirokawa, N., Sato-Yoshitake, R., Kobayashi, N., Pfister, K. K., Bloom, G. S., and Brady, S. T. (1991) Kinesin associates with anterogradely transported membranous organelles in vivo. *J. Cell Biol.* **114**, 295–302.
8. Ferreira, A., Niclas, J., Vale, R. D., Banker, G., and Kosik, K. S. (1992) Suppression of kinesin expression in cultured hippocampal neurons using antisense oligonucleotides. *J. Cell Biol.* **117**, 595–606.

9. McEwen, B. S. and Grafstein, B. (1968) Fast and slow components in axonal transport of protein. *J. Cell Biol.* **38,** 494–508.

10. Morin, P. J., Liu N., Johnson, R. J., Leeman, S. E., and Fine, R. E. (1991) Isolation and characterization of rapid transport vesicle subtypes from rabbit optic nerve. *J. Neurochem.* **56,** 415–427.

11. Morin, P. J., Johnson, R. J., and Fine, R. E. (1993) Kinesin is rapidly transported in the optic nerve as a membrane associated protein. *Biochim. Biophys. Acta.* **1146,** 275–281.

12. Ingold, A., Cohn, S. A., and Scholey, J. M. (1988) Inhibition of kinesin driven microtubule motility by monoclonal antibodies to kinesin heavy chains. *J. Cell Biol.* **107,** 2657–2667.

13. Laemmli, U. K. (1970) Cleavage of structural proteins during the assembly of the head of bacteriophage T4. *Nature* **277,** 680–685.

14. Amaratunga, A., Morin, P. J., Kosik, K. S., and Fine, R. E. (1993) Inhibition of kinesin synthesis and rapid anterograde axonal transport in vivo by an antisense oligonucleotide. *J. Biol. Chem.* **268,** 17,427–17,430.

15. James, D. E., Lederman, L., and Pilch, P. F. (1987) Purification of insulin-dependent exocytic vesicles containing the glucose transporter. *J. Biol. Chem.* **262,** 11,817–11,824.

16. Aizawa, H., Sekine, Y., Takemura, R., Zhang, Z., Nangaku, M., and Hirokawa, N. (1992) Kinesin family in murine central nervous system. *J. Cell Biol.* **119,** 1069–1076.

7

c-*Myb* in Smooth Muscle Cells

Michael Simons and Robert D. Rosenberg

1. c-*Myb*

The proto-oncogene c-*myb* is a cellular homolog of a viral oncogene v-*myb* found in two independently derived avian acute leukemia viruses: avian myeloblastosis virus (AMV) and E26 leukemia virus *(1)*. The *myb* gene is highly conserved in eukaryotes, and it usually consists of 15 exons spanning over 35 kb of genomic DNA *(1)*. In humans, c-*myb* gene locus has been mapped to chromosome 6 (6q22–23). However, *myb* mRNA expressed in thymus contains transcripts originating from chromosome 17 (17q25), suggesting that an intermolecular recombinant event may be involved in the formation of the mature protein *(2)*.

The *myb* gene generates a number of alternatively spliced mRNAs with a major (4 kb) as well as several minor (4.3 and 3.4 kb) transcripts *(1)*. These transcripts in turn produce at least two protein isoforms, 75 and 89 kDa, detectable in both normal and tumor cell lines *(3)*. c-*Myb* is a DNA binding protein with a DNA binding region consisting of 3 imperfect 50–52 amino acid repeats at the 5' end of the molecule possessing typical helix-loop-helix motif *(4,5)*. The presence of two of these repeats (R2 and R3) is needed for DNA binding *(5,6)*. Interestingly, as a result of N-terminal deletion, most of the first DNA binding domain is missing in v-*myb* sequences. The presence of *myb* binding domain is sufficient for transcriptional activation as demonstrated by trans-activation studies *(7)*, and c-*myb* has been shown to control transcription of a number of eukaryotic genes, including mim-1 *(8)* and c-*myc* *(9)* by binding to specific (PyACC^G/$_T$G) DNA binding site *(10)*.

The proto-oncogene was originally thought to be synthesized only by myeloid cells, but recently its expression has been detected in a number of other cell types, including chick embryo fibroblasts *(11)*, small-cell lung carci-

From: *Methods in Molecular Medicine: Antisense Therapeutics*
Edited by: S. Agrawal Humana Press Inc., Totowa, NJ

noma *(12)* and colon carcinoma *(13)* cells, neuroectodermal cells *(14)*, embryonal stem (ES) cells *(15)*, and smooth muscle cells *(16)*. The proto-oncogene's expression is growth-dependent, occurring at low levels in quiescent cells, increasing as cells begin to proliferate, and peaking in the late G_1 phase of the cell cycle *(11)*. Despite numerous investigations, c-*myb*'s function has not been well defined. The oncogene appears to be involved in regulation of growth and development with abundant expression noted in erythroid, myeloid, and lymphoid progenitors. c-*Myb*'s expression rapidly declines once cells begin to differentiate *(17)*, whereas overexpression of the oncogene blocks this process *(18)*. Cellular growth regulation by c-*myb* is suggested by the observation that oncogene's expression in fibroblasts results (directly or indirectly) in appearance of mRNA for DNA polymerase α, histone H3, and PCNA, and cell entry into the *S* phase *(19)*. Furthermore, c-*myb*'s expression is required for proliferation of a variety of cell types, including vascular smooth muscle cells *(13,14,16,20,21)* as well as for a number of other cellular processes, including activation of T-lymphocytes *(22)* and CD-4 gene expression *(23)*.

In this chapter, we will review the information regarding c-*myb*'s expression and function in vascular smooth muscle cells, and examine some of the molecular pathways involved in the oncogene's control of cell cycle and proliferation in these cells. The reader is referred to several recent reviews for consideration of other sides of c-*myb*'s function *(1,17,24,25)*.

2. c-*Myb*'s Function in Vascular Smooth Muscle Cells

A substantial body of evidence points to c-*myb*'s expression and important functional role in vascular smooth muscle. Proto-oncogene's expression in bovine *(26)*, and rat *(20)* aortic smooth muscle cells has been demonstrated by Northern analysis *(20,26)*, its cDNA has been cloned from a bovine vascular smooth muscle cell library *(26)*, and c-*myb* protein has been demonstrated in immortalized rat aortic smooth muscle cells *(20,27)* by immunofluorescence and immunoprecipitation (Fig. 1). Proto-oncogene's expression in smooth muscle cells parallels its expression in other cell types: It occurs at low levels in quiescent cells, increases by the time of G_1/S interface, and then declines to baseline levels *(26,27)*. However, unlike hematopoietic cells, at no point in the cell cycle in smooth muscle cells does there appear to be an increased rate of c-*myb* gene transcription, suggesting that changes in mRNA/protein levels are owing to changes in message and/or protein half-life *(26)*.

The proto-oncogene appears to be involved in control of intracellular ionized calcium ($[Ca^{2+}]_i$) in smooth muscle and other cell types. The intracellular concentration of ionized calcium has long been recognized as an important element in the control of cell growth. Transient increases in $[Ca^{2+}]_i$ have been documented during mitosis and anaphase *(28,29)*, and are thought to be

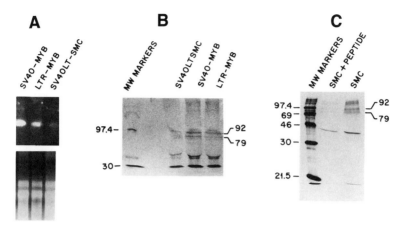

Fig. 1. c-*Myb* expression in smooth muscle cells. **(A)** Northern analysis of c-*myb*-transfected and untransfected SV40LT-SMC. Northern analysis of LTR-MYB (first lane), SV40-MYB (second lane), and untransfected SV40LT-SMC (third lane) probed with radiolabeled c-*myb* probe; the corresponding ethidium-bromide-stained gel is shown below. **(B)** SDS gel electrophoretic analysis of immunoprecipitated radiolabeled c-*myb* protein from c-*myb*-transfected and untransfected SV40LT-SMC. SDS gel electrophoresis on a 6% polyacrylamide gel of ^{14}C labeled mol-wt markers (lane 1), ^{35}S-labeled c-*myb* protein immunocaptured from untransfected SV40LT-SMC (lane 2), ^{35}S-labeled c-*myb* protein immunocaptured from SV40-MYB (lane 3), and ^{35}S-labeled c-*myb* protein immunocaptured from LTR-MYB (lane 4). **(C)** Characterization of protein immunocaptured with c-*myb* antisera. SDS gel electrophoresis on an 8% polyacrylamide gel of ^{14}C-labeled mol-wt markers (lane 1), ^{35}S-labeled c-*myb* protein immunocaptured from untransfected SV40LT-SMC in the presence of 10 μg of peptide against which the c-*myb* antibody was raised (lane 2), and ^{35}S-labeled c-*myb* protein immunocaptured from untransfected SV40LT-SMC in the absence of added peptide (lane 3). LTR-MYB: c-*myb*-transfected smooth muscle cells under a control of LTR promoter, SV40-MYB: c-*myb*-transfected smooth muscle cells under a control of SV40 promoter, SV40LT-SMC: SV40 large T immortalized smooth muscle cell line (from ref. *27*).

required for breakdown of nuclear envelope and mitotic spindles, chromosomal condensation, and activation of the contractile ring *(30–33)*. However, relatively little information exists regarding intracellular ionized calcium's role in progression of G_1 phase with earlier work, suggesting that the cation may be important during this part of the cycle *(34,35)*. With these results in mind, we thought to study whether c-*myb*-induced changes in $[Ca^{2+}]_i$ can account for its effect on cell proliferation.

To this end, we measured $[Ca^{2+}]_i$ in c-*myb*-transfected vascular smooth muscle cells as well as in the course of the cell cycle in wild-type cells *(27)*.

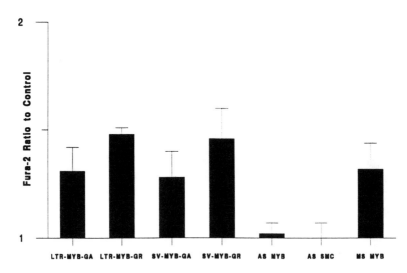

Fig. 2. The concentrations of intracellular calcium in c-*myb*-transfected smooth muscle cells determined with Fura-2. The concentrations of intracellular calcium in c-*myb* transfected SV40LT-SMC determined with Fura-2. LTR-MYB in growth arrest (LTR-MYB-GA) and 24 h after serum stimulation (LTR-MYB-GR); SV40-MYB in growth arrest (SV-MYB-GA) and 24 h after serum stimulation (LTR-MYB-GR); LTR-MYB in growth arrest treated with 25 μ*M* phosphorothioate antisense c-*myb* oligonucleotide (AS MYB) or 2 bp mismatch phosphorothioate c-*myb* oligonucleotide (MS MYB); untransfected SV40LT-SMC in growth arrest treated with 25 μ*M* phosphorothioate antisense c-*myb* oligonucleotide (AS SMC). All Fura-2 ratio measurements are normalized to control Fura-2 ratios in untransfected SV40LT-SMC determined simultaneously. The data are presented as mean ± SE (from ref. *27*). Abbreviations as in Fig. 1.

Stable transfection of full-length c-*myb* cDNA under control of two different promoters (SV40 and LTR) results in a twofold rise in the intracellular $[Ca^{2+}]_i$ levels as measured by Fura-2 imaging (Fig. 2). Note that the increase in $[Ca^{2+}]_i$ concentration is independent of the promoter used and the growth state of the cells, suggesting that the expression of the message itself is responsible for this effect. Further confirmation of this relationship between c-*myb* mRNA levels and intracellular calcium concentration comes from the observation that antisense (but not missense) c-*myb* oligonucleotides in an amount sufficient to reduce the c-*myb* mRNA level to baseline (wild-type cell) levels return $[Ca^{2+}]_i$ to normal levels (Fig. 2). Since vascular smooth muscle cells do not posses extensive intracellular calcium storage, it is likely that extracellular calcium entry is responsible for elevation in $[Ca^{2+}]_i$ seen in c-*myb* transfectants. This can be confirmed by placing c-*myb*-transfected cells in a calcium-free medium: Within 5 min $[Ca^{2+}]_i$ in c-*myb*-transfectants returns to wild-type levels *(27)*.

Since c-*myb* is able to increase $[Ca^{2+}]_i$ in stably transfected cells, it becomes important to determine whether physiologic changes in c-*myb* levels in the course of cell cycle can also affect intracellular calcium concentration. To this effect, we have measured $[Ca^{2+}]_i$ in synchronized smooth muscle cells at different stages of the cell cycle (Fig. 3). We find that as the cells progress through the G_1 phase, $[Ca^{2+}]_i$ begins to rise, reaching its peak at the G_1/S interface and then declines to baseline levels (Fig. 3). To demonstrate that changes in c-*myb* levels are responsible for the observed change in $[Ca^{2+}]_i$, we determined intracellular calcium levels during the cell cycle in the presence and absence of antisense c-*myb* oligonucleotides in synchronized smooth muscle cells (Fig. 4). The data show that an increase in c-*myb* mRNA levels precedes a rise in $[Ca^{2+}]_i$ and that suppression of c-*myb* mRNA with antisense oligonucleotides prevents the late G_1 rise in $[Ca^{2+}]_i$. The observed effect of c-*myb* on $[Ca^{2+}]_i$ is not limited to smooth muscle cells, since c-*myb* transfection of fibroblasts results in similar alterations.

Since inhibition of late G_1 c-*myb* expression results in suppression of the rise in intracellular calcium concentration normally seen at the G_1/S interface, and since calcium may be involved in cell-cycle progression *(34,35)*, it is interesting to determine whether these changes in c-*myb* and $[Ca^{2+}]_i$ levels affect the S-phase entry. To this end, we measured S-phase entry of cultured immortalized rat aortic smooth muscle cells, as well as primary baboon vascular smooth muscle cells in the presence and absence of antisense c-*myb* oligonucleotides. In both cases, there was severe inhibition of S-phase by antisense c-*myb* that could be overcome by addition of calcium ionophore in a concentration sufficient to raise $[Ca^{2+}]_i$ to levels normally seen at G_1/S interface (Fig. 5).

The experiments summarized above demonstrate that an increase in c-*myb* expression whether occurring physiologically during the cell-cycle progression or secondary to stable overexpression results in increased intracellular calcium concentration. Furthermore, late G_1 c-*myb*-dependent rise in intracellular calcium concentration appears necessary for the S-phase entry of cultured smooth muscle cells *(36)*. This increase in $[Ca^{2+}]_i$ appears dependent on the presence of a sufficient concentration of extracellular calcium in the tissue-culture medium. It should be noted that the late G_1 changes in $[Ca^{2+}]_i$ described above are fundamentally different from previously observed transient calcium spikes in cycling cells *(29)*, since these alterations in calcium levels persist for an extended (~6 h) time span suggesting that fundamental changes in cellular calcium handling occur during this time period.

Given these profound effects of c-*myb*'s expression on intracellular calcium levels and cell-cycle progression, it would appear likely that c-*myb* is critically involved in regulation of smooth muscle cell growth and proliferation. In the

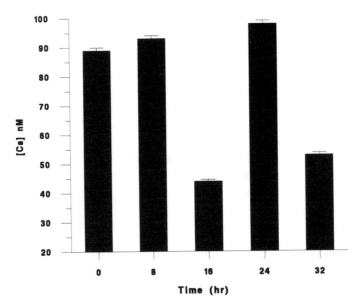

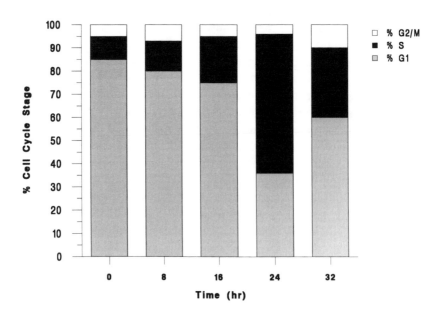

Fig. 3. Intracellular calcium levels during cell-cycle progression in vascular smooth muscle cells. **(Top)** The concentrations of intracellular calcium in SV40LT-SMC determined with Fura-2 following serum stimulation (mean ± SE). **(Bottom)** Cell-cycle distribution determined in parallel samples (from ref. *27*).

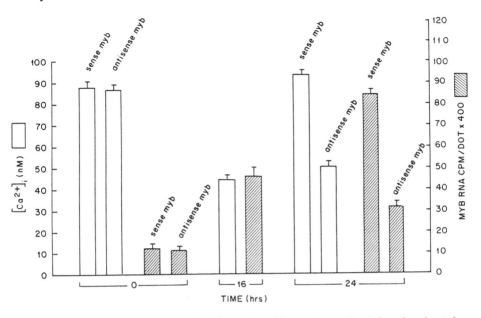

Fig. 4. Effect of antisense c-*myb* oligonucleotide on elevated calcium levels at the G_1/S interface. The concentrations of intracellular calcium (light bars) and the levels of c-*myb* mRNA (dark bars) in SV40LT-SMC at 0, 16, and 24 h after serum stimulation. The missense or antisense phosphorothioate c-*myb* oligonucleotides were added to cultures at a concentration of 25 μ*M* at the time of serum stimulation (24-h timepoint). The data are displayed as mean ± SE (from ref. *27*).

following sections, we will discuss the use of antisense oligonucleotides to study the role of c-*myb* in these processes under in vitro and in vivo conditions.

3. In Vitro Studies of Vascular Smooth Muscle Cell Proliferation

The vascular smooth muscle cells normally exist in a relatively quiescent state and display a well-defined set of contractile proteins, including α smooth muscle actin and myosin. However, once subjected to arterial injury or a disease process, smooth muscle cells undergo phenotypic transformation, acquiring the ability to proliferate. Indeed, smooth muscle cell proliferation underlies a number of important disease states, including atherosclerosis, hypertension, and diabetes mellitus, and is also responsible for postangioplasty restenosis and failure of coronary bypass grafts *(37–41)*. Therefore, the ability to control this process may provide a useful tool for treatment of a number of disease states. However, despite numerous attempts, no satisfactory therapy has emerged. Part of the reason for this failure may lie with the fact that a multitude of factors including numerous growth factors, and mitogens *(42)*, stretch and mechanical

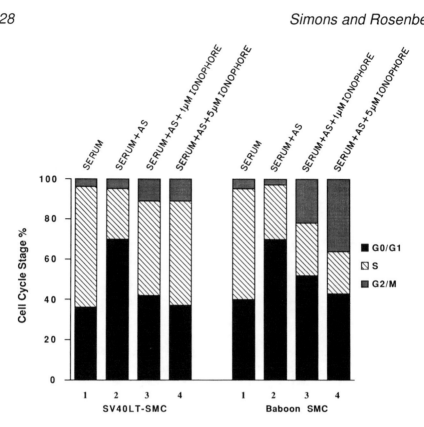

Fig. 5. Effect of calcium ionophore on cell-cycle progression in smooth muscle cells. Growth arrested SV40LT-SMC **(left)** or primary baboon smooth muscle cells **(right)** were stimulated with serum in the absense of any additives (first bar), in the presence of 25 μM antisense c-*myb* oligonucleotides (second bar), or in the presence of antisense oligonucleotides and 1 μM (third bar) or 5 μM (fourth bar) of calcium inophore 4-bromo A23187. The cell cycle distribution of cells in the population was determined by flow cytometry.

injury *(43)*, or simply loss of endothelial covering with subsequent platelet accumulation and thrombosis *(44,45)*, may activate smooth muscle proliferation. Therefore, interventions designed to deal with any one of these triggers while ignoring others are likely to fail. However, regardless of the mode of activation, once stimulated, smooth muscle cells must pass through a number of steps in what has been termed "common final pathway" in order to proliferate. These steps include gene and gene products responsible for cell cycle-control, as well as other events critical for cell division.

Previously presented information regarding c-*myb*'s function and its role in regulation of cellular calcium homeostasis and cell-cycle progression suggests that it may well be one of such critical genes. Indeed, experiments indicate that c-*myb* plays an important role in regulation of proliferation. Thus, antisense c-*myb* oligonucleotides have been shown to inhibit T-cell proliferation in response to phytohemagglutinin by blocking cells in the late G_1 phase *(46)* as well as proliferation of myeloid leukemic cells *(21)*. Therefore, we set out to test whether antisense oligonucleotide suppression of c-*myb* expression would result in inhibition of smooth muscle cell proliferation.

We carried out these studies using a well-described immortalized rat vascular smooth muscle cell line *(47)*, as well as a number of primary rat and mouse smooth muscle cell cultures using phosphorothioate oligonucleotides. To maximize the effect of antisense suppression, cultured smooth muscle cells were growth-arrested, and various antisense and control of oligonucleotide sequences were added at the time of serum stimulation. Such an approach allows for maximum effect of antisense sequences, since there is virtually no c-*myb* mRNA or protein in growth-arrested cells. The studies demonstrated that administration of c-*myb* antisense, but not control sense or scrambled oligonucleotides results in suppression of smooth muscle cell proliferation under these culture conditions *(20)*. Similar effect of antisense c-*myb* oligonucleotides was demonstrated on bovine vascular smooth muscle cells *(16)*. The inhibition of cell proliferation occurred in a dose-dependent manner (Fig. 6) and was accompanied by a decrease in the oncogene's mRNA and protein levels *(20)*. The antisense oligonucleotides were equally effective in immortalized as well as in primary smooth muscle cell cultures *(16,20)*.

Although careful selection of antisense sequences in theory allows unique specificity of effect, in practice, a number of steps should be taken to ensure that the observed biological consequence (or phenotype) of antisense oligonucleotides is indeed owing to suppression of intended sequences and not to a nonspecific interaction. Antisense oligonucleotide administration can potentially disrupt synthesis of unwanted messages owing to the existence of partial sequence matches between oligonucleotides and other mRNAs. Such occurrence of nonspecific degradation of mRNA sequences by synthetic oligonucleotides has been demonstrated under in vitro conditions *(48)* as well as in *Xenopous* oocytes *(49,50)*. Further nonspecificity may result from oligonucleotides binding to intercellular *(51)* or extracellular *(52)* proteins, thus interfering with their function. Finally, generation of double-stranded RNA molecules secondary to formation of RNA–oligonucleotide hybrids may result in activation of cellular enzymes, such as p68 kinase *(53)*, that can affect synthesis of a number of molecules.

Thus, a phenotype arising from oligonucleotide administration can be the result of unintended suppression of different genes and proteins rather than that of the targeted sequence. Therefore, a number of controls are necessary for

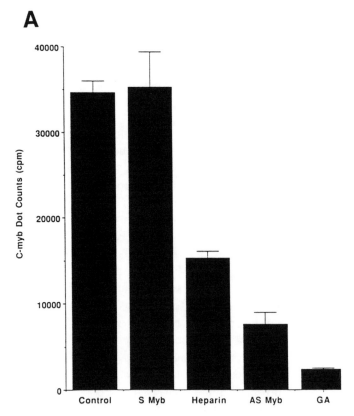

Fig. 6. Effects of antisense c-*myb* oligonucleotides on cell proliferation and mRNA levels. **(A)** The levels of c-*myb* message are displayed as numbers of cpm/dot normalized for large T RNA level. Control—untreated cells; S-*myb*—cells treated with 25 μ*M* sense c-*myb* phosphorothioate oligonucleotide; Heparin—cells treated with 100 μg/mL heparin; AS *myb*—cells treated with 25 μ*M* antisense c-*myb* phosphorothioate oligonucleotide; GA—growth-arrested cells (96 h in 0.5% FBS-DMEM) (from ref. *20*, used with permission).

demonstration of sequence specificity of effect. The use of multiple nonoverlapping antisense sequences minimizes the chance of an observed biological effect being the result of nonspecific interactions, since it is unlikely that two or more overlapping sequences would interact with the same unintended mRNAs or proteins producing identical effect. To assure that the observed phenotype is not the result of nonspecific chemical toxicity of the oligonucleotide, "scrambled" (oligonucleotide of the same base composition as an antisense sequence, but synthesized in a different order) and "missense" sequences (oligonucleotides with 2–3 nucleotide mismatches) should demon-

B

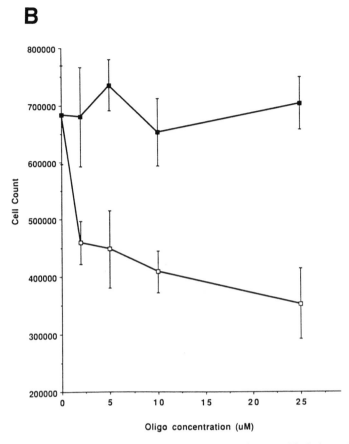

Fig. 6. *(continued)* **(B)** The antisense (light boxes) and sense (dark boxes) c-*myb* phosphorothioate oligonucleotides have been added at varying concentrations (μM) to growth-arrested SV40LT-SMC prior to serum stimulation, and cell counts were obtained 72 h later. The data are displayed as mean ± SD.

strate no biological effect and should not affect the targeted gene's mRNA and protein levels. Suppression of an unrelated gene's mRNA not expected to alter cell phenotype is used to demonstrate that formation of RNA–oligonucleotide hybrids *per se* does not lead to nonspecific degradation of multiple messages, thus producing unintended biological effects.

All of these controls were utilized in antisense c-*myb* studies. Two different antisense oligonucleotides produced a similar degree of cell proliferation suppression and reduced c-*myb*'s expression to similar levels *(16)*. Sense and missense (2-bp mismatch) oligonucleotides demonstrated no effect on cell proliferation or the proto-oncogene's expression, and antisense oligonucle-

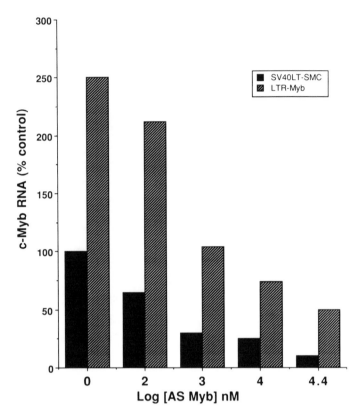

Fig. 7. Effect of antisense c-*myb* oligonucleotides on c-*myb* mRNA levels in c-*myb*-transfected and wild-type cells. Dot-blot determined c-*myb* mRNA levels (normalized for GAPDH) in the wild-type (SV40LT-SMC cells, dark bars) and c-*myb*-transfected (LTR-*Myb* cells, gray bars) in the presence of variable amounts of antisense c-*myb* oligonucleotides.

otides directed against thrombomodulin, a protein not thought to be involved in cell proliferation, had no effect on cell growth but fully suppressed mRNA levels *(20)*. However, to confirm further the specificity of antisense c-*myb* oligonucleotides in terms of their effect on cell growth, we demonstrated increased requirement for antisense oligonucleotides in smooth muscle cells stably transfected with c-*myb* expression construct (Fig. 7).

4. In Vivo Studies of Vascular Smooth Muscle Cell Proliferation

Since c-*myb* appears to be important for smooth muscle cell proliferation in vitro, it is interesting to determine whether the protooncogene is similarly important under in vivo conditions. To study the effect of antisense c-*myb* oli-

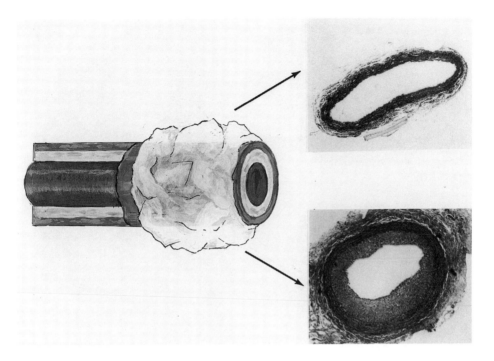

Fig. 8. Extravascular delivery of antisense *c-myb* oligonucleotides. Exposed portions of rat carotid artery were covered with oligonucleotide containing polaxamer gel (Pluronics® F127), and the extent of neointimal proliferation in antisense **(Top)** and control **(Bottom)** arteries was morphometrically determined 2 wk later.

gonucleotides in vivo, we chose to work with a rat carotid artery injury model. In this model, left carotid artery of the rat is exposed via a midline cervical incision, suspended on ties, and stripped of its adventitia *(54)*. A 2F Fogarty catheter is then introduced through the external carotid artery, advanced to the aortic arch, and withdrawn with a balloon firmly inflated to the entry point. The entire procedure is then repeated three times. This injury is reproducibly followed by the initiation of medial smooth muscle cell proliferation (reaching its peak after 72 h) and formation of neointima composed almost exclusively of vascular smooth muscle cells *(43,44)*. The process is complete by 2 wk after injury, and there are relatively little further changes in the neointimal area and extent of smooth muscle cell proliferation.

The oligonucleotides were mixed with a polaxamer polymer (Pluronics® F127) at concentration of 1 mg/mL, and 200 µL of oligonucleotide solution were applied to the adventitial aspect of the artery immediately after balloon injury (Fig. 8). The polaxamer gel used in our study has a unique property of

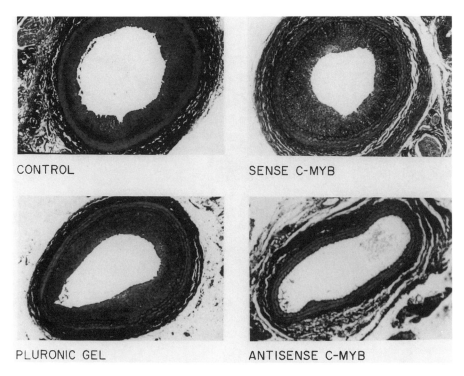

CONTROL   SENSE C-MYB

PLURONIC GEL   ANTISENSE C-MYB

Fig. 9. Effect of antisense and sense c-*myb* oligonucleotides on neointimal formation in rat carotid arteries subjected to balloon angioplasty. Representative cross-sec-
tions from the carotid artery of an untreated rat **(A)**, a rat treated with Pluronic gel
containing 200 µg of sense oligonucleotide **(B)**, a rat treated with Pluronic gel **(C)**, and
a rat treated with Pluronic gel containing 200 µg of antisense oligonucleotide **(D)**
(Mason trichrome, magnification ×80) (from ref. *55*, used with permission).

being liquid at 4°C and instantly gelling at body temperature. Once applied to
the artery in vivo, the gel is rapidly resorbed providing relatively short (<2 h)
bolus delivery of oligonucleotides. However, even this short exposure to
antisense oligonucleotide results in a significant and prolonged inhibition
of c-*myb* mRNA expression in the injured arterial wall *(55)*. Arteries exposed
to antisense, but not sense oligonucleotides or polaxamer gel alone, demonstrated little neointimal formation 2 wk after arterial injury (Fig. 9). Morpho-
metric analysis showed that antisense c-*myb* oligonucleotides inhibited
smooth muscle cell-mediated neointimal formation, but did not affect smooth
muscle cells in the arterial media (Fig. 10). This observation is consistent
with known c-*myb* expression and biological function in proliferating, but not
quiescent cells.

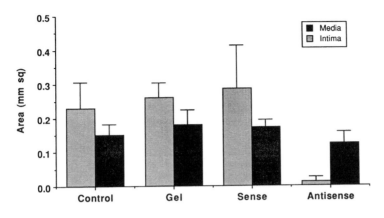

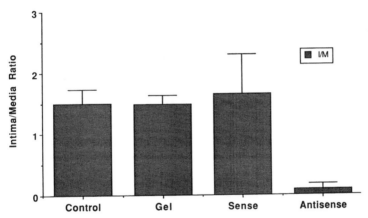

Fig. 10. Intimal and medial cross-sectional areas, as well as intimal/medial ratios of untreated and treated rat carotid arteries. **(Top)** Mean cross-sectional areas of the intimal (light stippled bars) and medial (dark stippled bars) regions of rat carotid arteries, which were untreated, treated with pluronic gel, treated with pluronic gel containing 200 μg of sense oligonucleotide, and treated with pluronic gel containing 200 μg of antisense oligonucleotide. **(Bottom)** Ratio of intimal to medial areas (stippled bars) for the same four groups of animals. The data are provided as mean ± SD (from ref. 55, used with permission).

At the time of sacrifice, rat carotid arteries were further examined for the extent of inhibition of neointimal formation. Interestingly, the antisense effect of c-*myb* oligonucleotides was localized to the site of gel application: Although the entire length of carotid artery was balloon-injured during the study, the gel containing oligonucleotides was applied only to the exposed (distal) portion of the artery, and only this arterial segment showed reduction of neointimal for-

Table 1
In Vivo Antisense c-*Myb* Studies[a]

Oligonucleotide	Delivery	% Suppression, I/M ratio
Antisense 1	Pluronics	93
Antisense 2	Pluronics	90
Antisense 1	EVA	85
Missense	Pluronics	0
Sense	Pluronics	0
Scrambled	EVA	3

[a]Effect of local extravascular administration of c-*myb* oligonucleotides on neointimal proliferation in a rat carotid injury model 2 wk after injury. Oligonucleotides were delivered using either Pluronics® or ethylenvinylacetate (EVA) polymers, and the extent of neointimal formation was determined as a ratio of neointimal area to the medial area using automated computer-guided edge detection system. Antisense 1: GTGTCGGGGTCTCCGGGC, nucleotides 4–21 of the mouse sequence; Antisense 2: GCTGTGTCGGGGTCTCCG, nucleotides 22–39 of the mouse sequence; Scrambled: CTGCGTTGGCGTG CGGCG; Sense: GCCCGGAGACCCCGACAC.

mation *(55)*. These observations suggest that very localized effect can be achieved with appropriate antisense oligonucleotide administration, thus minimizing potential systemic toxicity. As in the case of in vitro studies, specificity of antisense effect was demonstrated by the lack of inhibition of neointimal growth in arteries treated with missense or scrambled oligonucleotides (Table 1).

These experiments do not define the temporal extent of antisense oligonucleotide activity in the arterial wall. Indeed, earlier studies have shown that the suppression of SMC migration/proliferation for as little as 2–3 d is sufficient to affect neointimal formation 2 wk later *(56)*. Furthermore, neointimal formation in a rat carotid artery injury model is critically dependent on both medial cell proliferation as well as migration, and early suppression of these processes may be enough for a prolonged effect. Therefore, the action of antisense nucleotides could have been limited to only a few days. However, it is also possible that the oligonucleotide can persist in the vessel wall for an extended period of time. Indeed, we have recovered undegraded phosphorthioate oligonucleotides from arterial tissue up to 7 d after delivery, and a number of investigators have demonstrated extended survival of oligonucleotides in different tissues *(57–59)*.

Subsequent to this study, the same local extravascular antisense oligonucleotide delivery was used to study the in vivo role of a number of other genes, including c-*myc* *(60)*, PCNA *(61)*, and cdc2/cdk2 *(62)*. In all of these studies, extravascular polaxamer gel delivery was associated with prolonged suppression of neointimal formation in a rat carotid model, thus suggesting that inhibition of any of the final common pathway genes may be sufficient for effective control of cell proliferation. The studies of antisense oligonucleotides in prevention of smooth muscle cell proliferation after arterial injury have now been extended to other animal model and delivery systems *(63–66)* with one group of investigators demonstrating inhibitory effect of antisense c-*myb* oligonucleotides in porcine carotid artery injury model *(65)*.

The studies described in this chapter suggest an important role for c-*myb* in control of smooth muscle cell growth and proliferation. However, it remains to be seen whether antisense oligonucleotides will be proven to be useful therapeutic agents.

References

1. Shen-Ong, G. L. C. (1990) The myb oncogene. *Biochim. Biophys. Acta* **1032,** 39–52.
2. Vellard, M., Soret, J., Viegas-Pequignot, E., Galibert, F., Cong, N. V., Dutrillaux, B., and Perbal, B. (1991) c-Myb protooncogene: evidence for intermolecular recombination of coding sequences. *Oncogene* **6,** 505–514.
3. Dudek, H. and Reddy, E. P. (1989) Identification of two translational products for c-myb. *Oncogene* **4,** 1061–1066.
4. Frampton, J., Gibson, T. J., Ness, S. A., Doderlein, G., and Graf, T. (1991) Proposed structure for the DNA-binding domain of the c-myb oncoprotein based on model building and mutational analysis. *Prot. Eng.* **4,** 891–901.
5. Gabrielsen, O. S., Sentenac, A., and Fromageot, P. (1991) Specific DNA binding by c-Myb: evidence for a double helix-turn-helix-related motif. *Science* **253,** 1140–1143.
6. Howe, K. M., Reakes, C. F., and Watson, R. J. (1990) Characterization of the sequence-specific interaction of mouse c-myb protein with DNA. *EMBO J.* **9,** 161–169.
7. Nishina, Y., Nakagoshi, H., Imamoto, F., Gonda, T. J., and Ishii, S. (1989) Transactivation by the c-myb proto-oncogene. *Nucleic Acid Res.* **17,** 107–117.
8. Dudek, H., Tantravahi, R. V., Rao, V. N., Reddy, E. S., and Reddy, E. P. (1992) Myb and Ets proteins cooperate in transcriptional activation of the mim-1 promoter. *Proc. Natl. Acad. Sci. USA* **89,** 1291–1295.
9. Evans, J. L., Moore, T. L., Kuehl, W. M., Bender, T., and Ting, J. P. (1990) Functional analysis of c-Myb protein in T-lymphocytic cell lines shows that it transactivates the c-myc promoter. *Mol. Cell Biol.* **10,** 5747–5752.
10. Ramsay, R. G., Ishii, S., and Gonda, T. J. (1992) Interaction of the myb protein with specific DNA binding sites. *J. Biol. Chem.* **267,** 5656–5662.
11. Thompson, C. B., Challoner, P. B., Neiman, P. E., and Groudine, M. (1986) Expression of the c-myb proto-oncogene during cellular proliferation. *Nature* **319,** 374–380.
12. Griffin, C. A. and Baylin, S. B. (1985) Expression of the c-myb oncogene in human small cell lung carcinoma. *Cancer Res.* **45,** 272–275.

13. Melani, C., Rivoltini, L., Parmiani, G., Calabretta, B., and Colombo, M. P. (1991) Inhibition of proliferation by c-myb antisense oligodeoxynucleotides in colon adenocarcinoma cell lines that express c-myb. *Cancer Res.* **51**, 2897–2901.

14. Raschella, G., Negroni, A., Skorski, T., Pucci, S., Nieborowska-Skorska, M., Romeo, A., and Calabretta, B. (1992) Inhibition of proliferation by c-myb antisense RNA and oligodeoxynucleotides in transformed neuroectodermal cell lines. *Cancer Res.* **52**, 4221–4226.

15. Dyson, P. J., Poirier, F., and Watson, R. J. (1989) Expression of c-myb in embryonal carcinoma cells and embryonal stem cells. *Differentiation* **42**, 24–27.

16. Brown, K. E., Kindy, M. S., and Sonenshein, G. E. (1992) Expression of the c-myb proto-oncogene in bovine vascular smooth muscle cells. *J. Biol. Chem.* **267**, 4625–4630.

17. Luscher, B. and Eisenman, R. N. (1990) New light on Myc and Myb. Part II. Myb. *Genes Dev.* **4**, 2235–2241.

18. Clarke, M. F., Kukowska-Latallo, J. F., Westin, E., Smith, M., and Prochownik, E. V. (1988) Constitutive expression of a c-myb cDNA blocks Friend murine erythroleukemia cell differentiation. *Mol. Cell Biol.* **8**, 884–892.

19. Travali, S., Ferber, A., Reiss, K., Sell, C., Koniecki, J., Calabretta, B., and Baserga, R. (1991) Effect of the myb gene product on expression of the PCNA gene in fibroblasts. *Oncogene* **6**, 887–894.

20. Simons, M. and Rosenberg, R. D. (1992) Antisense nonmuscle myosin heavy chain and c-myb oligonucleotides suppress smooth muscle cell proliferation in vitro. *Circ. Res.* **70**, 835–843.

21. Anfossi, G., Gewirtz, A. M., and Calabretta, B. (1989) An oligomer complementary to c-myb-encoded mRNA inhibits proliferation of human myeloid leukemia cell lines. *Proc. Natl. Acad. Sci. USA.* **86**, 3379–3383.

22. Lipsick, J. S. and Boyle, W. J. (1987) c-myb protein expression is a late event during T-lymphocyte activation. *Mol. Cell Biol.* **7**, 3358–3360.

23. Siu, G., Wurster, A. L., Lipsick, J. S., and Hedrick, S. M. (1992) Expression of the CD4 gene requires a Myb transcription factor. *Mol. Cell Biol.* **12**, 1592–1604.

24. Gewirtz, A. M. and Calabretta, B. (1991) Role of the c-myb and c-abl proto-oncogenes in human hematopoiesis. *Ann. NY Acad. Sci.* **62**, 63–73.

25. Weston, K. M. (1990) The myb genes. *Semin. Cancer Biol.* **1**, 371–382.

26. Reilly, C. F., Kindy, M. S., Brown, K. E., Rosenberg, R. D., and Sonenshein, G. E. (1989) Heparin prevents vascular smooth muscle cell progression through the G_1 phase of the cell cycle. *J. Biol. Chem.* **264**, 6990–6995.

27. Simons, M., Morgan, K. G., Parker, C., Collin, E., and Rosenberg, R. D. (1993) Proto-oncogene c-myb mediates an intracellular calcium rise during the late G_1 phase of the cell cycle. *J. Biol. Chem.* **268**, 627–632.

28. Keith, C. H., Ratan, R., Maxfield, F. R., Bajer, A., and Shelanski, Z. (1985) Local cytoplasmic calcium gradients in living mitotic cells. *Nature* **316**, 848–850.

29. Poenie, M., Alderton, J., Steinhardt, R., and Tsien, R. (1986) Calcium rises abruptly and briefly throughout the cell at the onset of anaphase. *Science* **233**, 886–889.

30. Steinhardt, R. A. and Alderton, J. (1988) Intracellular free calcium rise triggers nuclear envelope breakdown in the sea urchin embryo. *Nature* **332**, 364–366.

31. Twigg, J., Patel R., and Whitaker, M. (1988) Translational control of InsP3-induced chromatin condensation during the early cycles of sea urchin embryos. *Nature* **332**, 366–369.

32. Sisken, J. E., Silver, R. B., Barrows, G. H., and Grasch, S. D. (1985) Studies on the role of Ca^{++} ions in cell division with the use of fluorescent probes and quantitative video intensification microscopy, in *Advances in Microscopy* (Cowden, R. R. and Harrison, F. W., eds.), Liss, New York, pp. 73–87.

33. McIntosh, J. R. and Koonce, M. P. (1989) Mitosis. *Science* **246**, 622–628.

34. Paul, D. and Ristom, H. J. J. (1979) Cell cycle control by Ca^{++} ions in mouse 3T3 cells and in transformed 3T3 cells. *J. Cell. Physiol.* **98**, 31–40.

35. Pardee, A. B., Dubrow, R., Hamlin, J. L., and Kletzien, R. F. (1978) Animal cell cycle. *Ann. Rev. Biochem.* **47**, 715–750.

36. Simons, M., Hideao, A., Salzman, E. W., and Rosenberg, R. D. (1995) c-Myb affects intracellular calcium handling in vascular smooth muscle cells. *Am. J. Physiol.: Cell Physiol.* **37**, C856–C868.

37. Garratt, K. N., Edwards, W. D., Kaufmann, U. P., Vliestra, R. E., and Holmes, D. R., Jr. (1991) Differential histopathology of primary atherosclerotic and restenotic lesions in coronary arteries and saphenous vein bypass grafts: analysis of tissue obtained from 73 patients by directional atherectomy. *J. Am. Coll. Cardiol.* **17**, 442–448.

38. Dilley, R. J., Mcgeachie, J. K., and Prendergast, F. J. (1988) A review of the histological changes in vein-to-artery grafts, with particular reference to intimal hyperplasia. *Arch. Surg.* **123**, 691–696.

39. Schwartz, S. M. and Reidy, M. A. (1987) Common, mechanisms of proliferation of smooth muscle in atherosclerosis and hypertension. *Human Pathol.* **18**, 240–247.

40. Schwartz, S. M., Campbell, G. R., and Campbell, J. H. (1986) Replication of smooth muscle cells in vascular disease. *Circ. Res.* **58**, 427–444.

41. Kocher, O., Gabbiani, F., Gabbiani, G., Reidy, M. A., Cokay, M. S., Peters, H., and Huttner, I. (1991) Phenotypic features of smooth muscle cells during the evolution of experimental carotid artery intimal thickening. *Lab. Invest.* **65**, 459–470.

42. Ross, R. (1986) The pathogenesis of atherosclerosis: an update. *N. Engl. J. Med.* **314**, 488–500.

43. Clowes, A. W., Clowes, M. M., Fingerle, J., and Reidy, M. A. (1989) Kinetics of cellular proliferation after arterial injury. V. Role of acute distension in the induction of smooth muscle proliferation. *Lab. Invest.* **60**, 360–364.

44. Clowes, A. W., Reidy, M. A., and Clowes, M. M. (1983) Kinetics of cellular proliferation after arterial injury. I. Smooth muscle growth in the absence of endothelium. *Lab. Invest.* **49**, 327–333.

45. Fingerle, J., Johnson, R., Clowes, A. W., Majesky, M. W., and Reidy, M. A. (1989) Role of platelets in smooth muscle cell proliferation and migration after vascular injury in rat carotid artery. *Proc. Natl. Acad. Sci. USA* **86**, 8412–8416.

46. Gewirtz, A. M., Anfossi, G., Venturelli, D., Valpreda, S., Sims, R., and Calabretta, B. (1989) G1/S transition in normal human T-lymphocytes requires the nuclear protein encoded by c-myb. *Science* **245**, 180–183.

47. Reilly, C. F. (1990) Rat vascular smooth muscle cells immortalized with SV40 large T antigen possess defined smooth muscle cell characteristics including growth inhibition by heparin. *J. Cell Physiol.* **142**, 342–351.

48. Giles, R. V. and Tidd, D. M. (1992) Increased specificity for antisense oligodeoxynucleotide targeting of RNA cleavage by RNase H using chimeric methylphosphonodiester/phosphodiester structures. *Nucleic Acid Res.* **20**, 763–770.

49. Woolf, T. M., Jennings, C. G., Rebagliati, M., and Melton, D. A. (1990) The stability, toxicity and effectiveness of unmodified and phosphorothioate antisense oligodeoxynucleotides in Xenopus oocytes and embryos. *Nucleic Acid Res.* **18**, 1763–1769.

50. Woolf, T. M., Melton, D. A., and Jennings, C. G. (1992) Specificity of antisense oligonucleotides in vivo. *Proc. Natl. Acad. Sci. USA* **89**, 7305–7309.

51. Block, L. C., Griffin, L. C., Latham, J. A., Vermass, E. H., and Toole, J. J. (1992) Selection of single-stranded DNA molecules that bind and inhibit human thrombin. *Nature* **355**, 564–566.

52. Guvakova, M. A., Yakubov, L. A., Vlodavsky, I., Tonkinson, J. L., and Stein, C. A. (1995) Phosphothioate oligodeoxynucleotides bind to basic fibroblast growth factor, inhibit its binding to cell-surface receptors, and remove it from low affinity binding sites on extracellular matrix. *J. Biol. Chem.* **270**, 2620–2627.

53. Offerman, M. K. and Medford, R. M. (1993) Induction of VCAM-1 gene expression by double-stranded RNA occurs by a p68 kinase-dependent pathway in endothelial cells. *Clin. Res.* **41**, 262a (abstr).

54. Edelman, E. R., Nugent, M. A., Smith, L. T., and Karnovsky, M. J. (1992) Basic fibroblast growth factor enhances the coupling of intimal hyperplasia and proliferation of vasa vasorum in injured rat arteries. *J. Clin. Invest.* **89**, 465–473.

55. Simons, M., Edelman, E. R., DeKeyser, J. L., Langer, R., and Rosenberg, R. D. (1992) Antisense c-myb oligonucleotides inhibit intimal arterial smooth muscle cell accumulation in vivo. *Nature* **359**, 67–70.

56. Guyton, J. R., Rosenberg, R. D., Clowes, A. W., and Karnovsky, M. J. (1980) Inhibition of rat arterial smooth muscle cell proliferation by heparin. In vivo studies with anticoagulant and nonanticoagulant heparin. *Circ. Res.* **46**, 625–634.

57. Agrawal, S., Temsamani, J., and Tang, J. Y. (1991) Pharmacokinetics, biodistribution, and stability of oligodeoxynucleotide phosphorothioates in mice. *Proc. Natl. Acad. Sci. USA* **88**, 7595–7599.

58. Temsamani, J., Tang, J. Y., and Agrawal, S. (1992) Capped oligodeoxynucleotide phosphorothioates. Pharmacokinetics and stability in mice. *Ann. NY Acad. Sci.* **660**, 318–320.

59. Goodarzi, G., Watabe, M., and Watabe, K. (1992) Organ distribution and stability of phosphorothioated oligodeoxyribonucleotides in mice. *Biopharm. Drug Dispos.* **13**, 221–227.

60. Bennett, M. R., Anglin, S., McEwan, J. R., Jagoe R., Newby, A. C., and Evan, G. I. (1994) Inhibition of vascular smooth muscle cell proliferation in vitro and in vivo by c-myc antisense oligodeoxynucleotides. *J. Clin. Invest.* **93**, 820–828.

61. Simons, M., Edelman, E. R., and Rosenberg, R. D. (1994) Antisense PCNA oligonucleotides inhibit intimal hyperplasia in a rat carotid injury model. *J. Clin. Invest.* **93**, 2351–2356.

62. Abe, J., Zhou, W., Taguchi, J., Takuwa, N., Miki, K., Okazaki, H., Kurokawa, K., Kumada, M., and Takuwa, Y. (1994) Suppression of neointimal smooth muscle cell accumulation in vivo by antisense cdc2 and cdk2 oligonucleotides in rat carotid artery. *Biochem. Biophys. Res. Comm.* **198**, 16–24.

63. Edelman, E. R., Simons, M., Sirois, M. G., and Rosenberg, R. D. (1995) c-Myc in vasculoproliferative disese. *Circ. Res.* **76**, 176–182.

64. Morishita, R., Gibbons, G. H., Ellison, K. E., Nakajima, M., Zhang, L., Kaneda, Y., Ogihara, T., and Dzau, V. J. (1993) Single intraluminal delivery of antisense cdc2 kinase and proliferating-cell nuclear antigen oligonucleotides results in chronic inhibition of neointimal hyperplasia. *Proc. Natl. Acad. Sci. USA* **90**, 8474–8478.

65. Azrin, M. A., Mitchel, J. F., Pedersen, C., Curley, T., Bow, L. M., Alberghini, T. V., Waters, D. D., and McKay, R. G. (1994) Inhibition of smooth muscle cell proliferation in vivo following local delivery of antisense c-myb oligonucleotides during angioplasty. *J. Am. Coll. Cardiol.* **23**, 396A.

66. Shi, Y., Fard, A., Vermani, P., and Zalewski, A. (1994) C-Myc antisense oligomers reduce neointimal formation in porcine coronary arteries. *J. Am. Coll. Cardiol.* **23**, 395A.

8

Antisense Therapy of Hepatitis B Virus Infection

In Vivo Analyses in the Duck Hepatitis B Virus Model

**Wolf-Bernhard Offensperger,
Silke Offensperger, and Hubert E. Blum**

1. Acquired Genetic Diseases and Antisense Oligonucleotide Therapy

Infectious diseases in general and viral infections in particular can be viewed as acquired genetic diseases *(1,2)*. At the molecular level, clinical signs and symptoms of viral infections are frequently caused by the expression or overexpression of the acquired genes. Based on this basic concept, such acquired genetic diseases should be amenable to treatment by a specific block of gene expression. Gene expression can be blocked at different levels by the following strategies: sense strategy, antigene strategy, ribozymes, antisense strategy, and interfering peptides or proteins (Fig. 1).

The antisense strategy has been explored for the treatment of malignant and viral diseases. The principle of this concept is the sequence-specific binding of an antisense oligonucleotide to the target mRNA, resulting in a translational arrest with or without degradation of the transcripts *(3–7)*. The simplicity, specificity, and elegance of this antisense strategy make it extremely attractive.

Antisense strategies have been employed in a variety of eukaryotic systems both to understand normal gene function and to block gene expression therapeutically in vitro *(8–10)*. Various oncogenes have been targeted with antisense oligonucleotides, resulting in suppression of cell growth and differentiation *(11–20)*. Viruses are particularly interesting targets for gene-inhibition therapy because they carry genetic information that is unique and distinct from the host cell. The antisense strategy has been successfully explored in vitro for the treatment of several viral infections: influenza virus *(21)*, Rous sarcoma virus *(22,23)*, human T-cell leukemia virus type I *(20)*, human immunodeficiency

From: *Methods in Molecular Medicine: Antisense Therapeutics*
Edited by: S. Agrawal Humana Press Inc., Totowa, NJ

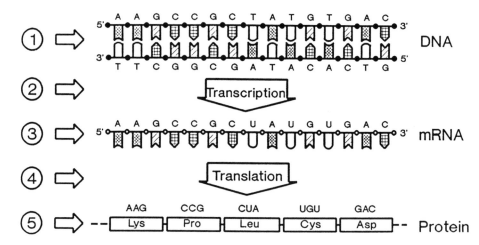

Fig. 1. Principle of gene expression and strategies aimed at block of gene expression: (1) sense strategy based on the binding of regulatory proteins, e.g., transcription factors, by oligonucleotides, resulting in a block of transcription; (2) antigene strategy based on triple helix formation between oligonucleotides and double-stranded DNA, resulting in a block of transcription; (3) ribozymes resulting in specifically targeted degradation of mRNA, resulting in a block of translation; (4) antisense strategy based on binding of oligonucleotides to mRNA, resulting in a block of translation; (5) functional inactivation of proteins by binding to other proteins or peptides synthesized intracellularly after transduction of the specific coding sequences.

virus *(24–27)*, vesicular stomatitis virus *(28)*, herpes simplex virus *(29)*, bovine papilloma virus type 1 *(30)*, Sendai virus *(31)*, and hepatitis B virus *(32–34)*. In the following, we describe antisense oligonucleotide studies aimed at the termination of duck hepatitis B virus (DHBV) infection in vitro and in vivo.

2. Hepatitis B Virus and Related Viruses

Infection with the hepatitis B virus (HBV) is endemic throughout much of the world with an estimated 300 million persistently infected people. HBV infection is associated with a wide spectrum of clinical presentations ranging from a carrier state without evidence of disease to acute, fulminant, or chronic hepatitis and liver cirrhosis. Further, chronic HBV infection is associated with the development of hepatocellular carcinoma (HCC), worldwide a leading cause of death from cancer *(35)*.

The structure and biology of HBV have been characterized in great detail *(36)*. HBV belongs to a group of hepatotropic DNA viruses (hepadnaviruses) that include the hepatitis viruses of the woodchuck *(37)*, ground squirrel *(38)*, Pekin duck *(39)*, and heron *(40)*. HBV is a small virus of about 42 nm diameter

("Dane particle") and is composed of a lipid bilayer envelope containing hepatitis B surface antigen (HBsAg) and an internal nucleocapsid structure ("core"). The nucleocapsid consists of the hepatitis B core antigen (HBcAg) and the viral DNA genome of about 3.2-kbp length with an associated DNA polymerase/reverse transcriptase. The diagnosis of HBV infection is made by serologic detection of viral antigens (HBsAg, hepatitis B e antigen [HBeAg]) or antibodies (anti-HBc, anti-HBe, anti-HBs) or by the direct demonstration of viral DNA using molecular hybridization techniques.

The hepadnaviral genomes are of similar size and structure, and replicate asymmetrically via reverse transcription of an RNA intermediate *(41)*, a replication strategy central to the life cycle of retroviruses. The genome has four open reading frames. Three of these encode for viral proteins whose structures and functions have been well characterized: the HBsAgs, the HBcAg and its cryptic antigenic determinant HBeAg, as well as the DNA polymerase/reverse transcriptase. The fourth open reading frame, which is conserved among all mammalian hepadnaviruses, encodes for a small protein X, whose biological function for the viral life cycle and the pathobiology of HBV has not been firmly established. Recent evidence suggests that the X-protein has transactivating potential and appears to induce liver cancer in transgenic mice. The factors determining the clinical course of HBV infection have not been clearly defined. Recently, mutations in various viral genes have been identified in patients with acute fulminant and chronic HBV infection *(42)*.

Studies of the life cycle and biology of HBV were greatly facilitated by the different animal models of HBV infection mentioned above. Using primary duck hepatocytes, hepadnaviral replication could be demonstrated in vitro after infection with DHBV DNA-positive serum *(43)*. Primary hepatocytes can be prepared either from uninfected ducks, followed by infection in vitro, or from DHBV-infected ducks. Thus, it is possible to study systematically different strategies aimed at termination of hepadnaviral replication. Using the primary duck hepatocyte system, nucleoside analogs *(44)*, phosphonoformate *(45)*, and supercoiled DNA-active compounds *(46)* were shown to inhibit DHBV replication. In vivo studies showed a marked, but transient inhibition of DHBV replication by nucleoside analogs *(47,48)*, or phosphonoformate *(49)*.

3. DHBV-Specific Antisense Oligodeoxynucleotides In Vitro

Phosphorothioate-modified oligodeoxynucleotides were synthesized using standard cyanoethylphosphoramidites and the sulfur transfer reagent 3H-1.2-benzodithiol-3-one 1.1-dioxide *(50)*. The genetic organization of the DHBV genome and the position of the nine antisense oligodeoxynucleotides tested are shown in Fig. 2: Four antisense oligodeoxynucleotides are located in the pre-S/ S-region (AS 1–4), one at the start of the polymerase region (AS 5), and four in

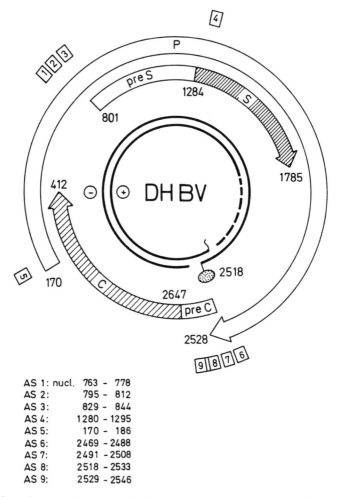

AS 1: nucl. 763 - 778
AS 2:   795 - 812
AS 3:   829 - 844
AS 4:   1280 - 1295
AS 5:   170 - 186
AS 6:   2469 - 2488
AS 7:   2491 - 2508
AS 8:   2518 - 2533
AS 9:   2529 - 2546

Fig. 2. Genetic organization of DHBV genome and location of nine antisense oligodeoxynucleotides.

the pre-C/C-region (AS 6–9). To isolate primary hepatocytes, about 10-d-old DHBV-infected Pekin ducklings were anesthesized, the livers were perfused *in situ* with collagenase (0.5 mg/mL in Williams' medium E buffered with 20 mM HEPES, pH 7.4, and 2.5 mM CaCl$_2$). The hepatocytes were seeded at a density of 2×10^5 cells/cm^2 on Primaria tissue-culture dishes using Williams medium E supplemented with 20 mM HEPES, pH 7.4, 5 mM glutamine, 0.066 μM insulin, 10 mM dexamethasone, 100 μg/mL penicillin, 100 μg/ mL streptomycin, and 1.5% dimethylsulfoxide. The infected hepatocytes were kept in culture with daily medium change containing the respective antisense oligo-

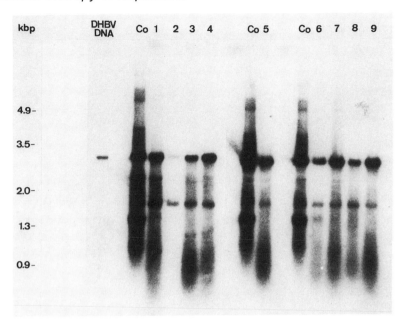

Fig. 3. Effect of antisense oligodeoxynucleotides on DHBV DNA replication in vitro. Co: untreated control cells. Numbers on top indicate antisense oligodeoxynucleotides described in Fig. 2. Size markers: *Hind*III-cut λ DNA and cloned DHBV DNA of 3.0-kbp length.

deoxynucleotide at a final concentration of 1.5 μ*M*. After 10 d, the cells were harvested and DNA was analyzed by Southern blot hybridization (Fig. 3). AS 2 and 6 showed a very strong inhibitory activity in comparison to the other antisense oligos, which all showed a decrease of intracellular viral replicative intermediates. AS 2, directed against the start of the pre-S region, led to a nearly complete inhibition of viral replication with only residual covalently closed circular DNA (ccc DNA) molecules left. Quantifying these effects by liquid scintillation counting, the hybridization signal was 3.7% after incubation with AS 2 and 9.4% after incubation with AS 6 as compared to the control cells (100%). Toxic effects could be excluded because trypan blue exclusion demonstrated identical viability of control cells and cells treated with the antisense oligodeoxynucleotides; in addition, the total RNA content of hepatocytes cultured in the presence or absence of the antisense oligos was nearly identical.

To demonstrate the specificity of the inhibition by AS 2 and AS 6, the respective sense oligodeoxynucleotides were synthesized. As shown in Fig. 4, the sense oligos did not significantly affect viral replication, demonstrating the specificity of the inhibition by the antisense oligos. To test whether the virus

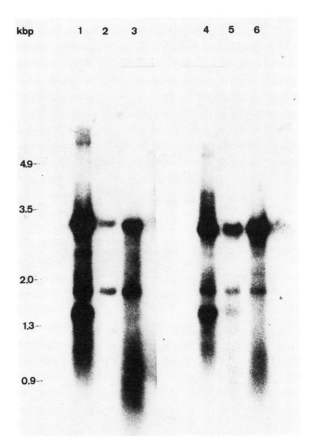

Fig. 4. Effect of AS 2 (lane 2), sense oligodeoxynucleotide 2 (lane 3), AS 6 (lane 5), and sense oligodeoxynucleotide 6 (lane 6) on DHBV replication in vitro. Untreated control cultures in lanes 1 and 4, respectively.

was definitely eliminated after this short-term treatment in vitro, cells that showed complete viral inhibition after a 10-d incubation with AS 2 were kept in culture for another 3 or 6 d in the absence of AS 2 (Fig. 5). It can be seen that 6 d after stopping the antisense treatment, viral replication re-emerges, suggesting that the residual cccDNA forms serve as template for viral reactivation.

4. DHBV-Specific Antisense Oligodeoxynucleotides In Vivo

For these experiments 1-d-old ducklings were infected with DHBV by iv injection of DHBV DNA-positive duck serum. Two weeks later, the treatment with AS 2 was started with daily iv application for 10 d. Then the ducks were

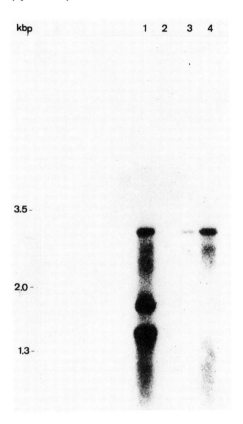

Fig. 5. DHBV replication after termination of treatment with AS 2 in vitro: lane 1: control cells; lane 2: inhibition of viral replication after a 10-d therapy with AS 2; lane 3: DHBV replication 3 d after termination of therapy with AS 2; lane 4: DHBV replication 6 d after termination of therapy with AS 2.

sacrificed, and the liver DNA was analyzed for the presence of viral replicative intermediates by Southern blot. As shown in Fig. 6, treatment with AS 2 resulted in a dose-dependent inhibition of viral replication in vivo with a nearly complete elimination of viral DNA forms at a daily dose of 20 µg/g body wt. These results demonstrate for the first time the effective in vivo treatment of a viral infection by antisense oligodeoxynucleotides.

To determine the effect of AS 2 on viral gene expression in vivo duck hepatitis B surface antigen (DHBsAg) from serum and duck hepatitis B core antigen (DHBcAg) from liver were analyzed by Western blot using polyclonal antibodies. Fig. 7 demonstrates the in vivo inhibition of viral gene expression with disappearance of viral pre-S- and S-antigens from serum and viral pre-C- and C-antigens from liver.

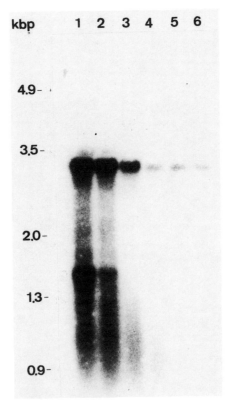

Fig. 6. Effect of AS 2 on DHBV DNA replication in vivo in DHBV-infected ducks. Lane 1: control duck; lanes 2–6: ducks treated for 10 d with daily iv of AS 2 at a concentration of 5 µg (lane 2), 10 µg (lane 3), and 20 µg/g body wt (lanes 4–6).

To detect possible side effects of this in vivo therapy several clinicochemical parameters, including alanine aminotransferase, aspartate aminotransferase, γ-glutamyl transpeptidase, cholinesterase, total protein, and albumin, were measured in serum. No differences were observed between the control ducks and the ducks treated with AS 2.

The specificity of the in vivo inhibition of DHBV replication and gene expression was analyzed using two control phosphorothioate oligodeoxynucleotides: the sense oligo complementary to AS 2 and a random oligonucleotide. As shown in Fig. 8, neither oligo causes a significant decrease in viral replicative intermediates. These data demonstrate the specificity of action of the antisense oligodeoxynucleotides in vivo.

The effect of AS 2 and control oligodeoxynucleotides on DHBV infection was tested in vivo. Two DHBV-negative ducklings were pretreated with AS 2

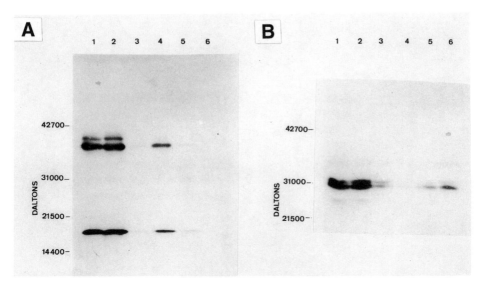

Fig. 7. (A) Western blot analysis of sera obtained from DHBV-infected ducks treated with AS 2 in vivo using a polyclonal antibody against native DHBsAg. Lane 1: control duck; lanes 2–6: ducks treated for 10 d with AS 2 at a concentration of 5 µg (lane 2), 10 µg (lane 3), and 20 µg (lanes 4–6)/g body wt. **(B)** Western blot analysis of liver extracts obtained from DHBV-infected ducks treated with AS 2 in vivo using a polyclonal antibody against DHBcAg. The numbers on top of the lanes correspond to those in (A).

starting 3 d after hatching. Twelve hours later, they were infected by iv injection of DHBV-DNA positive serum. The ducklings were treated over 10 d with a daily dose of 20 µg AS 2/g body wt. At the end of the experiment, viral DNA in liver was analyzed by Southern blot hybridization, and viral antigens in serum were analyzed by Western blot analyses (Fig. 9). In contrast to untreated ducks and to ducks treated with the sense or random oligodeoxynucleotides, no viral DNA could be detected in AS 2-treated ducks, demonstrating the successful prevention of the DHBV infection.

5. Summary and Perspectives

The data demonstrate that DHBV replication and gene expression in liver cells can be blocked by antisense oligodeoxynucleotides in vivo. The action is antisense-specific, i.e., the corresponding sequence of sense polarity does not affect viral replication or gene expression. In addition, the effect of the antisense oligodeoxynucleotides is sequence-dependent with the strongest inhibition observed for an oligonucleotide binding to a DHBV mRNA region 5' to the pre-S gene. Apart from these positive in vivo antisense oligodeoxynucleotide

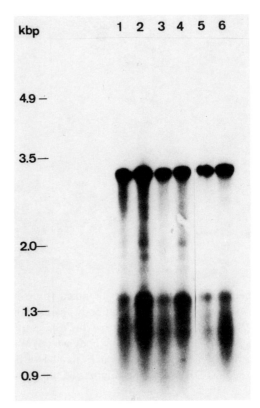

Fig. 8. Effect of sense or random oligodeoxynucleotides on DHBV DNA replication in vivo. Lanes 1 and 2: two control ducks; lanes 3 and 4: two ducks treated with the sense oligo 2 at a concentration of 20 μg/g body wt; lanes 5 and 6: two ducks treated with the random oligo at a concentration of 20 μg/g body weight.

results, several issues remain to be resolved before this strategy is potentially clinically useful for the treatment of HBV infection:

1. Definition of optimal target sequences: In in vitro assays, one of nine antisense oligodeoxynucleotides directed against different targets in the DHBV genome proved very effective. Possible factors involved in this phenomenon are the hybridization efficiency depending on secondary and tertiary structures, the significance of the targeted region for viral replication, the steric hindrance by proteins that may prevent oligonucleotide annealing, and the intracellular distribution of antisense oligodeoxynucleotides (51).

2. Definition of optimal oligodeoxynucleotide properties: Criteria that must be met by a therapeutic oligonucleotide are easy and large-scale synthesis, in vivo stability, uptake into the target cells, specific delivery to the target cells, interaction with their intracellular targets, and target sequence-specific interaction (52).

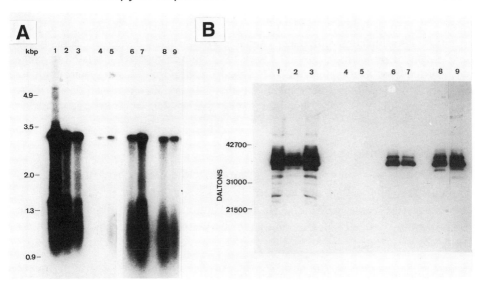

Fig. 9. In vivo effect of oligodeoxynucleotides on DHBV infectivity. The treatment of the ducklings with the oligodeoxynucleotides was started 12 h before infection with DHBV-positive serum. **(A)** Southern blot analysis: lanes 1–3: three control ducklings; lanes 4 and 5: two ducklings treated with AS 2; lanes 6 and 7: two ducklings treated with sense oligo 2; lanes 8 and 9: two ducklings treated with the random oligo. **(B)** Western blot analysis of the sera using a polyclonal antibody against DHBpre-sAg. The numbers above the lanes correspond to those in (A).

Clearly, more detailed analyses of pharmacokinetic properties of the oligonucleotides and their side effects are necessary. Further, modified oligodeoxynucleotides, e.g., peptide nucleic acids *(53)* or phosphorothioate analogs containing C-5 propyne pyrimidines *(54)*, may prove more efficient than unmodified oligodeoxynucleotides.

3. Transient effect of antisense oligodeoxynucleotides: In vitro and in vivo experiments showed the reactivation of DHBV replication after cessation of antisense oligodeoxynucleotide treatment, presumably because of the resistance of the cccDNA to this therapeutic strategy. Two possibilities to solve this problem are under investigation in our laboratory. Long-term treatment with AS 2 could eliminate all viral DNA forms, and triple helix-forming oligonucleotides could inhibit DHBV transcription and directly attack viral cccDNA *(55)*.

4. Targeted delivery of antisense nucleotides. To reduce the amount of oligonucleotide necessary to inhibit viral replication, the oligonucleotides should be targeted to liver cells. This could be achieved by encapsidation of the antisense oligodeoxynucleotides into liposomes or immunoliposomes *(56,57)* or by coupling to an asialoglycoprotein for which liver cells have a specific and unique receptor *(34,58)*. However, with this strategy, extrahepatic viruses in lympho-

cytes, spleen, pancreas, and so forth, could escape therapy, resulting in a reservoir for reinfection of liver cells and reactivation of the disease. To address this potential problem, adenovirus-based or transferrin-receptor-mediated delivery systems could be explored *(59–61)*. Finally, a retroviral delivery system could be used to target a construct to liver cells that will produce antisense DNA or RNA intracellularly *(62,63)*, thereby reducing the problem of oligonucleotide size and stability.

Acknowledgments

The study was supported by a grant from the Deutsche Forschungsgemeinschaft. The advice and help of F. Eckstein (Göttingen, Germany), S. Beaucage (Bethesda, MD), and H. Will (Hamburg, Germany) as well as the technical assistance of B. Hockenjos, P. Kary, and E. Schiefermayr are gratefully acknowledged.

References

1. Anderson, W. F. (1992) Human gene therapy. *Science* **256,** 808–813.
2. Morgan, R. A. and Anderson, W. F. (1993) Human gene therapy. *Annu. Rev. Biochem.* **62,** 191–217.
3. Agrawal, S. (1992) Antisense oligonucleotides as antiviral agents. *Trends Biotechnol.* **10,** 281–287.
4. Helene, C. (1991) Rational design of sequence-specific oncogene inhibitors based on antisense and antigene oligonucleotides. *Eur. J. Cancer* **27,** 1466–1471.
5. Helene, C. and Toulme, J. J. (1990) Specific regulation of gene expression by antisense, sense and antigene nucleic acids. *Biochim. Biophys. Acta* **1049,** 99–125.
6. Uhlmann, E. and Peyman, A. (1990) Antisense oligonucleotides: a new therapeutic principle. *Chem. Rev.* **90,** 544–583.
7. Wickstrom, E. (1992) Strategies for administering targeted therapeutic oligodeoxynucleotides. A review. *Trends Biotechnol.* **10,** 281–287.
8. Calabretta, B. (1991) Inhibition of protooncogene expression by antisense oligodeoxynucleotides: biological and therapeutic implications. *Cancer Res.* **51,** 4505–4510.
9. Ma, D. D. F. and le Doan, T. (1994) Antisense oligonucleotide therapies: are they the "magic bullets"? *Ann. Intern. Med.* **120,** 161–163.
10. Stein, C. A. and Cohen, J. S. (1988) Oligodeoxynucleotides as inhibitors of gene expression: a review. *Cancer Res.* **48,** 2659–2668.
11. Becker, D., Meier, C. B., and Herlyn, M. (1989) Proliferation of human malignant melanomas is inhibited by antisense oligodeoxynucleotides targeted against basic fibroblast growth factor. *EMBO J.* **8,** 3685–3691.
12. Harel-Bellan, A., Ferris, D. K., Vinocour, M., Holt, J. T., and Farrar, W. L. (1988) Specific inhibition of c-*myc* protein biosynthesis using an antisense synthetic deoxyoligonucleotide in human T lymphocytes. *J. Immunol.* **140,** 2431–2435.

13. Holt, J. T., Redner, R. L., and Nienhuis, A. W. (1988) An oligomer complementary to c-myc mRNA inhibits proliferation of HL-60 promyelocytic cells and induces differentiation. *Mol. Cell. Biol.* **8,** 963–973.

14. McManaway, M. E., Neckers, L. M., Loke, S. L., Al-Nasser, A. A., Redner, R. L., Shiramizu, B. T., Goldschmidts, W. L., Huber, B. E., Bhatia, K., and Magrath, I. T. (1990) Tumour-specific inhibition of lymphoma growth by an antisense oligodeoxynucleotide. *Lancet* **335,** 808–811.

15. Melani, C., Rivoltini, L., Parmiani, G., Calabretta, B., and Colombo, M. P. (1991) Inhibition of proliferation by c-myb antisense oligodeoxynucleotides in colon adenocarcinoma cell lines that express c-myb. *Cancer Res.* **51,** 2897–2901.

16. Reed, J. C., Stein, C., Subasinghe, C., Haldar, S., Croce, C. M., Yum, S., and Cohen, J. (1990) Antisense-mediated inhibition of BCL2 protooncogene expression and leukemic cell growth and survival: comparisons of phosphodiester and phosphorothioate oligodeoxynucleotides. *Cancer Res.* **50,** 6565–6570.

17. Rivera, R. T., Pasion, S. G., Wong, D. T., Fei, Y. B., and Biswas, D. K. (1989) Loss of tumorigenic potential by human lung tumor cells in the presence of antisense RNA specific to the ectopically synthesized alpha subunit of human chorionic gonadotropin. *J. Cell Biol.* **108,** 2423–2434.

18. Saison-Behmoaras, T., Tocque, B., Rey, I., Chassignol, M., Thuong, N. T., and Helene, C. (1991) Short modified antisense oligonucleotides directed against Ha-ras point mutation induce selective cleavage of the mRNA and inhibit T24 cells proliferation. *EMBO J.* **10,** 1111–1118.

19. Szczylik, C., Skorski, T., Nicolaides, N. C., Manzella, L., Malaguarnera, L., Venturelli, D., Gewirtz, A. M., and Calabretta, B. (1991) Selective inhibition of leukemia cell proliferation by BCR-ABL antisense oligodeoxynucleotides. *Science* **253,** 562–565.

20. von Rüden, T. and Gilboa, E. (1989) Inhibition of human T-cell leukemia virus type I replication in primary human T cells that express antisense RNA. *J. Virol.* **63,** 677–682.

21. Kabanov, A. V., Vinogradov, S. V., Ovcharenko, A. V., Krivonos, A. V., Melik, N. N. S., Kiselev, V. I., and Severin, E. S. (1990) A new class of antivirals: antisense oligonucleotides combined with a hydrophobic substituent effectively inhibit influenza virus reproduction and synthesis of virus-specific proteins in MDCK cells. *FEBS Lett.* **259,** 327–330.

22. Chang, L. J. and Stoltzfus, C. M. (1987) Inhibition of Rous sarcoma virus replication by antisense RNA. *J. Virol.* **61,** 921–924.

23. Zamecnik, P. C. and Stephenson, M. L. (1978) Inhibition of Rous sarcoma virus replication and cell transformation by a specific oligodeoxynucleotide. *Proc. Natl. Acad. Sci. USA* **75,** 280–284.

24. Agrawal, S., Goodchild, J., Civeira, M. P., Thornton, A. H., Sarin, P. S., and Zamecnik, P. C. (1988) Oligodeoxynucleoside phosphoramidates and phosphorothioates as inhibitors of human immunodeficiency virus. *Proc. Natl. Acad. Sci. USA* **85,** 7079–7083.

25. Matsukura, M., Shinozuka, K., Zon, G., Mitsuya, H., Reitz, M., Cohen, J. S., and Broder, S. (1987) Phosphorothioate analogs of oligodeoxynucleotides: inhibitors of replication and cytopathic effects of human immunodeficiency virus. *Proc. Natl. Acad. Sci. USA* **84,** 7706–7710.

26. Renneisen, K., Leserman, L., Matthes, E., Schröder, H. C., and Müller, W. E. (1990) Inhibition of expression of human immunodeficiency virus-1 in vitro by antibody-targeted liposomes containing antisense RNA to the env region. *J. Biol. Chem.* **265,** 16,337–16,342.

27. Zamecnik, P. C., Goodchild, J., Taguchi, Y., and Sarin, P. S. (1986) Inhibition of replication and expression of human T-cell lymphotropic virus type III in cultured cells by exogenous synthetic oligonucleotides complementary to viral RNA. *Proc. Natl. Acad. Sci. USA* **83,** 4143–4146.

28. Lemaitre, M., Bayard, B., and Lebleu, B. (1987) Specific antiviral activity of a poly-lysine-conjugated oligodeoxyribonucleotide sequence complementary to vesicular stomatitis virus N protein mRNA initiation site. *Proc. Natl. Acad. Sci. USA* **84,** 648–652.

29. Smith, C. C., Aurelian, L., Reddy, M. P., Miller, P. S., and Ts'o, P. O. P. (1986) Antiviral effect of an oligo (nucleoside methylphosphonate) complementary to the splice junction of herpes simplex virus type 1 immediate early pre-mRNAs 4 and 5. *Proc. Natl. Acad. Sci. USA* **83,** 2787–2791.

30. Bergman, P., Ustav, M., Moreno-Lopez, J., Vennstroem, B., and Pettersson, U. (1986) Replication of the bovine papillomavirus type 1 genome; antisense transcripts prevent episomal replication. *Gene* **50,** 185–193.

31. Gupta, K. C. (1987) Antisense oligodeoxynucleotides provide insight into mechanism of translation initiation of two Sendai virus mRNAs. *J. Biol. Chem.* **262,** 7492–7496.

32. Blum, H. E., Galun, E., von Weizsäcker, F., and Wands, J. R. (1991) Inhibition of hepatitis B virus by antisense oligodeoxynucleotides (letter). *Lancet* **337,** 1230.

33. Goodarzi, G., Gross, S. C., Tewari, A., and Watabe, K. (1990) Antisense oligodeoxyribonucleotides inhibit the expression of the gene for hepatitis B virus surface antigen. *J. Gen. Virol.* **71,** 3021–3025.

34. Wu, G. Y. and Wu, C. H. (1992) Specific inhibition of hepatitis B viral gene expression in vitro by targeted antisense oligonucleotides. *J. Biol. Chem.* **267,** 12,436–12,439.

35. Wands, J. R. and Blum, H. E. (1991) Primary hepatocellular carcinoma. *N. Engl. J. Med.* **325,** 729–731.

36. Tiollais, P., Pourcel, C., and Dejean, A. (1985) The hepatitis B virus. *Nature* **317,** 489–495.

37. Summers, J., Smolec, J. M., and Snyder, R. (1978) A virus similar to hepatitis B virus associated with hepatitis and hepatoma in woodchucks. *Proc. Natl. Acad. Sci. USA* **75,** 4533–4537.

38. Marion, P. L., Oshiro, L. S., Regnery, D. C., Scullard, G. H., and Robinson, W. S. (1980) A virus in Beechey ground squirrels that is related to hepatitis B virus of humans. *Proc. Natl. Acad. Sci. USA* **77,** 2941–2945.

39. Mason, W. S., Seal, G., and Summers, J. (1980) Virus of Pekin ducks with structural and biological relatedness to human hepatitis B virus. *J. Virol.* **36**, 829–836.

40. Sprengel, R., Kaleta, E. F., and Will, H. (1988) Isolation and characterization of a hepatitis B virus endemic in herons. *J. Virol.* **62**, 3832–3839.

41. Summers, J. W. and Mason, W. S. (1982) Replication of the genome of a hepatitis B-like virus by reverse transcription of an RNA intermediate. *Cell* **29**, 403–415.

42. Carman, W., Thomas, H., and Domingo, E. (1993) Viral genetic variation: hepatitis B virus as a clinical example. *Lancet*, **341**, 349–353.

43. Tuttleman, J. S., Pugh, J. C. and Summers, J. W. (1986) In vitro experimental infection of primary duck hepatocyte cultures with duck hepatitis B virus. *J. Virol.* **58**, 17–25.

44. Suzuki, S., Lee, B., Luo, W., Tovell, D., Robins, M. J., and Tyrrell, D. L. (1988) Inhibition of duck hepatitis B virus replication by purine 2',3'- dideoxynucleosides. *Biochem. Biophys. Res. Commun.* **156**, 1144–1151.

45. Fourel, I., Gripon, P., Hantz, O., Cova, L., Lambert, V., Jacquet, C., Watanabe, K., Fox, J., Guillouzo, C., and Trepo, C. (1989) Prolonged duck hepatitis B virus replication in duck hepatocytes cocultivated with rat epithelial cells: a useful system for antiviral testing. *Hepatology* **10**, 186–191.

46. Civitico, G., Wang, Y. Y., Luscombe, C., Bishop, N., Tachedjian, G., Gust, I., and Locarnini, S. (1990) Antiviral strategies in chronic hepatitis B virus infection: II. Inhibition of duck hepatitis B virus in vitro using conventional antiviral agents and supercoiled-DNA active compounds. *J. Med. Virol.* **31**, 90–97.

47. Hirota, K., Sherker, A. H., Omata, M., Yokosuka, O., and Okuda, K. (1987) Effects of adenine arabinoside on serum and intrahepatic replicative forms of duck hepatitis B virus in chronic infection. *Hepatology* **7**, 24–28.

48. Kassianides, C., Hoofnagle, J. H., Miller, R. H., Doo, E., Ford, H., Broder, S., and Mitsuya, H. (1989) Inhibition of duck hepatitis B virus replication by 2',3'-dideoxycytidine. A potent inhibitor of reverse transcriptase. *Gastroenterology* **97**, 1275–1280.

49. Sherker, A. H., Hirota, K., Omata, M., and Okuda, K. (1986) Foscarnet decreases serum and liver duck hepatitis B virus DNA in chronically infected ducks. *Gastroenterology* **91**, 818–824.

50. Iyer, R. P., Philips, L. R., Egan, W., Regan, J. B., and Beaucage, S. L. J. (1990) The automated synthesis of sulphur-containing oligodeoxyribonucleotides using 3H-1,2-benzodithiol-3-one 1,1-dioxide as a sulphur-transfer reagent. *Org. Chem.* **55**, 4693–4698.

51. Leonetti, J. P., Mechti, N., Degols, G., Gagnor, C., and Lebleu, B. (1991) Intracellular distribution of microinjected antisense oligonucleotides. *Proc. Natl. Acad. Sci. USA* **88**, 2702–2706.

52. Stein, C. A. and Cheng, Y. C. (1993) Antisense oligonucleotides as therapeutic agents—is the bullet really magical? *Science* **261**, 1004–1012.

53. Hanvey, J. C., Peffer, N. J., Bisi, J. E., Thomson, S. A., Cadilla, R., Josey, J. A., Ricca, D. J., Hassman, C. F., Bonham, M. A., Au, K. G., et al. (1992) Antisense and antigene properties of peptide nucleic acids. *Science* **258**, 1481–1485.

54. Wagner, R. W., Matteucci, M. D., Lewis, J. G., Gutierrez, A. J., Moulds, C., and Froehler, B. C. (1993) Antisense gene inhibition by oligonucleotides containing C-5 propyne pyrimidines. *Science* **260**, 1510–1513.

55. Duval-Valentin, G., Thuong, N. T., and Helene, C. (1992) Specific inhibition of transcription by triple helix-forming oligonucleotides. *Proc. Natl. Acad. Sci. USA* **89**, 504–508.

56. Felgner, P. L., Gadek, T. R., Holm, M., Roman, R., Chan, H. W., Wenz, M., Northrop, J. P., Ringold, G. M., and Danielsen, M. (1987) Lipofection: a highly efficient, lipid-mediated DNA-transfection procedure. *Proc. Natl. Acad. Sci. USA* **84**, 7413–7417.

57. Leonetti, J. P., Machy, P., Degols, G., Lebleu, B., and Leserman, L. (1990) Antibody-targeted liposomes containing oligodeoxyribonucleotides complementary to viral RNA selectively inhibit viral replication. *Proc. Natl. Acad. Sci. USA* **87**, 2448–2451.

58. Liang, T. J., Makdisi, W. J., Sun, S., Hasegawa, K., Zhang, Y., Wands, J. R., Wu, C. H., and Wu, G. Y. (1993) Targeted transfection and expression of hepatitis B viral DNA in human hepatoma cells. *J. Clin. Invest.* **91**, 1241–1246.

59. Cotten, M., Wagner, E., Zatloukal, K., Phillips, S., Curiel, D. T., and Birnstiel, M. L. (1992) High-efficiency receptor-mediated delivery of small and large (48 kilobase) gene constructs using the endosome-disruption activity of defective or chemically inactivated adenovirus particles. *Proc. Natl. Acad. Sci. USA* **89**, 6094–6098.

60. Cristiano, R. J., Smith, L. C., and Woo, S. L. (1993) Hepatic gene therapy: adenovirus enhancement of receptor-mediated gene delivery and expression in primary hepatocytes. *Proc. Natl. Acad. Sci. USA* **90**, 2122–2126.

61. Wagner, E., Zenke, M., Cotten, M., Beug, H., and Birnstiel, M. L. (1990) Transferrin-polycation conjugates as carriers for DNA uptake into cells. *Proc. Natl. Acad. Sci. USA* **87**, 3410–3414.

62. Kay, M. A., Baley, P., Rothenberg, S., Leland, F., Fleming, L., Ponder, K. P., Liu, T., Finegold, M., Darlington, G., Pokorny, W., and Woo, S. L. C. (1992) Expression of human alpha 1-antitrypsin in dogs after autologous transplantation of retroviral transduced hepatocytes. *Proc. Natl. Acad. Sci. USA* **89**, 89–93.

63. Ledley, F. D., Darlington, G. J., Hahn, T., and Woo, S. L. (1987) Retroviral gene transfer into primary hepatocytes: implications for genetic therapy of liver-specific functions. *Proc. Natl. Acad. Sci. USA* **84**, 5335–5339.

9

Antisense-Mediated Inhibition of Protein Synthesis

Rational Drug Design, Pharmacokinetics, Intracerebral Application, and Organ Uptake of Phosphorothioate Oligodeoxynucleotides

Wolfgang Brysch, Abdalla Rifai, Wolfgang Tischmeyer, and Karl-Hermann Schlingensiepen

1. Introduction

There is hardly any class of drugs for which the term "rational drug design" is more appropriate than for the currently developing antisense therapeutics. The specificity of the hybridization reaction and the surprisingly efficient uptake of synthetic oligonucleotide derivatives provide a new class of selective protein synthesis inhibitors. Concurrently with the development of the antisense technology, elucidation of the pathogenetic role of individual proteins for certain diseases is rapidly progressing, most notably in the fields of cancer research and virology. Consequently, the first clinical trials are being conducted with antisense therapeutics for the treatment of viral diseases and neoplasms.

Despite the success of different antisense studies, in particular, with the very stable phosphorothioate oligodeoxynucleotides (S-ODN), several properties of these compounds can be improved. This chapter focuses on the quantification, time-course, and specificity of antisense-mediated inhibition of protein synthesis. The issue of S-ODN specificity was addressed by targeting two genes, c-*jun* and *jun*B, with similar induction properties, high sequence homology, and functional similarity, in the same cell line. Both antisense S-ODN specifically inhibited synthesis of the respective target protein, but not of the other Jun family member.

Antisense oligonucleotides are generally designed to inhibit protein synthesis by hybridization to mRNA. However, other RNA molecules may also pro-

From: *Methods in Molecular Medicine: Antisense Therapeutics*
Edited by: S. Agrawal Humana Press Inc., Totowa, NJ

vide target sequences. Thus, as a novel model system for the rational design of an antisense oligonucleotide, we developed a sequence complementary to the human methionine transfer RNA (Met-tRNA) and targeted it to the anticodon region. This anti-Met-tRNA sequence should mask the anticodon region and make it unavailable for recognition of AUG codons. As predicted, this anti-Met-tRNA S-ODN inhibited *de novo* protein synthesis and cell proliferation. In contrast, S6 kinase activity, a potent regulator of the protein translation rate, was strongly increased. Activation of this enzyme would be expected as a compensatory feedback mechanism and confirms the specificity of the anti-Met-tRNA S-ODN activity.

In a further set of experiments, cellular uptake of phosphorothioate oligonucleotides was studied in culture and in vivo. For in vivo experiments, oligonucleotides were injected intravenously or intracerebrally. For iv application, a poly-dT S-ODN was used, which may be regarded as the universal antisense sequence against all poly-A RNAs. As a control, we used a poly-dA S-ODN. Surprisingly, some significant differences in the pharmacokinetics of the two sequences became evident. Furthermore, they also differed with respect to organ uptake. Oligo-T S-ODN uptake was highest in the liver, whereas Oligo-A uptake was most pronounced in the kidney.

Following intracerebral application, wide distribution and efficient neuronal and glial uptake of the S-ODN injected into rat brain could be observed. The efficient neuronal uptake and intraneuronal protein suppression are encouraging with respect to the development of antisense pharmaceuticals for the treatment of neurological disorders, possibly including such prevalent diseases as Alzheimer type neuronal degeneration and various types of mental illness.

1.1. Molecular Basis for Antisense Therapeutics

The first experiment demonstrating inhibition of protein translation by antisense oligodeoxynucleotides was reported in 1978 by Stephenson and Zamecnik *(1)*. The same authors showed that the inhibition of translation of Rous sarcoma virus mRNA could decrease viral replication and cell transformation *(2)*. Selective inhibition of protein synthesis thus opened the possibility to suppress expression of those proteins that play a significant role in particular disease processes. However, for effective therapy, antisense oligonucleotides have to be more stable than either single-stranded DNA or single-stranded RNA, since both are rapidly degraded by nucleases in serum and inside the cell *(3)*. Therefore, the discovery by Eckstein and coworkers of phosphorothioate DNA, a chemical derivative that is very resistant to nuclease digestion (reviewed in ref. *4*), marked a significant step forward in the development of antisense-based therapeutics (reviewed in ref. *5*). Chemically synthesized S-ODN have been by far the most effective antisense molecules to date. Thus, the majority of clinical studies with antisense molecules that are currently planned or under way use S-ODN compounds (*see* Chapters 2, 11, and 14).

1.1.1. Identification of the Molecular Pathophysiology of Diseases

Effective antisense therapy is based on the identification of the pathophysiological role of individual protein products in a disease state and requires information about the encoding gene's nucleotide sequence. In recent years, an ever-increasing number of human genes were sequenced, and many of these genes had been traced by their suspected involvement in disease processes. In cancer research, the relevance of different oncogenes for the development of neoplasms could be elucidated *(6)*. With respect to infectious diseases, a large array of genes, ranging from viral to protozoal, could be sequenced.

Since not only gene sequences, but also the function of many protein products, could be identified—now increasingly by the use of loss of function analysis with the antisense technique—the etiologic relevance of a variety of genes for pathological processes became clear. The antisense technology now potentially allows a causal therapeutic approach and enables us to tackle diseases at their molecular roots.

1.1.2. Molecular Basis of the Antisense Approach

Antisense drugs have several properties that theoretically make them almost ideal candidates for pharmaceutical compounds. Hybridization of two complementary nucleic acid sequences by Watson-Crick base pairing is a highly selective and efficient process. The structure of the DNA double helix and the replication of DNA both depend on the process of hybridization, just as does RNA transcription and protein translation. The latter, for example, involves specific recognition of messenger RNA (mRNA) base triplets by the respective anticodon triplet of the transfer RNA (tRNA). The effectiveness and specificity of these biological processes, founded on the specificity of base pairing, are a strong incentive to use the hybridization process for the design of specific antisense pharmaceuticals. Furthermore, in contrast to antibodies, which interfere with extracellular and transmembrane proteins, antisense oligonucleotides can potentially inhibit the synthesis of any protein, regardless of its final cellular or extracellular localization.

Finally, one of the major attractions of the antisense approach is that it allows the rational design of drugs. The conventional approach to the development of pharmaceuticals often requires the testing of 2000–10,000 test substances, leading to one or several lead substances for further development *(7)*. In contrast, knowledge of a potential target RNA sequence, together with data on its predicted secondary structure, as well as certain desirable or undesirable sequence motifs, allows one to design an antisense drug, taking into account stacking energies and the likelihood of hybridization with related or nonrelated RNA sequences.

The design and use of antisense molecules are based on several assumptions, e.g., prediction of certain effects following hybridization to the target sequence and, furthermore, to be effective as a therapeutic agent, an antisense drug must fulfill several requirements.

1.2. Design of and Requirements for Antisense Oligonucleotide Drugs

1.2.1. Cellular Uptake; Availability at Target Localization

Antisense drugs need to be internalized by the cells involved in a particular pathological process. Furthermore, inside the cell a significant proportion of the oligonucleotide must be available in the same cellular compartment as the target RNA to allow hybridization to occur.

1.2.2. Effective Protein Synthesis Suppression

The hybridization of the antisense drug with its target RNA must lead to a significant biological effect. In the case of mRNA targeting, this means to reduce substantially or even shut off expression of a protein that is involved in a particular pathological process.

1.2.3. Selectivity

Ideally, inhibition of protein synthesis should be completely selective and leave expression of other proteins unaffected. Furthermore, it is desirable that the interaction of the antisense molecules with other molecules, e.g., proteins, carbohydrate moieties, lipids, second messenger molecules, and so forth, does not lead to adverse biological effects.

1.2.4. Rational Design and Prediction of Pharmacological Effects

Rational drug design is one of the major attractions of the antisense pharmacological approach. It is based on the assumption that hybridization to a target nucleic acid sequence will lead to predicted effects. As an exemplary approach to such a design of an antisense compound, we will present data on an S-ODN construct complementary to Met-tRNA. As a model system for the design of an antisense oligonucleotide, tRNA offers several advantages. One is the shortness of the sequence, which allows accurate prediction of secondary structure with the algorithm described by Zuker and Stiegler *(8)*. For several yeast tRNAs, this correlates very precisely with crystal structure analysis *(9)*. Second, the region of the tRNA anticodon offers a well-defined target sequence. Hybridization of the antisense S-ODN with this anticodon target region should lead to a predictable effect, i.e., interference with binding to the respective mRNA codon. In the case of Met-tRNA, this should then interfere with the initiation of protein translation and inhibit protein synthesis in general if the assumptions for the molecular design were correct.

1.2.5. Purity of Compound, Homogeneity of Product, and Low Toxicity

Like any other drug, antisense pharmaceuticals need to meet certain standards regarding purity and homogeneity of the compound and the definition of GMP standards. Furthermore, unwanted toxic side effects need to be as low as possible.

1.2.6. Pharmacology: Effective Drug Levels at the Site of Pathology

To achieve therapeutic effects, antisense molecules have to reach the target site in sufficient concentrations after systemic or local application. A general trend toward achieving some organ specificity of a particular drug has led to a rapidly growing subfield of pharmacology, the development of drug-delivery systems, which attempts to meet these goals. Once developed, most of these site-specific delivery systems will be usable for antisense pharmaceuticals. However, organ specificity is a general goal of drug delivery that is not required by antisense drugs in particular. The specificity of drugs that are currently used like that of receptor blockers, e.g., histamine H2-antagonists, does not depend on the specificity of their delivery systems for a particular organ, but rather on the selectivity of their interaction with a certain receptor. Similarly, the selectivity of antisense drugs will largely depend on their specificity for single RNA molecules and on the levels at which the tRNA is expressed in different cell populations. In the future, site-specific delivery systems may give additional specificity to certain antisense drugs. Furthermore, there may be one useful approach to enhance uptake at sites that do not show strong uptake of systemically applied oligonucleotides.

2. Materials and Methods
2.1. Oligonucleotide Synthesis, Labeling, and Analysis
2.1.1. Synthesis

S-ODN were synthesized and highly purified as described previously *(10,11)*. Briefly, oligonucleotides were synthesized on an Applied Biosystems model 394 DNA synthesizer using β-cyanoethyl phosphoramidites and a 30-min sulfurization step with 0.5 g elemental sulfur/4.8 mL CS_2/4.8 mL pyridine/0.4 mL triethylamine. In some of the experiments, sulfurization was achieved using $0.5M$ tetraethylthiuram disulfide in CH_3CN (15 min). Oligonucleotide sequences were: c-*jun*: TGCAGTCATAGAAC; *jun*-B: AGTTTGTAGTCGTG; p53: CAACAGTGAGGGAC; *erb*B-2: CATGGTGC

TCACTG; Met-tRNA: CTCTGGGTTATGGG; randomized control S-ODN: GTCCCTATACGAAC; Oligo-T: TTTTTTTTTTTTTT; Oligo-A: AAAA AAAAAAAAAA. Oligodeoxynucleotides were purified by reverse-phase HPLC on a Waters Delta-Pak C_{18} column and detritylated with 80% acetic acid, extracted with diethyl ether, lyophilized, and resuspended in water. S-ODN were further purified by anion-exchange chromatography on a Waters Protein-Pak DEAE 8HR column. Gradient: 10–100% B, linear over 45 min, hold for 10 min; buffer A: 25 mM Tris-HCl, 1 mM EDTA, pH 8.0, 10% acetonitrile; buffer B = 2.5M NaCl in A. S-ODN eluted around 90% B. S-ODN were precipitated from the eluate with 2 × vol 100% EtOH, resuspended in sterile water and dialyzed against 0.1M Tris, pH 8.0.

2.1.2. Analysis of Oligonucleotide Homogeneity

Homogeneity with respect to substitution of oxygen residues with thioate residues was analyzed as previously described *(10)* by ^{31}P-NMR spectroscopy using a Bruker spectrometer at 162 MHz and 22°C. Homogeneity with respect to length was analyzed using a capillary electrophoresis system (Beckmann Instruments) following the manufacturer's instructions using a polyacrylamide gel capillary and Tris-borate buffer. Samples of 10 µg/mL were loaded into the sample vial. Following an equilibration run, the sample was injected for 2 s and run for 60 min. The absorbance was read at 254 nm. Results were collated and analyzed using System Gold (Beckmann Instruments).

2.1.3. Oligonucleotide Labeling

The radiolabeling reaction of oligonucleotides was performed via a 5' C_6-amino-linker (Glen Research). The amino moiety was then radiolabeled using either ^{125}I-Bolton-Hunter reagent or ^{35}S-Bolton-Hunter reagent (Amersham). Briefly, 200 µCi of ^{125}I- or ^{35}S-Bolton-Hunter reagent were dried in a gentle stream of nitrogen. Oligonucleotides were then dissolved in 30 µL of double-distilled dimethylformamide and added to the dried reagent. Following incubation on ice for 1 h, 30 µL of borate-buffered saline were added to the tube. The tubes containing the reaction mixture were placed on a rotator at 4°C overnight. The reaction was stopped by adding 10 µL of 1M Tris buffer. Unbound ^{125}I or ^{35}S was removed by centrifugation in a Biospin 6 column (Bio-Rad, Richmond, CA). The oligonucleotides were precipitated with ethanol/magnesium chloride, pelleted, and vacuum-dried.

FITC labeling was performed using a C_3-amino-linker (Glen Research). Briefly, 100 nmol of 5' C_3-amino-modified oligonucleotide were diluted in 400 µL 1M carbonate-bicarbonate buffer (pH 9.0) to which 5 mg of fluorescein isothiocyanate in 400 µL 50% dimethylformamide were added. The labeling reaction was allowed to proceed at 37°C for 2 h in the dark. Labeled oligonucleotides were separated from unincorporated product by reverse-phase HPLC on a C_{18} column, and the product peak was lyophilized and processed as previously described *(10,11)*.

2.2. Cell-Culture and Proliferation Assays

Cells were maintained in RPMI medium (Gibco), supplemented with 100 U/mL penicillin, 100 µg/mL streptomycin, 10% FCS. For ^{3}H-thymidine incorporation, 2500 cells/well were plated into 96-well microtiter plates. S-ODN was added at 2-μM concentration 1 h after seeding. At the times indicated, cells were incubated with 0.15 µCi ^{3}H-thymidine/well for 6 h. Cells were lysed by freezing, spotted onto glass filters, washed, and the amount of incorporated tritium was determined by liquid scintillation counting.

2.3. Western Blots

SK-Br-3 human mammary carcinoma cells were kept in RPMI/2% FCS for 3 d. Cells were seeded into 260-mL culture flasks at a density of 3×10^6 in 30 mL of RPMI medium supplemented with 5% FCS and 2 μM S-ODN. At the time-points indicated, cells were harvested, pelleted, and lysed by the addition of lysis buffer consisting of $0.1M$ Tris-buffer, pH 6.8, 20% glycerin, and 2% sodium dodecyl sulfate (SDS). Proteins were separated by electrophoresis on a 10% SDS-polyacrylamide (SDS-PAGE) gel and blotted onto Immobilon-P membrane (Millipore). Blots were probed with either rabbit polyclonal anti-c-JUN antibody (Oncogene Science, PC 07), rabbit polyclonal anti-JUNB antibody (Santa Cruz, N-17), mouse monoclonal anti-p53 antibody (Oncogene OP 09), or rabbit polyclonal anti-ErbB-2 antibody (Oncogene PC 04). For detection, either goat antirabbit IgG–alkaline-phosphatase conjugate (Boehringer Mannheim) or goat antimouse IgG–alkaline-phosphatase conjugate (DAKO GmbH D486) was used as the second antibody. Detection was performed by chemiluminescence with CSPD (Tropix) according to manufacturer's instructions. Band intensity was quantified by densitometric analysis of scanned blots using the "NIH Image" software package.

2.4. Protein Concentration and Kinase Activity Measurements

To determine protein concentration for the Western blotting procedure, cell lysates were equilibrated to pH 11.2, and 10 µL of cell lysate were incubated with 200 µL of BCA-reagent (Pierce) in 96-well microtiter plates. The color reaction was measured at 562 nm in an automated microtiter plate reader. Known concentrations of bovine serum albumin (Sigma) were used to titrate a standard curve.

De novo protein synthesis was determined according to standard techniques. Briefly, cells were incubated in 96-well microtiter plates with 0.5 µCi of ^{35}S-methionine/well for 4 h. The amount of trichloroacetic acid-insoluble radioactivity was determined in a liquid scintillation counter.

S6 kinase activity was determined *in situ* in 96-well microtiter plates. Cells were seeded at a density of 25,000/mL. S-ODN were added 1 h after plating. Cells were permeabilized with digitonin, washed, and incubated with 100 mM S6-specific substrate peptide (RRLSSLRA) and 2 µCi of γ-[^{32}P]-dATP (3000 Ci/mmol). The substrate peptide labeling procedure was stopped by the addition of trichloroacetic acid. Incorporation of ^{32}P into the substrate peptide was measured by spotting aliquots of the supernatant on phosphocellulose paper and counting in a liquid scintillation counter.

2.5. Intravenous and Intracerebral Injection: Distribution and Pharmacokinetics

2.5.1. Oligonucleotide Injection

For injections into rat brain, adult rats were implanted under deep barbiturate anesthesia with a microinjection guide-cannula targeted to the dorsolateral hippocampus (2.8 mm posterior and 2.2 mm lateral of the bregma). On d 8 after implantation, 1 µL of a 2-mM solution of either FITC- or ^{35}S-labeled randomized control S-ODN was injected 10 h before sacrifice. A second dose of 1 µL was applied 2 h before decapitation. For intravenous (iv) injection into mice, 2 × 10^6 µCi of ^{125}I-labeled S-ODN, diluted in 200 µL of borate-buffered saline were injected into the tail vein.

2.5.2. Serum Clearance

From mice injected with ^{125}I-oligonucleotide, blood samples were taken at 1, 2.5, 5, 10, 20, 30, 45, and 60 min after injection. Each sample was mixed with 200 µL of 10 mM EDTA to prevent blood clotting. Excess MgCl$_2$ was added to each sample, followed by addition of 0.6 mL of ethanol. Samples were kept on ice and then centrifuged at 12,000g for 20 min. Supernatants of each sample were counted in a liquid scintillation counter.

2.5.3. Organ Uptake after IV Application in Mice

On sacrifice, mice were perfused transcardially with 10–15 mL of sterile saline. Organs were removed, and the incorporated radioactivity of each organ was determined in a liquid scintillation counter.

2.5.4. Uptake into Rat Brain After Intracerebral Injection

Rats were killed by decapitation. Brains were removed from the skull, submerged in embedding medium, and frozen on liquid nitrogen. Parasagittal sections were prepared on a cryostat. Sections from brains injected with ^{35}S-oligonucleotide were processed for autoradiography. Sections from animals injected with FITC-labeled oligonucleotide were examined and photographed using a fluorescence microscope (Zeiss).

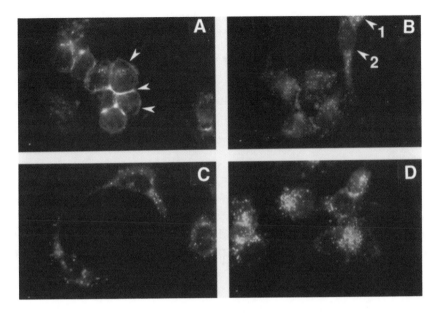

Fig. 1. Uptake of FITC-labeled control S-ODN into SK-Br-3 mammary carcinoma cells at different time-points after addition of phosphorothioate oligonucleotides to cell culture medium. **(A)** After 5 h S-ODN are mainly associated with cell membrane (arrows). **(B)** After 10 h, two adjacent cells (arrows 1 and 2) show different levels of fluorescence. Strong labeling of cells is seen after **(C)** 24 h and **(D)** 48 h.

3. Results

3.1. Cellular Uptake

3.1.1. Uptake in Cell Culture

Cellular uptake of FITC-labeled control S-ODN at 2 mM concentration was followed over a 48-h period in human SK-Br-3 mammary carcinoma cells. This epithelial-derived cell line was also used to determine the effectiveness and specificity of different antisense sequences described below. After 5 h, the majority of the fluorescence signal was associated with the cell membranes. Also, a moderate intracellular fluorescence signal in a punctate appearance could be observed at this time-point (Fig. 1A). After 10 h, little membrane associated fluorescence remained, but the punctate fluorescence signal had strongly increased (Fig. 1B). Also at this time-point, marked differences in labeling intensity could be seen in different cells of the same preparation (arrows 1 and 2, Fig. 1B). These differences had disappeared by 24 h, and the intensity of the punctate labeling signal progressively increased until 48 h (Fig. 1C,D). Furthermore, at the latter two time-points, a diffuse labeling signal could be observed, which largely spared the nuclear region. Similar results were obtained with two further S-ODN sequences (data not shown).

3.1.2. Uptake into Neurons In Vivo After Brain Injections

After injection of FITC-labeled or ^{35}S-labeled S-ODN into the brain, marked uptake into glial cells and heavy labeling of neurons was observed. A detailed description is given below under organ uptake into the brain.

3.2. Effective Protein Synthesis Suppression

Effective and selective suppression of translation of the target mRNA is the key feature by which therapeutic antisense oligonucleotides will be judged. The effectiveness and time-course of protein suppression depends not only on cellular oligonucleotide uptake and stability, but also on various biological parameters, including concentration and half-life of the mRNA and the encoded protein. To study these variations, expression of two different proteins, p53 and ErbB-2, was blocked in human SK-Br-3 mammary carcinoma cells under identical culture conditions.

3.2.1. Inhibition of p53 Synthesis

At 10 h after application of 2 μM anti-p53 S-ODN, expression of the p53 protein in SK-Br-3 cells was reduced to 60% of controls and to <10% after 24 h. Levels had returned to approx 80% after 48 h (Fig. 2A).

3.2.2. Inhibition of ErbB-2 Synthesis

In contrast to the results with anti-p53 S-ODN, application of anti-c-*erb*B-2 S-ODN in the same cell line led to a different time-course of protein suppression. The amount of ErbB-2 protein was only marginally reduced 10 h after anti-c-*erb*B-2 S-ODN treatment of SK-Br-3 cells. Subsequently, levels progressively decreased to <15% after 48 h (Fig. 2B).

3.3. Selectivity of Gene Suppression by Antisense S-ODN for c-jun or junB

The specificity of gene expression is difficult to evaluate. A variety of controls have to be performed. The use of control probes, either sense, randomized, or mismatch containing sequences is of obvious importance. Furthermore, effects on the expression of proteins other than the protein to be suppressed have to be studied. We chose to evaluate the expression of two members of the same gene family, c-*jun* and *jun*B, in response to treatment with S-ODNs targeted to their respective mRNAs. The two immediate early genes share high sequence homology *(12)*. Western blot analysis revealed that antisense S-ODN complementary to c-*jun* mRNA reduced c-JUN protein expression by more than 90%, whereas JUNB protein levels were not decreased; rather, the amount

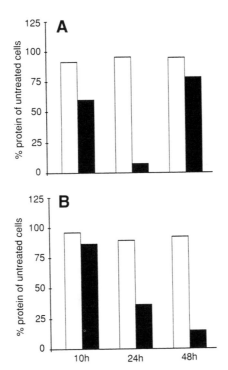

Fig. 2. Time-course of antisense-mediated protein suppression: Densitometric analysis of Western blots from SK-Br-3 cell lysates. **(A)** Relative amounts of p53 protein after single treatment with anti-p53 S-ODN (black bars) and randomized control S-ODN (white bars). Inhibition of p53 protein expression was maximal after 24 h. **(B)** Relative amounts of ErbB-2 protein after single treatment with anti-c-*erb*B-2 S-ODN (black bars) and randomized control S-ODN (white bars). In contrast to p53, the decrease of ErbB-2 protein gradually continued over the 48 h period following administration of anti c-*erb*B-2 S-ODN.

of JUNB increased. In contrast, antisense S-ODN specific for *jun*B mRNA significantly reduced JUNB expression, but only marginally affected c-JUN protein levels (Fig. 3). Blots that were stained for total protein showed no differences in the overall amount or gross distribution of protein bands between untreated, control, and antisense-treated cells (data not shown).

3.4. Rational Design:
The Anticodon Region of Met-tRNA as a Target

The same SK-Br-3 cell line that was used to test the effectiveness and specificity of the antisense sequences targeted to c-*jun*, *jun*B, p53, and c-*erb*B-2 mRNAs was used to determine whether a sequence complementary to Met-

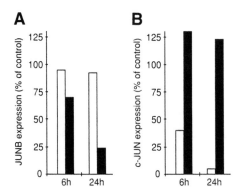

Fig. 3. Specificity of suppression of two highly homologous proteins, c-JUN and JUNB. White bars: SK-Br-3 cells treated with anti-c-*jun* S-ODN. Black bars: SK-Br-3 cells treated with anti-*jun*B S-ODN. **(A)** Densitometric quantification of a Western blot probed with anti-JUNB antibody. **(B)** Densitometric quantification of a Western blot probed with anti-c-JUN antibody. Each antisense S-ODN selectively inhibits expression of its target gene, but not of the related gene.

tRNA would lead to the predicted inhibition of protein translation. Anti-Met-tRNA S-ODN treatment of SK-Br-3 cells at 2-μM concentration inhibited protein synthesis to <20% of controls after 48 h (Fig. 4B). Analysis of the consequences on cell proliferation of this decrease in protein synthesis showed a reduction in thymidine incorporation to 35% of controls after 2 d (Fig. 4C).

Phosphorylation of the S6 protein subunit of the ribosome by S6 kinase is a rate-limiting step in protein translation *(13)*. To analyze whether the activity of this enzyme shows a compensatory increase, S6 phosphorylating activity was determined. Cells treated with 2 μM anti-Met-tRNA S-ODN showed a more than twofold increase in S6 protein kinase activity compared to control S-ODN-treated cells (Fig. 4D).

3.5. Homogeneity

Homogeneity of phosphate backbone modification was analyzed by ^{31}P-NMR spectroscopy. Sulfurization of virtually all phosphor linkages (Fig. 5, trace 1) was routinely achieved with a 30-min sulfurization step using elemental sulfur, as described in Section 2. Using other sulfurization reagents, such as TEDT, and shortening the time of the sulfurization step often led to incomplete backbone modification with residual P-O linkages as high as 10% (Fig. 5, trace 2). The homogeneity of S-ODN preparations with respect to their length was evaluated using capillary gel electrophoresis (Fig. 6). By combining reverse-phase and anion-exchange chromatography, a purity >98% was routinely achieved.

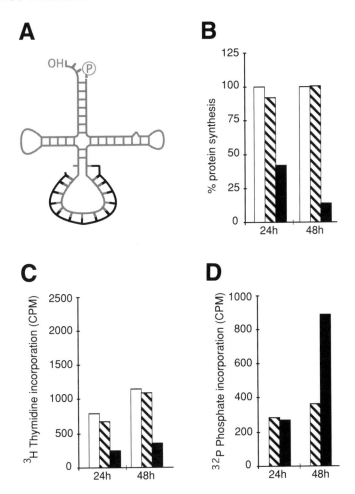

Fig. 4. Effects of masking the Met-tRNA anticodon in SK-Br-3 cells. **(A)** Design of the antisense Met-tRNA S-ODN. The predicted secondary structure of Met-tRNA is depicted in gray. The complementary oligonucleotide sequence is shown in black. **(B–D)** Effects of control S-ODN and anti-Met-tRNA S-ODN on (B) total cellular protein synthesis, (C) ^{3}H-thymidine incorporation, and (D) S6 kinase activity. White bars: untreated control cells; hatched bars: cells treated with randomized control S-ODN; black bars: effects of anti-Met-tRNA S-ODN.

3.6. Effects of Chemical Purity

3.6.1. Purification Grades

To evaluate the influence of purity on cell proliferation of different purification grades of phosphorothioate oligonucleotides, SK-Br-3 cells were treated

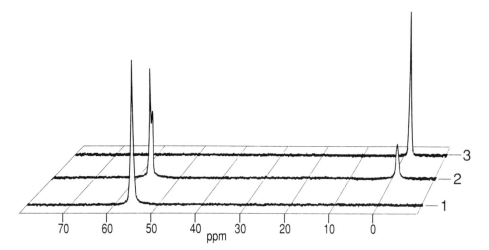

Fig. 5. Assessment of the degree of phosphate-backbone modification in different oligonucleotide preparations by ^{31}P-NMR spectroscopy. Trace 1 = fully substituted (>99%) phosphorothioate oligonucleotide. Trace 2 = partially substituted ($\approx$92%) phosphorothioate as a result of incomplete sulfurization during synthesis. Trace 3 = unmodified phosphodiester DNA oligonucleotide.

with 2 μM randomized control S-ODN of varying purities. S-ODN for these experiments had been sulfurized using the TEDT reagent. S-ODN that had not undergone HPLC purification strongly inhibited cell proliferation (Fig. 7A, hatched bars). In contrast, S-ODN that had undergone RP-HPLC strongly increased ^{3}H-thymidine uptake into SK-Br-3 cells. To a lesser extent, this was also the case after treatment of S-ODN with organic extraction following the HPLC. Treatment of cells with randomized control S-ODN that underwent further purification by dialysis resulted in thymidine incorporation rates that were not significantly different from those in untreated control cells (Fig. 7A, gray bars). Analysis of all S-ODN fractions by PAGE showed that before HPLC purification, oligonucleotides contained about 18% nonfull-length fragments. In contrast, all HPLC-purified preparations looked identical in PAGE analysis (>98% full length, data not shown).

3.6.2. Counterion

As with many other pharmacological substances, the salt form of the preparation can alter the biological behavior and effect of the substance itself. To test whether this also applies to S-ODN, SK-Br-3 cells were treated with randomized control S-ODN both in the form of an ammonium salt (which is usually obtained after reverse-phase HPLC purification) and as a sodium salt. Cell

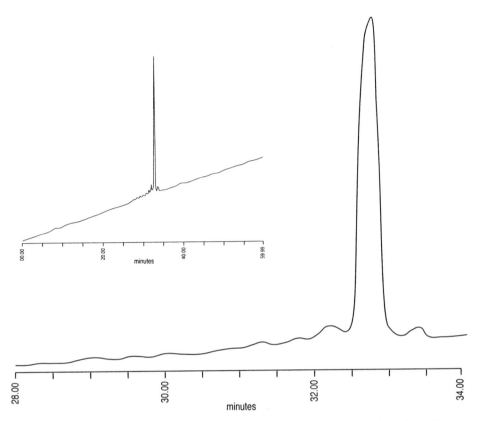

Fig. 6. Product peak from capillary gel electrophoresis of purified phosphorothioate oligonucleotide. Full length product >98%. Inset: complete trace.

proliferation rates were determined using ^{3}H-thymidine incorporation assays at 24 and 48 h after addition of 2 μM of the respective S-ODN to the culture medium. Compared to untreated cells, the S-ODN-Na$^+$ form had no measurable effect on cell proliferation, whereas the ammonium salt inhibited incorporation of ^{3}H-thymidine into SK-Br-3 cells by 30 and 20% at 24 and 48 h respectively (Fig. 7B).

3.7. Pharmacology In Vivo

3.7.1. Serum Clearance, Liver and Kidney Uptake

To investigate the pharmacokinetics and organ distribution of S-ODN in mice, they were injected with ^{125}I-labeled Oligo-A and Oligo-T S-ODNs. During the first phase, Oligo-T and Oligo-A were both rapidly eliminated from the serum with $t_{1/2}$ of 2.9 and 2.4 min, respectively. During this early phase serum

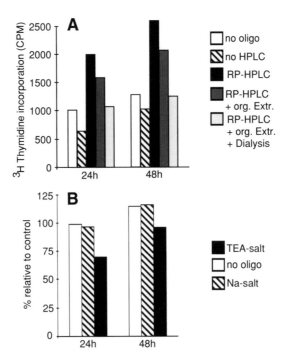

Fig. 7. Effects of oligonucleotide purity on proliferation of SK-Br-3 cells. **(A)** Cells treated with 2 μ*M* S-ODN, of different purification grades: white bars: untreated control cells; hatched bars: treatment with S-ODN that were just deprotected and lyophilized; black bars: application of S-ODN, subjected to RP-HPLC; dark gray bars: S-ODN purification with RP-HPLC followed by organic extraction; light gray bars: S-ODN purification with RP-HPLC + organic extraction + dialysis. **(B)** Effect of counterion on SK-Br-3 cell proliferation: White bars: no S-ODN; gray bars: S-ODN-Na+-salt; black bars: S-ODN-ammonium salt.

clearance amounted to 81% of Oligo-T and 94% of Oligo-A. During the second, slower phase of elimination, the remaining Oligo-T and Oligo-A were cleared form the serum, and clearance was faster for Oligo-T ($t_{1/2} = 31.5$ min) than for Oligo-A ($t_{1/2} = 48.1$ min) (Fig. 8).

3.7.2. Organ Uptake

Two organs showed particularly strong uptake of S-ODN during the first phase of clearance from serum. Elimination from serum of oligo-T was mediated mostly by the liver, where 37% of Oligo-T was localized within 10 min, compared to 10% in the kidney. Surprisingly, Oligo-A showed a reciprocal finding with uptake into the kidneys of 26% and uptake into the liver of only 15% (Fig. 9A).

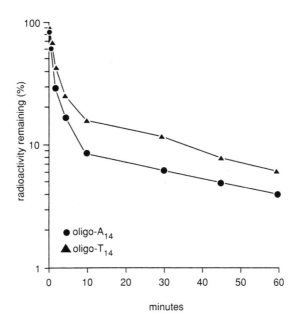

Fig. 8. Serum clearance of S-ODN over 60 min following iv injection. Circles: Oligo-A$_{14}$ S-ODN; triangles: Oligo-T$_{14}$ S-ODN.

At 60 min after injection, the difference in liver uptake was still pronounced, but the amounts localized in the kidney were similar at this time-point. Although a large proportion of both S-ODNs is localized either in the kidney or within the liver, significant uptake is also seen in the gastrointestinal tract, the skin, and the carcass, and to a lesser extent in spleen and lung (Fig. 9B).

3.7.3. Organ Uptake and Cellular Localization After Intracerebral Injection

After injection of FITC-labeled or ^{35}S-labeled S-ODN into rat brain, strong uptake was seen immediately surrounding the injection site (Fig. 10). The localized twofold injection of only 1 µL of ^{35}S-labeled oligonucleotide, at the dorsolateral end of the hippocampus, led to strong labeling in most of the pyramidal neuronal layer of the entire hippocampus. Furthermore, strong labeling was also observed in fiber tract systems and those parts of the cerebral cortex that surround the tract of the injection needle (Fig. 10A,B).

In animals injected with FITC-conjugated S-ODN, many of the fluorescence-labeled cells surrounding the tip of the injection needle probably represent glial cells that have migrated to and/or undergone proliferation at the site of injection. Furthermore, in the hippocampal pyramidal cell layer, heavy labeling of large pyramidal neurons could be observed (Fig. 10C, white

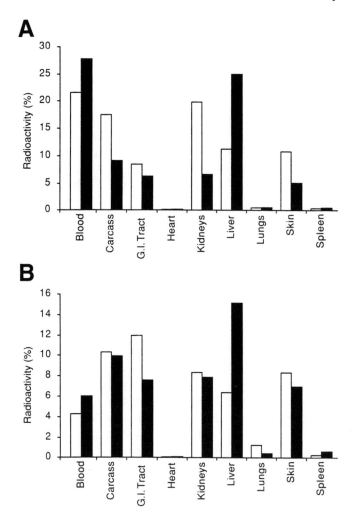

Fig. 9. Organ distribution of ^{125}I-labeled phosphorothioate oligonucleotides in mice as percentage of total administered dose. White bars: Oligo-T_{14} S-ODN; Black bars: Oligo-A_{14} S-ODN. **(A)** Organ distribution 10 min after iv injection of S-ODN. **(B)** Distribution 60 min after injection.

arrows). Several smaller fluorescence-labeled cells may represent interneurons or subpopulations of glial cells.

4. Discussion

The phosphorothioate analogs of DNA fulfill most of the requirements for an antisense drug described in Section 1. Cellular uptake of fluorescein-labeled

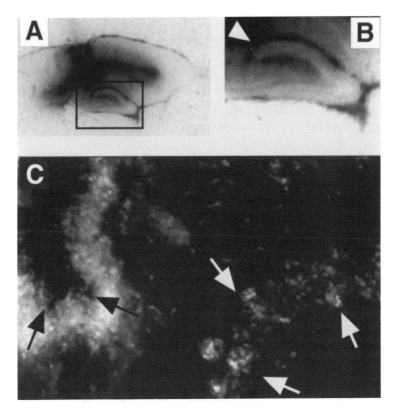

Fig. 10. Uptake of S-ODN into rat brain after intracerebral injection. **(A)** Sagittal section of rat brain after two injections of ^{35}S-labeled S-ODN (10 and 2 h before decapitation). The tip of the injection cannula was targeted to the dorsolateral hippocampus. The autoradiographic signal shows wide distribution of oligonucleotide in the pyramidal layer of the hippocampus and in the granule cell layer of the dentate gyrus (boxed region). A strong labeling signal is also seen over cortical regions surrounding the injection site and in fiber tract systems. **(B)** Higher magnification of the boxed region from (A). The white arrowhead points to the tip of the injection cannula. **(C)** Labeling of pyramidal neurons (white arrows) of the dorsolateral hippocampus in the vicinity of the injection site 26 h after a twofold injection of FITC-labeled S-ODN. The tip of the injection site is marked by black arrows. The labeling signal surrounding the injection site most likely represents labeling of glial cells that have proliferated around the injection site.

S-ODN was very efficient in SK-Br-3 cells. The results were very similar to our uptake studies using bromo-deoxyuridine (Br-dU)-labeled S-ODN (data not shown), indicating that the fluorescence residue plays a minor role, if any, in the uptake of S-ODN. The early phases of the uptake have been previously described in other cell types, and a punctate appearance of oligonucleotide distribution for unmodified phosphodiester as well as for phosphorothioate oligodeoxynucleotides has been noted (reviewed in ref. *5*). Our data at the later time-points 24 and 48 h after S-ODN application suggest that accumulation of the fluorescence moiety progresses until this time. Analysis of S-ODN integrity at these time-points is currently under way. Despite the punctate distribution of much of the fluorescence signal, the belief that the final intracellular localization of ODN and S-ODN is in late endosomes and lysosomes appears not to be true. Neither differential centrifugation of subcellular structures nor pH-quenched FITC fluorescence of cells treated with monesin and FITC-labeled oligodeoxynucleotide substantiated the hypothesis of a localization in these cellular compartments *(5)*. Further studies will be needed to determine the subcellular localization and mode of entry into the cytoplasm of charged oligonucleotide derivatives.

With the antisense sequences used in this study, we found sequence-specific suppression of proteins by targeting the encoding mRNAs. Synthesis of four very different proteins, ErbB-2, p53, c-JUN, and JUNB, was efficiently inhibited by specific antisense S-ODN oligonucleotides. All antisense sequences were able to decrease the respective proteins by at least 75% compared to untreated cells at a concentration of 2 μM. The maxima of suppression, however, occurred at different time-points, indicating that the efficiency of antisense S-ODN should always be judged in terms of a time–response curve. These differences are probably at least partly owing to differences in the expression patterns of the target genes, i.e., suppression of a rapidly regulated cytoplasmic protein like p53 shows a different time-course compared to a receptor protein like ErbB-2. These data suggest that evaluation of biological effects and target protein levels during similar time windows is helpful to determine accurately the functional significance of antisense-mediated protein suppression.

Analysis of the specificity of antisense effects is particularly important, since nonsequence specific effects of S-ODN have been described in particular with respect to the inhibition of viral replication and the activity of viral enzymes (reviewed in ref. *5*). As a stringent control experiment for sequence specificity, we targeted two highly homologous mRNAs, encoded by members of the same gene family. Antisense S-ODN complementary to c-*jun* mRNA reduced c-JUN, but not JUNB protein levels, whereas the reverse finding was seen with anti-*jun*B S-ODN. The c-*jun* and *jun*B genes have very similar induction properties. Both are immediate early genes that are induced within minutes in response to stimuli that alter cellular programs *(14,15)*, and the two encoded transcription factors play a central role in coordinating a cell's genetic response to extracellular stimulation *(10)*. Taking into account the similarity of induction properties, their key role in orchestration of genetic responses, and the high sequence homology, they can serve as an excellent paradigm for comparative analysis of S-ODN effects. In summary, the example shown here demonstrates that carefully designed and tested S-ODN can act selectively to suppress just one of two closely related genes with similar biological characteristics and high sequence homology.

The protein translation inhibitory effect of antisense oligonucleotides is based on their hybridization to the target mRNA resulting in a steric block of the translation process and/or cleavage of the mRNA at the RNA:DNA hybrid site by RNaseH (reviewed in ref. *16*). Second, targeting mRNA splice regions of pre-mRNA has been a useful approach *(17)*. Taking Met-tRNA as a model system, we show here that an antisense effect can also be mediated through the masking of other sequences on the target molecule. Masking the Met-tRNA anticodon region led to a marked inhibition of total cellular protein synthesis. Similarly, thymidine incorporation also decreased, as would be expected as a consequence of the reduction of protein synthesis. In contrast, S6 kinase activity, a potent regulator of the protein translation rate *(13)*, was strongly increased, representing most likely a compensatory activation of feedback loops. This increase in the activity of one enzyme further demonstrates that the general decrease in *de novo* protein synthesis is a specific effect of the anti-Met-tRNA S-ODN.

Beyond the model system of tRNA anticodon targeting, antisense oligonucleotides that mask certain sites on RNA molecules serve as an example for a novel antisense strategy. Thus, RNA sites that regulate expression posttranscriptionally like the stable stem-loop structure in the TGF-β 1 5'-untranslated region (UTR) *(18,19)*, can serve as valuable target sites for antisense molecules (W. B. and K. H. S., unpublished observations).

Proliferation assays of cells treated with batches of randomized control S-ODN of different purities revealed that nonsequence-specific effects of oligonucleotides may be the result of impurities in the particular batch. These impurities produced not only proliferation-inhibiting effects, but could in some cases also stimulate cell growth. Generally, oligonucleotide preparations are analyzed by methods focusing on homogeneity rather than purity. PAGE and capillary electrophoresis, HPLC, and even mass spectroscopy yield information about the length and integrity of the oligonucleotides, but small molecules that are byproducts of the chemical synthesis are often overlooked. As demonstrated by our results, such contaminants can produce profound biological effects that may falsely be attributed to the antisense oligonucleotide itself. The suppression of cell proliferation by S-ODN, which had not been HPLC-purified, could mainly be owing to the hybridization of short oligonucleotide fragments to a large number of unrelated mRNAs. On the other hand, stimulation of cell proliferation by partially purified S-ODN was apparently owing to traces of synthesis reagents. This stimulatory effect was only seen when TEDT reagents were used for sulfurization of the phosphate backbone. These control experiments emphasize the importance of purity of an oligonucleotide preparation used for in vivo experiments. Too often the term "pure" is falsely applied for oligonucleotides with only the knowledge of backbone homogeneity and full-length product homogeneity.

The pharmacology of intravenously injected radiolabeled S-ODN showed rapid serum clearance in agreement with the study of Agrawal and coworkers *(20)*. In addition to the time-points observed in this study, which started at 5 min, we obtained additional samples at 1 and 2.5 min. These data reveal that the majority of S-ODN are cleared from the serum during the first 5 min. The organ uptake is highest in kidney and liver after 10 min and 1 h. After 24 h, again the highest concentration of S-ODN is found in these two organs (data not shown). This finding is in agreement with the above-mentioned study *(20)*. Surprisingly, however, there was a pronounced difference in the serum clearance and the organ uptake between Oligo-T and Oligo-A. Among the factors influencing this difference may be differential binding to macromolecules in the serum and on the cell surfaces, and differences in diffusion and cellular uptake. Further studies are warranted to define in more detail the influence of base composition on pharmacokinetics and organ distribution.

Direct injection of a very small volume of S-ODN into rat brain led to a surprisingly wide distribution. Neuronal uptake in the hippocampus, the dentate gyrus, and the cerebral cortex was very efficient. Furthermore we found selective suppression of *jun* gene expression in rat brain neurons in vivo *(21)*. These findings demonstrate that the physiological and pathophysiological relevance of gene expression in the central nervous system (CNS) can be conveniently studied with antisense phosphorothioate oligodeoxynucleotides. Classical pharmacological methods can be used to achieve penetrance through the blood–brain barrier of S-ODN after systemic application.

The wide distribution of S-ODN after localized injection was unexpected and may be of great significance for the potential treatment of diseases localized in the CNS, including tumors, e.g., medulloblastomas in children and gliomas *(11,22)* in adults, as well as neurodegenerative disorders, e.g., Alzheimer's disease.

5. Conclusion

Phosphorothioate antisense oligodeoxynucleotides can be effective and highly specific inhibitors of gene expression. Increasing knowledge about the sequence and biological role of genes makes the antisense technique one of the most rational and intriguing approaches to drug design.

However, even though the design of a specific antisense molecule by adhering to the rules of Watson-Crick base pairing may seem straightforward, a number of other factors, like choice of the target sequence within the gene, homogeneity and purity of the oligonucleotide preparation, and its pharmacology have to be carefully considered in order to achieve the desired effects and to minimize nonsequence-specific side effects. The biological rationale behind the inhibition of a particular gene will ultimately determine whether a given antisense drug is therapeutically effective.

References

1. Stephenson, M. and Zamecnik, P. C. (1978) Inhibition of Rous sarcoma viral RNA translation by a specific oligodeoxyribonucleotide. *Proc. Natl. Acad. Sci. USA* **75,** 285–288.
2. Zamecnik, P. C. and Stephenson, M. (1978) Inhibition of Rous sarcoma virus replication and cell transformation by a specific oligodeoxynucleotide. *Proc. Natl. Acad. Sci. USA* **75,** 280–284.
3. Campbell, J., Bacon, T., and Wickstrom, E. (1990) Oligodeoxynucleoside phosphorothioate stability in subcellular abstracts, culture media, sera and cerebrospinal fluid. *J. Biochem. Biophys. Methods* **20,** 259–267.
4. Eckstein, F. (1983) Phosphorothioate analogues of nucleotides—tools for the investigation of biochemical processes. *Angewandte Chemie* **22,** 423–506.
5. Stein, C. A. and Cheng, Y.-C. (1993) Antisense oligonucleotides as therapeutic agents—is the bullet really magical? *Science* **261,** 1004–1012.
6. Bishop, J. M. (1991) Molecular themes in oncogenesis. *Cell* **64,** 235–248.
7. Griffin, J. P., O'Grady, J., and Wells, F. O. (eds.) (1993) *The Textbook of Pharmaceutical Medicine.* The Queen's University of Belfast, Belfast.
8. Zuker, M. and Stiegler, P. (1981) Optimal computer folding of large RNA sequences using thermodynamics and auxiliary information. *Nucleic Acids Res.* **9,** 133–148.
9. Schevitz, R. W., Podjarny, A. D., Krishnamachari, N., Hughes, J. J., Sigler P. B., and Sussman, J. L. (1979) Crystal structure of an eukaryotic initiator tRNA. *Nature* **278,** 188–190.
10. Schlingensiepen, K.-H., Schlingensiepen, R., Kunst, M., Klinger, I., Gerdes, W., Seifert, W., and Brysch W. (1993) Opposite functions of Jun-B and c-Jun in growth regulation and neuronal differentiation. *Dev. Genet.* **14,** 305–312.
11. Jachimczak, P., Bogdahn, U., Schneider, J., Behl, C., Meixensberger, J., Apfel, R., Dörries, R., Schlingensiepen, K.-H., and Brysch W. (1993) TGF-β2-specific phosphorothioate antisense oligodeoxynucleotides may reverse cellular immunosuppression in malignant glioma. *J. Neurosurg.* **78,** 944–951.
12. Vogt, P. K. and Bos, T. J. (1990) *jun*: oncogene and transcription factor. *Adv. Cancer Res.* **55,** 1–35.
13. Erikson, R. L. (1991) Structure, expression and regulation of protein kinases involved in the phosphorylation of ribosomal S6. *J. Biol. Chem.* **266,** 6007–6009.

14. Bravo, R. (1990) Genes induced during the G0/G1 transition in mouse fibroblasts. *Semin. Cancer Biol.* **1,** 37–46.

15. Sheng, M. and Greenberg, M. E. (1990) The regulation and function of c-fos and other immediate early genes in the nervous system. *Neuron* **4,** 477–485.

16. Toulmé, J.-J. and Hélène, C. (1988) Antimessenger oligodeoxyribonucleotides: an alternative to antisense RNA for artificial regulation of gene expression. *Gene* **72,** 51–58.

17. Becker, D., Meier, C. B., and Herlyn, M. (1989) Proliferation of human melanomas is inhibited by antisense oligodeoxynucleotides against basic fibroblast growth factor. *EMBO J.* **8,** 3685–3691.

18. Kim, S. J., Park, K., Koeller, D., Kim, K. Y., Wakefield, L. M., Sporn, M. B., and Roberts, A. B. (1992) Post-transcriptional regulation of the human transforming growth factor-beta 1 gene. *J. Biol. Chem.* **267,** 13,702–13,707.

19. Romeo, D. S., Park, K., Roberts, A. B., Sporn, M. B., and Kim, S. J. (1993) An element of the transforming growth factor-beta 1 5'-untranslated region represses translation and specifically binds a cytosolic factor. *Mol. Endocrinol.* **7,** 759–766.

20. Agrawal, S., Temsamani, J., and Tang, J. Y. (1991) Pharmacokinetics, biodistribution, and stability of oligodeoxynucleotide phosphorothioates in mice. *Proc. Natl. Acad. Sci. USA* **88,** 7595–7599.

21. Behl, C., Bogdahn, U., Winkler, J., Meixensberger, J., Schlingensiepen, K.-H., and Brysch, W. (1993) Autocrine growth regulation in neuroectodermal tumors as detected with oligodeoxynucleotide antisense molecules. *Neurosurgery* **33,** 679–684.

22. Schlingensiepen, K.-H., Wollnik, F., Kunst, M., Schlingensiepen, R., Herdegen, T., and Brysch, W. (1994) The role of *jun* transcription factor expression and phosphorylation in neuronal differentiation, neuronal cell death and in plastic adaptations in vivo. *Cell. Mol. Neurobiol.* **14,** 487–505.

10

Antisense *rel* A in Cancer

Jose R. Perez, Kimberly A. Higgins-Sochaski, Jean-Yves Maltese, and Ramaswamy Narayanan

1. Introduction

NF-κB is a pleiotropic transcription factor that participates in the induction of various cellular and viral genes (for a review, *see* refs. *1* and *2*). The principal form of this active complex is composed of two polypeptides, p65 (recently renamed *rel* A) and p50 (recently renamed NFKB1) *(3,4)*. Both of these subunits belong to the *rel* family of transcription factors with homology in the amino terminus *(5,6)*. Since the original identification of these factors, several others have been identified in this family of transcription factors: P49, c-*rel*, p100, *rel* B, *dorsal*, Bcl-3, and p105 *(1)*. The NF-κB heterodimer is associated in the cytoplasm with the IκB subunit, which sequesters this factor in an inactive state *(7)*. On activation by various stimuli, the IκB subunit separates from the active NF-κB heterodimer, and the active complex is translocated to the nucleus where it binds to its nuclear DNA target sequence (Fig. 1). The principal cause for disassociation of IκB from NF-κB is phosphorylation *(8)*, but it may involve proteolysis of IκB *(9)*. Several autoregulatory loops have been proposed for the involvement of NF-κB in the regulation of the inhibitor IκB. Furthermore, NF-κB has been shown to regulate the transcription of IκB and NFKB1 *(10–12)*.

Many of the previous studies involving the subunits of NF-κB have utilized the expression of these factors in host cells by transient transfection, and have studied their role in the transactivation of a reporter construct *(13–16)* or the binding of these nuclear proteins to target sequences by electrophoretic mobility shift assays (EMSA) *(14,17–19)*. These experiments have presented an artificial situation in which to study these factors because their expression levels have been unnaturally high. It was our goal to develop systems in which to study the function of NF-κB at its endogenous level. To accomplish this, we

From: *Methods in Molecular Medicine: Antisense Therapeutics*
Edited by: S. Agrawal Humana Press Inc., Totowa, NJ

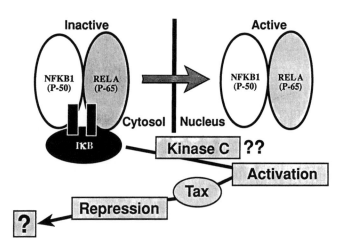

Fig. 1. NF-κB transcription factor. This figure depicts the basis for the function of the NF-κB transcription factor. The principal form of the NF-κB transcription complex is as a heterodimer composed of the two subunits: *rel* A and NFKB1. The inactive complex is sequestered in the cytoplasm associated with the inhibitor protein, IκB. On activation, the IκB protein separates from the active complex through a mechanism involving the phosphorylation and proteolysis of IκB. The active NF-κB complex is then translocated to the nucleus where it binds DNA in a sequence-specific manner and interacts with the transcriptional machinery to modulate the rate of transcription of target genes.

chose to inhibit the individual subunits of NF-κB using antisense technology (for a review, *see* ref. *20*). Antisense phosphorothioate (PS) oligonucleotides and stable antisense RNA expression vectors were utilized in these studies.

2. Antisense Inhibition of NF-κB Activity

The oligonucleotides used in these studies were PS oligonucleotides directed against the translation initiation site of the *rel* A and NFKB1 coding sequences. Also used in these studies was a double-stranded PS oligonucleotide that contained three tandem repeats of the NF-κB consensus sequence (Table 1).

Antisense- and sense-expressing constructs were generated using pMAM-NEO-CAT, a dexamethasone-inducible mammalian expression vector, as previously described *(21,22)*. A fragment of 350 bp, which encompassed the 5' end of the *rel* A cDNA, including the AUG initiation codon, was cloned into the *Nhe*I site in either the sense or the antisense orientation to express the appropriate RNA. This construct contained the *E. coli* neomycin-resistance gene, which allowed for G418 selection of stable transfected cells, the bacterial chloramphenicol acetyl transferase (CAT) gene for generation of chimeric CAT-antisense RNA, and a mouse mammary tumor virus long-terminal repeat (MMTV-LTR), which allowed for dexamethasone-mediated expression of the RNA product.

Table 1
PS Oligomers Used in this Study

Gene, species	Sequence, 5' to 3'
rel A S (m)	ACC <u>ATG</u> GAC GAT CTG TTT CCC CTC
rel A AS (m)	GAG GGG AAA CAG ATC GTC CAT GGT
rel A S (h)	GCC <u>ATG</u> GAC GAA CTG TTC CCC CTC
rel A AS (h)	GAG GGG GAA CAG TTC GTC CAT GGC
NFKB1 S (m)	ACC <u>ATG</u> GCA GAC GAT GAT CCC
NFKB1 AS (m)	GGG ATC ATC GTC TGC CAT GGT
NFKB1 S (h)	AGA <u>ATG</u> GCA GAA GAT GAT CCA
NFKB1 AS (h)	TGG ATC ATC TTC TGC CAT TCT
NF-κB DS	<u>GGG GAC TTT CC</u>G CTG <u>GGG ACT TTC C</u>AG <u>GGG GAC TTT CC</u>
NF-κB DSΔ	GTC TAC TTT CCG CTG TCT ACT TTC CAC GGT CTA CTT TCC

The sense (S) and antisense (AS) PS oligomers correspond to the 5' end of the cDNAs encompassing the AUG initiation codon (underlined). The NF-κB double-stranded sequence (DS) or the mutant (DSΔ) encompass three tandem repeats of NF-κB binding sites. The single-stranded sense and the antisense oligomers for these sequences were annealed to create double-stranded PS oligomers. NF-κB binding sites are underlined.

3. Inhibition of Cellular Adhesion

The use of antisense technology to inhibit the subunits of the NF-κB transcription factor led to an unexpected observation pointing to a possible role for the *rel* A subunit in cellular adhesion. The exposure of the trypsinized cells to the antisense *rel* A PS oligonucleotides resulted in a block of adhesion of the treated cells (Fig. 2). This effect was not observed when the cells were treated with *rel* A sense, NFKB1 (sense or antisense), double-stranded NF-κB, or many other oligonucleotides (data not shown). The block in cellular adhesion was found to be very sequence-specific and was seen in several transformed cell lines, including K-Balb, B-16, Rat-1, HT-29, SW480, and T47-D *(21,22)*. The human and murine *rel* A antisense oligonucleotides, which differ by only 3 bases in their sequences, did not show cross-species reactivity (Fig. 2). Although a low level of nonspecificity by diverse oligonucleotides was seen in this adhesion assay, the antisense *rel* A PS oligonucleotides consistently elicited a complete block of cellular adhesion.

3.1. Differential Involvement of rel A
in Adherent vs Suspension Cell Growth

The role of *rel* A and NF-κB was also examined in the HL-60 leukemia cell model system. Interestingly, the growth of undifferentiated HL-60 cells was not inhibited by the antisense *rel* A PS oligonucleotides *(23)*. Similarly, growth of several other nonadherent cells, including Jurkat, WEHI-3B, and primary

PBS   **REL A - S**

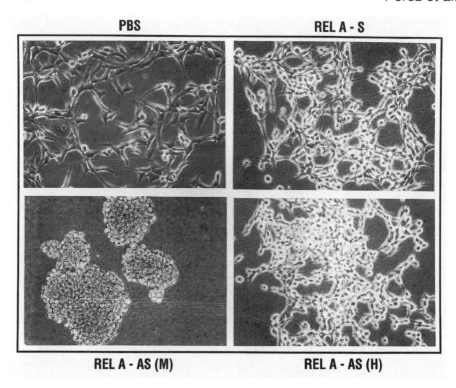

REL A - AS (M)   **REL A - AS (H)**

Fig. 2. Block of adhesion of the fibrosarcoma K-Balb cells to the substratum by murine antisense *rel* A PS oligonucleotides. The K-Balb cells were trypsinized and plated at 4×10^4 in the presence of vehicle (PBS), or 20 μM of murine sense *rel* A, murine antisense *rel* A, or human antisense *rel* A PS oligonucleotides. Photographs were taken after 48 h. Magnification, 45×. M, murine; H, human.

bone marrow cells, was also not inhibited by these antisense oligonucleotides, suggesting that *rel* A NF-κB might be important in the regulation of growth of adherent cells. However, treatment of the DMSO-differentiated HL-60 cells with antisense *rel* A PS oligonucleotides prior to treatment with 12-*O*-tetra-decanoylphorbol 13-acetate (TPA, 1 μg/mL) prevented the cells from attaching and spreading to the plate surface. Pretreatment of the HL-60 cells with *rel* A sense or NFKB1 (sense or antisense) oligonucleotides did not have an effect on adhesion and spreading on treatment with TPA. The block in cellular adhesion by *rel* A antisense oligonucleotides was subsequently found to correlate with a marked reduction in the level of a cell-surface protein, an integrin, CD 11b *(23)*. Eck et. al. *(24)* recently made similar observations implicating NF-κB with adhesion in the HL-60 cell system. These authors used an alternative approach to inhibit NF-κB activity, utilizing a double-stranded PS oligonucleotide composed of three tandem repeats of the NF-κB binding sequence. These oligonucleotides would in theory compete for the binding of the NF-κB complex to the cognate DNA binding sequence and hence function as a direct in vivo competitor by a nonantisense mechanism. These studies demonstrated a block in TPA-induced adhesion of HL-60 cells and the expression of the leukocyte integrin CD 11b. This finding confirms the link between NF-κB and cellular adhesion by blocking cellular adhesion through the disruption of normal NF-κB function. Interestingly, studies in our laboratory have not found this double-stranded NF-κB consensus sequence to be effective in blocking adhesion in adherent cell lines, although the nuclear NF-κB activity was very effectively inhibited in diverse cell lines, including cells that grow in suspension (data not shown). These results suggest that the *rel* A subunit of NF-κB, but not the NF-κB heterodimer, may regulate distinct cell adhesion molecules and integrins that are involved in the maintenance of cell adhesion.

Further corroboration of the antisense *rel* A PS oligonucleotides' effects on cell adhesion came from results of experiments involving stable transfectants of a rat pheochromocytoma cell line (PC-12) with a dexamethasone-inducible antisense *rel* A expression vector *(22)*. When clones transfected with the antisense construct were replated in the presence of the inducing agent, dexamethasone, the cells failed to attach to the substratum *(22)* Such a block of adhesion was not seen with control clones. These results helped overcome skepticism about the results of PS oligonucleotides.

4. In Vitro Inhibition of Growth

Cell-to-cell and cell-to-substratum adhesion play an important role in various disease states, including cancer, metastasis, and inflammation *(25,26)*. Hence, we reasoned that a block in cellular adhesion by *rel* A antisense PS oligonucleotides might contribute to inhibition of growth. The growth of tumor cell lines in vitro was monitored by their ability to form colonies in semisolid agar cultures. Inhibition of soft-agar colony formation by antisense *rel* A oligonucleotides was observed in B-16 (murine melanoma), HOS-MNNG (a chemically transformed human osteosarcoma), SW-480 (human colon carcinoma), and T47-D (human breast carcinoma) cell lines (Fig. 3). This inhibition of soft-agar colony formation was not observed with the use of either NFKB1 antisense oligonucleotides or the double-stranded NF-κB consensus PS oligonucleotides at the same concentration. This observation is indicative of a differential role in cell growth regulation for the transcription factor complex NF-κB vs the *rel* A subunit of this transcription factor.

The use of antisense expression constructs stably transfected into K-Balb cells again aided in the distinction between a true antisense mechanism and an

REL A-S   **REL A-AS**

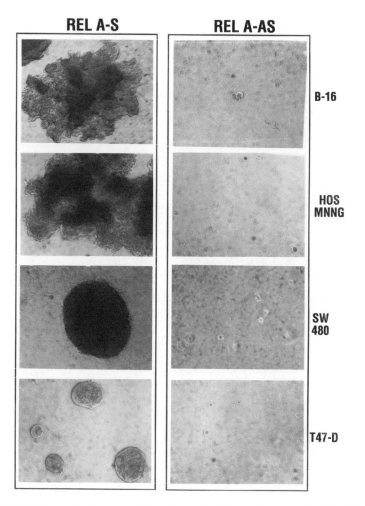

Fig. 3. Inhibition of soft-agar colony formation by antisense *rel* A PS oligonucleotides. Transformed cells, B-16 murine melanoma, SW-480 human colon carcinoma, HOS-MNNG human osteosarcoma, and T47-D human breast carcinoma cells were treated for 72 h with oligonucleotides (sense or antisense) before being trypsinized and seeded in 0.33% agar (10^3 cells/well for murine cells and 10^5 cells/well for human cell lines). Experiments were done in quadruplicate wells in the presence of 20 µ*M* oligonucleotides. Colonies were scored after 7 (B-16) or 14 (SW-480, HOS-MNNG, and T-47D) d. Photographs were taken on d 7 or 14. Magnification, 40×.

oligo-mediated, nonantisense-mediated effect. In the presence of the inducing agent dexamethasone, the antisense-transfected cells could not form colonies in soft agar (data not shown).

5. Inhibition of NF-κB Activity

The use of antisense oligonucleotides in a cellular system requires assays to verify the inhibition of the target mRNA and protein. In these studies, we employed two methods to verify the effectiveness of the antisense oligonucleotides to inhibit *rel* A. Reverse transcriptase polymerase chain reaction (RT-PCR) was used to determine whether the target mRNA was disrupted. The sequence-specific inhibition of *rel* A mRNA demonstrated an antisense effect that correlated with the block of cellular adhesion *(21–23)*. The subsequent disruption of protein function for the NF-κB subunits was examined by the use of an electrophoretic mobility shift assay (EMSA). The EMSA is used to detect the presence of DNA binding proteins in cellular extracts by coincubating a radiolabeled DNA consensus sequence probe with nuclear extracts and subsequently resolving the DNA/protein complexes on a nondenaturing gel. Nuclear DNA binding proteins are believed to be transcriptional regulatory proteins. They function by binding DNA in a sequence-specific manner and subsequently modulating the rate of transcription of their target genes though interaction with the transcriptional machinery. Hence, the binding activity of these factors could be used as a measure of their function. The treatment of K-Balb cells with *rel* A antisense oligonucleotides inhibited the nuclear NF-κB binding activity in a dose-dependent manner. At concentrations of 2 and 5 μ*M*, no effect was observed, but at 10- and 20-μ*M* concentrations of *rel* A antisense PS oligonucleotides, a partial and complete inhibition of nuclear NF-κB binding was observed, respectively (Fig. 4). In contrast, the control sense PS oligonucleotide did not cause an inhibition of nuclear NF-κB activity. The inhibition of NF-κB activity was also seen in the cytoplasmic extracts of the antisense-treated cells (not shown). The specificity of this binding was verified by competitively inhibiting the binding with a 50-fold excess of unlabeled oligonucleotide, and the position of the complex was verified by including a control reticulocyte lysate extract. This inhibition of nuclear binding activity by antisense *rel* A PS oligonucleotides was seen within 4–6 h of treatment, which correlated with the duration of the block in cell adhesion (data not shown). Several control PS oligonucleotides, including the human antisense *rel* A (which differs by only 3 bases from murine antisense *rel* A), did not elicit such a pronounced inhibition of nuclear NF-κB activity *(27)*. The authenticity of the heterodimeric complex was confirmed by antibody-based supershift experiments (not shown).

6. In Vivo Experiments

The inhibition of growth of adherent cells in vitro by antisense *rel* A PS oligonucleotides raised the possibility that inhibition of *rel* A NF-κB activity might contribute to the inhibition of tumor cell growth in vivo. Hence, we

Concentration μM   0   2   5   10   20

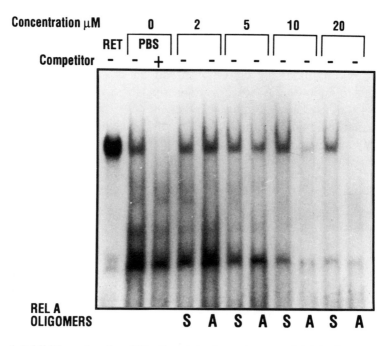

RET | PBS

Competitor   −   −   +   −   −   −   −   −   −   −   −

REL A
OLIGOMERS   S   A   S   A   S   A   S   A

Fig. 4. Inhibition of nuclear NF-κB activity by antisense *rel* A PS oligonucleotides. K-Balb cells were treated for 48 h with vehicle (PBS) or either sense *rel* A or antisense *rel* A PS oligonucleotides at 2-, 5-, 10-, and 20-μ*M* concentrations prior to the preparation of nuclear extracts. A ^{32}P-labeled oligonucleotide (1 ng probe, 2×10^5 cpm) corresponding to the NF-κB consensus binding sequence was incubated with nuclear extracts (10 μg) from treated cells, and the DNA/protein complexes were separated on a nondenaturing 0.5X TBE 4% acrylamide gel (80:1 acrylamide:bis). For the PBS sample, binding was competitively inhibited with a 50-fold excess of unlabeled oligonucleotide. Rabbit reticulocyte lysate was used as a positive control.

decided to test the in vivo effects of inhibition of *rel* A NF-κB activity using nude mouse tumorigenicity assays.

6.1. Use of Inducible Antisense RNA In Vivo

Prior to testing the antisense *rel* A PS oligonucleotides in vivo, we attempted to take advantage of the availability of K-Balb clones, which express dexamethasone-inducible, stable antisense *rel* A. Utilizing the identical vector involved in earlier studies of the tumor-suppressor gene deleted in colorectal cancer (DCC), we demonstrated the in vivo efficacy of antisense RNA expression in response to dexamethasone treatment of nude mice *(28)*. The *rel* A sense- or antisense-transfected K-Balb cells were injected sc in Balb/c nu mice, the mice were treated with or without dexamethasone (0.28 mg/L) in their drinking water, and the growth of the tumors was monitored. The *rel* A antisense clones formed smaller tumors than the *rel* A sense clones even in the absence of dexamethasone treatment. This is perhaps owing to low-level basal transcription from the antisense expression construct *(28)*. Oral administration of dexamethasone to mice injected with K-Balb *rel* A antisense clones inhibited the formation of tumors in more than 70% of treated animals. In addition, regression of established tumors in mice bearing antisense clones could be achieved by the subsequent oral administration of dexamethasone *(21)*. Such regression was not seen with control sense clones or in nude mice injected with the vector-transfected cells.

6.1.1. Efficacy of Antisense Oligonucleotides In Vivo

Encouraged by these results, we decided to test the efficacy of *rel* A antisense PS oligonucleotides in vivo. We chose two routes of oligonucleotide administration: sc injection and a sustained-release osmotic pump delivery. Experiments were done in Balb/c nu mice injected subcutaneously with K-Balb fibrosarcoma cells or B-16 murine melanoma cells. An sc injection schedule was followed with 2–3 injections/wk for 2 wk at 7, 35, or 70 mg/kg. Injection of oligonucleotides began 24 h after the mice were inoculated with the tumor cells. In the second route of delivery, the osmotic pumps (Alza Corp., Palo Alto, CA) were loaded to deliver 10, 20, or 40 mg/kg/d for 7–14 d and were implanted sc 24–48 h prior to inoculation. Localized toxicity at the site of delivery was observed by both routes at the highest dosage used. Use of the antisense *rel* A oligonucleotide was found to be effective in inhibiting tumor growth in vivo *(21)*. The K-Balb tumor cells were growth-inhibited in the nude mice when the animals were treated with the antisense *rel* A oligonucleotide by either route of administration *(21)*. A typical picture of inhibition of B-16 melanoma, which forms a readily visible pigmented tumor, is shown in Fig. 5. Several control oligonucleotides, including *rel* A sense and NFKB1 (sense and antisense), did not inhibit tumor cell growth in vivo *(21)*. While our studies were in progress, Kitajima et al. *(29)* demonstrated the inhibition of transplanted human T-cell leukemia virus (HTLV-1) Tax-transformed tumors in mice utilizing similar antisense *rel* A PS oligonucleotides. Interestingly, the use of Tax antisense PS oligonucleotides did not inhibit the growth of the tumor, suggesting that the *rel* A antisense effect may not be related to Tax function.

6.1.2. In Vivo Distribution of the rel A Antisense Oligonucleotides

In an experiment to examine the distribution of oligonucleotides in vivo, uniformly substituted ^{35}S-phosphorothioate *rel* A antisense oligonucleotides (1×10^7 cpm) were subcutaneously injected into Balb/c nu mice bearing K-Balb tumors 48 h prior to sacrifice. Several tissues were subsequently examined for

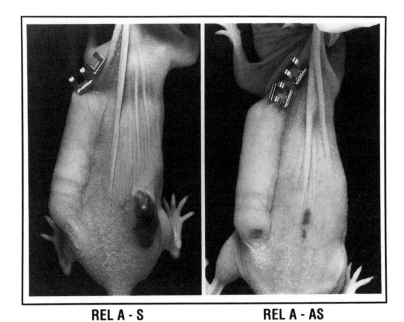

REL A - S   REL A - AS

Fig. 5. Inhibition of in vivo tumorigenicity of the B-16 murine melanoma by antisense *rel* A PS oligonucleotides. Five Balb/c nu mice per group were implanted with sc Alza osmotic pumps (Alza Corp., Palo Alto, CA) to deliver either sense *rel* A or antisense *rel* A PS oligonucleotides at a dosage of 10 mg/kg/d for 14 d. After 48 h, the B-16 murine melanoma cells (5×10^6 cells) were introduced sc on the contralateral side of the mouse. Representative mice from one of four independent experiments are shown. Photographs were taken on d 14 of treatment. Magnification, 0.11×.

uptake of the oligonucleotide by solubilizing samples of tissue in Solvable (NEN Research Products, Boston, MA) at 50°C for 3 h and counting the cpm/g weight of tissue. The greatest accumulations of labeled material was found in the kidneys and liver: approx 1.21×10^6 and 1.10×10^6 cpm/g, respectively (Fig. 6). Other tissues, such as the heart and the spleen, had lesser accumulations of radiolabeled material: 3.48×10^5 and 1.41×10^5 cpm/g, respectively. The implanted K-Balb tumor tissue also showed an accumulation of labeled material of approx 8.0×10^5 cpm/g. The brain had the lowest accumulation of radiolabeled material with counts of 1.0×10^4 cpm/g (Fig. 6). The lack of radiolabled material in the brain is thought to be owing to the impermeability of the oligonucleotide through the blood–brain barrier. In a previously published in vitro study *(29)*, it was demonstrated that a human T-cell leukemia virus (HTLV) Tax-transformed fibroblast cell line showed an enhanced uptake of radiolabled antisense oligonucleotides over the parental nontransformed

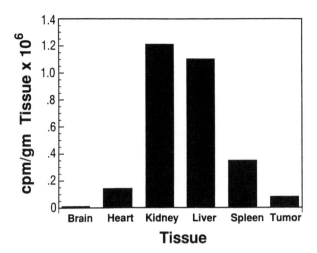

Fig. 6. In vivo distribution of ^{35}S radiolabeled antisense *rel* A PS oligonucleotide. Duplicate Balb/c nu mice were injected sc with uniformly ^{35}S-labeled *rel* A antisense oligonucleotide (1×10^7 cpm), and the animals were sacrificed 48 h later. Selected tissues (150 mg) were solubilized in Solvable (NEN, Boston) at 50°C for 3 h, and the average cpm/g tissue were determined. The values of cpm from each animal did not vary more than 5%.

cells. Our findings did not show an enhanced uptake of the oligonucleotides by the transformed tumor cell in vivo, but instead a predominant accumulation of the labeled material in the kidneys and liver.

7. Toxicology

Previous reports have demonstrated the in vivo pharmacological and the toxicological effects of phosphorothioate oligonucleotides *(30–36)*. These toxicity studies demonstrated mortality within 3 d of an ip injection of oligonucleotides at doses of 160 mg/kg and within minutes after administration of doses of 640 mg/kg.

A comprehensive 2-wk study of phosphorothioate oligonucleotide toxicity was performed using antisense *rel* A phosphorothioate *(37)*. In agreement with earlier studies, ip doses of up to 50 mg/kg were not toxic to the animal. However, doses of 150 mg/kg were lethal to 5 of 12 mice within 2–4 d. Pathological examination revealed that the principal cause of death in these animals was acute renal failure owing to cortical necrosis or acute tubular nephrosis. This observation might be predicted based on the high rate of clearing of oligonucleotide from the blood. Pharmacokinetic studies have shown that oligonucleotides introduced by iv or ip administration have a half-life in the blood of approx 20 min and accumulate maximally in organs, such as liver, kidney, lung, and spleen; once cleared, the oligonucleotides are excreted in the urine over the next 20–40 h *(31,34,35,38,39)*. Significant toxicity was also detected in the liver and bone marrow megakaryocytes. These findings correlated with the changes detected in the blood chemistry, including elevation in liver enzymes aspartate aminotransferase, and alanine aminotransferase, as well as a decrease in the platelet count with increased platelet size *(37)*. The results of the toxicology study correlated with the findings of the in vivo distribution of the radiolabeled oligonucleotide. The highest concentrations of the labeled oligonucleotide material were detected in the kidneys and the liver.

8. Complications of Oligonucleotide Therapy

PS oligonucleotides exhibit both sequence-specific (antisense) and sequence-selective (nonantisense) effects *(40)*. These effects complicate our ability to interpret antisense-based experiments. The precise nature of the nonantisense effects is not clear. PS oligonucleotides are polyanionic in nature, and may interact with cell-surface components and proteins (for a review, *see* ref. *20*). Clarifying these issues becomes essential for successful antisense therapeutics. To date, all the studies have dealt with efficacy of these potent molecules. We have inadvertently addressed some of these questions in our studies.

8.1. Sequence-Selective Effects

In our in vivo studies, we unexpectedly observed that the control sense *rel* A PS oligonucleotides, but not the antisense *rel* A antisense PS oligonucleotides, caused massive splenomegaly in mice *(41)*. This surprising result of the control sense *rel* A PS oligonucleotide treatment was caused by a sequence-specific stimulation of splenic cell proliferation both in vitro and in vivo. Primarily, the expanded population of cells was the surface antigen B220+, sIg+ B-cell population. Treatment of these splenic cells with the sense *rel* A PS oligonucleotides in vitro also stimulated an 11-fold increase in the amount of secreted IgM in the medium. To date three PS oligomers have been found to cause splenomegaly: *rel* A sense *(41)*, *Rev* AS *(42)*, and MCF *Env* gene AS *(43)*. There were no obvious sequence similarities among these three PS oligonucleotides. These results showed that PS oligonucleotides can exhibit sequence-specific nonantisense effects. Furthermore, our demonstration in these studies that the control sense *rel* A PS oligonucleotides causes enhanced proliferation of primary splenic B-cells, but not of transformed B-cell lines, indicates that the oligonucleotide-mediated growth stimulation is cell-type specific. Surprisingly, these three oligonucleotides caused an induction in nuclear NF-κB binding activity in primary splenocytes in vitro, but not in established cell lines *(41)*. However, the connection between splenomegaly and induction of nuclear NF-κB is yet to be resolved.

8.2. Sequence-Independent Effects

Charged oligonucleotides can interact with proteins and may contribute to some nonspecific effects *(40)*. We have recently obtained interesting insights into these interactions. In our efforts to elucidate the molecular mechanisms of antisense *rel* A PS oligonucleotides in regulating cell growth, we observed that the control sense *rel* A PS oligonucleotide, but not the antisense *rel* A PS oligonucleotide, caused the induction of a nuclear DNA binding activity *(27)*. Subsequent experiments revealed that this was activity of the transcription factor Sp-1. In addition to sense *rel* A PS oligonucleotides, several PS oligonucleotides (both in the sense and antisense orientation) were found to cause the induction of Sp-1 activity in diverse cell lines, including primary splenic cells in vitro. The induction of Sp-1 was also observed in vivo in splenic cells of mice 24 h after oligonucleotide administration. Surprisingly, inhibition of NF-κB activity prior to treatment with PS oligonucleotides completely prevented the induction of Sp-1 activity. In the context of the role of the Sp-1 transcription factor in regulation of the expression of various cellular and viral genes *(44–50)*, these results suggest that care should be exercised before proposing the use of PS oligonucleotides for human therapy.

9. Perspectives on Antisense *rel* A in Cancer

The studies described herein were undertaken to explore the usefulness of antisense techniques in cancer therapeutics. While investigating the role of the transcription factor NF-κB in gene regulation, we made a chance observation that suggested involvement of the *rel* A subunit of NF-κB in cellular adhesion. Since adhesion plays an important role in various disease processes, including cancer, we decided to study this unexpected observation in tumor models in vitro and subsequently in vivo. Our results indicate that it is possible to inhibit a key regulatory molecule, such as a transcription factor, without causing serious toxicity in the subject, at least in short-term treatment. Our results further demonstrate that the individual subunits of the heterodimeric NF-κB complex elicit distinct effects: Whereas the *rel* A subunit is critical for the growth of the transformed cell, the NFKB1 subunit has a limited role, if any. How does the inhibition of *rel* A NF-κB contribute to the block of tumorigenicity in vivo? Is there a correlation between the in vitro block of cellular adhesion and the in vivo effects? Do the *rel* A oligonucleotides exhibit a true antisense-mediated effect? Could there be a sequence-specific antisense effect as well as a sequence-specific, nonantisense effect associated with the *rel* A antisense oligonucleotides? We have begun systematic studies to address these issues. Our results should be taken in the context of sequence-specific and sequence-independent effects exhibited by the PS oligonucleotides. The antisense approach offers considerable promise as a molecular drug for the treatment of diverse diseases, and clinical trials are already underway. The eventual success of this novel mode of treatment will be predicated on our ability to achieve true antisense-mediated effects from a selected sequence. Investigation of the non-specific components associated with an oligomer is critical to our ability to evaluate the target, in addition to being more efficient in the long term. Perhaps the *rel* A NF-κB model can offer an avenue to clarifying some of these issues. This may enable us to bring antisense research to a true therapeutic reality.

Acknowledgments

We thank S. Agrawal for the *rel* A antisense oligonucleotides used in the toxicology studies, U. Sarmiento for toxicology support, C. A. Stein for synthesizing the ^{35}S-labeled oligonucleotides, and J. Narayanan for editorial assistance.

References

1. Grilli, M., Chiu, J. S., and Lenardo, M. J. (1993) NF-κB and Rel—participants in a multiform transcriptional regulatory system. *Int. Rev. Cytol.* **143,** 1–62.
2. Liou, H. C. and Baltimore, D. (1993) Regulation of the NF-κB/rel transcription factor and IκB inhibitor system. *Curr. Opin. Cell Biol.* **5,** 477–487.
3. Nabel, G. and Baltimore, D. (1987) An inducible transcriptional factor activates expression of the human immunodeficiency virus in T cells. *Nature* **335,** 683.
4. Lenardo, M. J. and Baltimore, D. (1989) NF-κB: a pleiotropic mediator of inducible and tissue-specific gene control. *Cell* **58,** 227–229.
5. Nolan, G. P., Ghosh, S., Liou, H. C., Tempst, P., and Baltimore, D. (1991) DNA binding and IκB inhibition of the cloned p65 subunit of NF-κB, a *rel*-related polypeptide. *Cell* **64,** 961–969.
6. Ruben, S. M., Dillon, P. J., Schreck, R., Henkel, T., Chen, C. H., Maher, M., Baeuerle, P. A., and Rosen, C. A. (1991) Isolation of a *rel*-related human cDNA that potentially encodes the 65-kD subunit of NF-κB. *Science* **251,** 1490–1493.
7. Baeuerle, P. A. and Baltimore, D. (1988) IKB: a specific inhibitor of the NFκB transcription factor. *Science* **242,** 540–546.
8. Ghosh, S. and Baltimore, D. (1990) Activation *in vitro* of NF-κB by phosphorylation of its inhibitor I-κ-B. *Nature* **344,** 678–682.
9. Henkel, T., Machleidt, T., Alkalay, I., Kronke, M., Ben-Nerlah, Y., and Baeuerle, P. A. (1993) Rapid proteolysis of IκB-α is necessary for activation of transcription factor NF-κB. *Nature* **365,** 182–184.
10. Brown, K., Park, S., Tanno, T., Franzoso, G., and Siebenlist, U. (1993) Mutual regulation of the transcriptional activator NF-κB and its inhibitor, IκB-α. *Proc. Natl. Acad. Sci. USA* **90,** 2532–2536.
11. Scott, M. L., Fujita, T., Liou, H. C., Nolan, G. P., and Baltimore, D. (1993) The p65 subunit of NF-κB regulates IκB by two distinct mechanisms. *Genes Dev.* **7,** 1266–1276.
12. Sun, S. C., Ganchi, P. A., Ballard, D. W., and Greene, W. C. (1993) NF-κB controls expression of inhibitor IκBα: evidence for an inducible autoregulatory pathway. *Science* **259,** 1912–1915.

13. Ruben, S. M., Narayanan, R., Klement, J. F., Chen, C. H., and Rosen, C. A. (1992) Functional characterization of the NF-κB p65 transcriptional activator and an alternatively spliced derivative. *Mol. Cell. Biol.* **12**, 444–454.

14. Duckett, C. S., Perkins, N. D., Kowalik, T. F., Schmid, R. M., Huang, E. S., Baldwin, A. S., Jr., and Nabel, G. J. (1993) Dimerization of NF-KB2 with RelA (p65) regulates DNA binding, transcriptional activation, and inhibition by an IκB-α (MAD-3). *Mol. Cell. Biol.* **13**, 1315–1322.

15. Stein, B., Cogswell, P. C., and Baldwin, A. S., Jr. (1993) Functional and physical associations between NF-κB and C/EBP family members: a rel domain-bZIP interaction. *Mol. Cell. Biol.* **13**, 3964–3974.

16. La Rosa, F. A., Pierce, J. W., and Sonenshein, G. E. (1994) Differential regulation of the c-*myc* oncogene promoter by the NF-κB rel family of transcription factors. *Mol. Cell. Biol.* **14**, 1039–1044.

17. Kunsch, C., Ruben, S. M., and Rosen, C. A. (1992) Selection of optimal κB/Rel DNA-binding motifs: interaction of both subunits of NF-κB with DNA is required for transcriptional activation. *Mol. Cell. Biol.* **12**, 4412–4421.

18. Ganchi, P. A., Sun, S. C., Greene, W. C., and Ballard, D. W. (1993) A novel NF-κB complex containing p65 homodimers: implications for transcriptional control at the level of subunit dimerization. *Mol. Cell. Biol.* **13**, 7826–7835.

19. Stein, B. and Baldwin, A. S., Jr. (1993) Distinct mechanisms for regulation of the interleukin-8 gene involve synergism and cooperativity between C/EBP and NF-κB. *Mol. Cell. Biol.* **13**, 7191–7198.

20. Stein, C. A., and Cheng, Y. C. (1993) Antisense oligonucleotides as therapeutic agents—is the bullet really magical? *Science* **261**, 1004–1012.

21. Higgins, K. A., Perez, J. R., Coleman, T. A., Dorshkind, K., McComas, W. A., Sarmiento, U. M., Rosen, C. A., and Narayanan, R. (1993) Antisense inhibition of the p65 subunit of NF-κB blocks tumorigenicity and causes tumor regression. *Proc. Natl. Acad. Sci. USA* **90**, 9901–9905.

22. Narayanan, R., Higgins, K. A., Perez, J. R., Coleman, T. A., and Rosen, C. A. (1993) Evidence for differential functions of the p50 and p65 subunits of NF-κB using a cell adhesion model. *Mol. Cell. Biol.* **13**, 3802–3810.

23. Sokoloski, J. A., Sartorelli, A. C., Rosen, C. A., and Narayanan, R. (1993) Antisense oligonucleotides to the p65 subunit of NF-κB block CD11b expression and alter adhesion properties of differentiated HL-60 granulocytes. *Blood* **82**, 625–632.

24. Eck, S. L., Perkins, N. D., Carr, D. P., and Nabel, G. J. (1993) Inhibition of phorbol ester-induced cellular adhesion by competitive binding of NF-κB in vivo. *Mol. Cell. Biol.* **13**, 6530–6536.

25. Albelda, S. M. and Buck, C. A. (1990) Integrins and other cell adhesion molecules. *FASEB J.* **4**, 2868–2880.

26. Hynes, R. O. (1992) Integrins: versatility, modulation, and signaling in cell adhesion. *Cell* **69**, 11–25.

27. Perez, J. R., Li, Y., Stein, C. A., Majumder, S., van Oorschot, A., and Narayanan, R. (1994) Sequence independent induction of Sp-1 transcription factor by phosphorothioate oligodeoxynucleotides. *Proc. Natl. Acad. Sci. USA* **91**, 5957–5961.

28. Narayanan, R., Schaapveld, R. Q., Cho, K. R., Vogelstein, B., Tran, P. B., Osborne, M. P., and Telang, N. T. (1992) Antisense RNA to the putative tumor-suppressor gene DCC transforms Rat-1 fibroblasts. *Oncogene* **7**, 553–561.

29. Kitajima, I., Shinohara, T., Bilakovics, J., Brown, D. A., Xu, X., and Nerenberg, M. (1992) Ablation of transplanted HTLV-1 *tax*-transformed tumors in mice by antisense inhibition of NF-κB. *Science* **258**, 1792–1795.

30. Agrawal, S., Goodchild, J., Civeira, M. P., Thornton, A. H., Sarin, P. S., and Zamecnik, P. C. (1988) Oligodeoxynucleoside phosphoramidates and phosphorothioates as inhibitors of human immunodeficiency virus. *Proc. Natl. Acad. Sci. USA* **85**, 7079–7083.

31. Agrawal, S., Temsamani, J., and Tang, J. Y. (1991) Pharmacokinetics, biodistribution, and stability of oligodeoxynucleotide phosphorothioates in mice. *Proc. Natl. Acad. Sci. USA* **88**, 7595–7599.

32. Goodchild, J., Agrawal, S., Civeira, M. P., Sarin, P. S., Sun, D., and Zamecnik, P. C. (1988) Inhibition of human immunodeficiency virus replication by antisense oligonucleotides. *Proc. Natl. Acad. Sci. USA* **85**, 5507–5511.

33. Goodchild, J., Byung, K., and Zamecnik, P. C. (1991) The clearance and degradation of oligodeoxynucleotides following intravenous injection into rabbits. *Antisense Res. Dev.* **1**, 153–160.

34. Iversen, P. (1991) *In vivo* studies with phosphorothioate oligonucleotides: pharmacokinetics prologue. *Anticancer Drug Des.* **6**, 531–538.

35. Iversen, P. (1993) *In vivo* studies with phosphorothioate oligonucleotides: rationale for systemic therapy, in *Antisense Research and Applications* (Crooke, S. T. and Lebleu, B., eds.), CRC, Boca Raton, FL, pp. 461–469.

36. Crooke, S. T. (1992) Therapeutic applications of oligonucleotides. *Annu. Rev. Pharmacol. Toxicol.* **32**, 329–376.

37. Sarmiento, U. M., Perez, J. R., Becker, J. M., and Narayanan, R. (1994) *In vivo* toxicologic effects of *rel* A antisense phosphorothioates in CD-1 mice. *Antisense Res. Dev.* **4**, 99–107.

38. Agrawal, S. (1991) Antisense oligonucleotides: a possible approach for chemotherapy of AIDS, in *Prospects for Antisense Nucleic Acid Therapy of Cancer and AIDS* (Wickstrom, E., ed.), Wiley-Liss, New York, pp. 143–215.

39. Crooke, R. M. (1993) *In vitro* and *in vivo* toxicology of first generation analogs, in *Antisense Research and Applications* (Crooke, S. T., and Lebleu, B. eds.), CRC, Boca Raton, FL, pp. 471–492.

40. Helene, C. and Toulme, J. J. (1990) Specific regulation of gene expression by antisense, sense and antigene nucleic acids. *Biochim. Biophys. Acta* **1049**, 99–125.

41. McIntyre, K. W., Lombard-Gilooly, K., Perez, J. R., Kunsch, C., Sarmiento, U. M., Larigan, J. D., Landreth, K. T., and Narayanan, R. (1993) A sense phosphorothioate oligonucleotide directed to the initiation codon of transcription factor NF-κB p65 causes sequence-specific immune stimulation. *Antisense Res. Dev.* **3**, 309–322.

42. Branda, R. F., Moore, A. L., Mathews, L., McCormack, J. J., and Zon, G. (1993) Immune stimulation by an antisense oligomer complementary to the *rev* gene of HIV-I. *Biochem. Pharmacol.* **45**, 2037–2043.

43. Mojcik, C. F., Gourley, M. F., Klinman, D. M., Krieg, A. M., Gmelig-Meyling, F., and Steinberg, A. D. (1993) Administration of a phosphorothioate oligonucleotide antisense to murine endogenous retroviral MCF *env* causes immune effects *in vivo* in a sequence-specific manner. *Clin. Immunol. Immunopathol.* **67**, 130–137.
44. Ishii, S., Xu, Y. H., Stratton, R. H., Roe, B. A., Merlino, G. T., and Pastan, I. (1985) Characterization and sequence of the promoter region of the human epidermal growth factor receptor gene. *Proc. Natl. Acad. Sci. USA* **82**, 4920–4924.
45. Jones, K. A., and Tjian, R. (1985) Sp1 binds to promoter sequences and activates herpes simplex virus "immediate-early" gene transcription in vitro. *Nature* **317**, 179–182.
46. Jones, K. A., Kadonaga, J. T., Luciw, P. A., and Tjian, R. (1986) Activation of the AIDS retrovirus promoter by the cellular transcription factor, Sp1. *Science* **232**, 755–759.
47. Melton, D. A. (1985) Injected anti-sense RNAs specifically block messenger RNA translation in vivo. *Proc. Natl. Acad. Sci. USA* **82**, 144–148.
48. Valerio, D., Duyvesteyn, M. G. C., Dekker, B. M. M., Weeda, G., Berkvens, T. M., van der Voorn, L., van Ormondt, H., and van der Eb, A. J. (1985) Adenosine deaminase: characterization and expression of a gene with a remarkable promoter. *EMBO J.* **4**, 437–443.
49. Yamaguchi, M., Hayashi, Y., and Matsukage, A. (1988) Mouse DNA polymerase β gene promoter: fine mapping and involvement of Sp1-like mouse transcription factor in its function. *Nucleic Acids Res.* **16**, 8773–8787.
50. Kim, S. J., Jeang, K. T., Glick, A. B., Sporn, M. B., and Roberts, A. B. (1989) Promoter sequences of the human transforming growth factor-β1 gene responsive to transforming growth factor-β1 autoinduction. *J. Biol. Chem.* **264**, 7041–7045.

11

Continuous Infusion of Antisense Phosphorothioate Therapeutics

Patrick L. Iversen, Bryan L. Copple,
Hemant K. Tewary, Eliel Bayever, and Michael R. Bishop

1. Introduction

Current therapy for acute myelogenous leukemia (AML) includes induction with Ara-C and an anthracycline, such as daunorubicin, idarubicin, or mitoxantrone. Unfortunately, most patients relapse from initial remission. Nearly one-fifth of early relapses experience treatment-related deaths. In addition, patients refractory to Ara-C die within months. Hence, new therapeutic agents must be identified capable of enhanced remission rates, diminished treatment-related mortality, or that can achieve remissions in refractory patients.

1.1. p53 Background

p53, a tumor-suppressor protein, is also a DNA-damage responsive protein. p53 protein levels are augmented in response to several DNA damaging agents, including radiation, mitomycin C, and hydrogen peroxide (1–7). Other agents that do not directly damage DNA, such as the antimetabolite cytosine arabinoside, do not induce p53 protein (2). Further, the use of halogenated pyrimidines as radiosensitizers, which produces direct DNA damage with minimal effect on cellular targets in the presence of ionizing radiation, causes the induction of GADD45, a downstream effector of p53 (8). Hence, p53 appears to play a very important role in the consequences of DNA-damaging agents, including arrest in the cell cycle at G1/S and G2/M or programmed cell death.

The p53 protein appears to have a role in the process of apoptosis. Wild-type p53 expression mediated by transfects into the cell lines of myeloid leukemia, colon tumors, and erythroleukemic cells results in apoptosis (9–11).

From: *Methods in Molecular Medicine: Antisense Therapeutics*
Edited by: S. Agrawal Humana Press Inc., Totowa, NJ

Cell-cycle checkpoints are essential for maintaining genomic integrity *(12,13)*. At least some of these signal transduction pathways, which arrest cells in G1/S and G2/M phases of the cell cycle in response to DNA damage, allow repair of DNA *(12,13)*, preventing the transfer of heritable mutations and disallowing the segregation of damaged chromosomes *(12,13)*. Recently, p53 has been determined to be in the G1 arrest pathway induced by DNA-damaging agents *(2,3,14)*. Cells lacking p53 or cells containing dominant negative forms of p53 do not arrest in G1 phase of the cell cycle in response to ionizing radiation *(3)*. In addition, restoration of wild-type p53 into p53-negative cell lines restores the G1 arrest, and the addition of dominant negative forms of p53 to cell lines containing wild-type p53 inhibits the G1 arrest *(14)*. Inhibition of p53 by antisense oligonucleotides (ODNs) can recover rat hepatocytes from UV-induced G1 arrest *(15)*. Recently, WAF-1 protein, which is induced by p53, has been tentatively determined to mediate the G1 arrest by blocking Cdk enzymes resulting in the inhibition of DNA synthesis *(16)*. Thus, p53 appears to play the role of checkpoint pathway protein in G1/S similar to that found in the RAD9 gene discovered in *Saccharomyces cerevisiae*, which induces a G2/M delay in response to ionizing radiation *(12,17,18)*.

Since p53 is involved in the G1/S checkpoint pathway, loss of p53 may be expected to induce radiosensitivity or chemosensitivity similar to observations made in yeast lacking RAD9 protein *(12,17,18)*. Cells from patients with Ataxia-telangiectasia show no induction of p53 after exposure to ionizing radiation and, subsequently, no G1 arrest in the cell cycle *(2)*. These cells also exhibit radiosensitivity when exposed to ionizing radiation (for review, *see* ref. *19*). Mouse fibroblasts lacking p53 and normal human fibroblasts treated with an antisense ODN to p53 show increased chemosensitivity to direct DNA-damaging agents *(20)*. These results clearly show that lack of response by p53 or loss of p53 can induce radiosensitivity and chemosensitivity. Other groups, however, have shown that loss of p53 induces chemoresistance and radioresistance *(21–24)*. Thus, whether loss of p53 causes increased or decreased sensitivity to DNA-damaging agents may be cell-type-specific.

1.2. OL(1)p53 Background

We have shown previously that antisense ODNs directed against p53 cause a decrease in human AML blast cell viability *(25)*. p53 is expressed at higher levels in AML blast stem cells than normal hematopoietic cells *(26,27)*, and p53 expressed in AML blast stem cells is predominantly wild-type *(28–30)*.

OL(1)p53, 5'-d(CCC TGC TCC CCC CTG GCT CC)-3' complementary to exon 10 region of the human p53 mRNA (mol wt = 6,625 g/mol) was supplied by Lynx Therapeutics, Inc. (Foster City, CA) as a sterile, apyrogenic, dry powder in vials each containing 50 mg. A radioactive OL(1)p53 was prepared by combining an H-phosphonate analog of OL(1)p53 prepared by Glenn Research (Sterling, VA) and shipped on solid support in the original synthesis column with elemental ^{35}S purchased from Amersham (Arlington Heights, IL) according to the method of Stein et al. *(31)*.

1.3. Goals

The studies described in this chapter encompass the broad goals of:

1. Evaluation of the pharmacokinetics of OL(1)p53 following iv administration to the mouse, rat, monkey, and human;
2. Comparison of the pharmacokinetic behavior of OL(1)p53 between the species;
3. Establishment of relationships between dose and the various pharmacokinetic parameters measured and comparison of these observations with the availability of OL(1)p53 at target cells; and
4. Evaluation of the toxicity and observations related to efficacy of OL(1)p53 as a function of exposure.

These studies would then provide insight into the basis of how accurately investigators may predict human exposures. These studies all involve the phosphorothioate, OL(1)p53, as a model. Finally, comparison of the cellular concentration of OL(1)p53 achieved in vivo compared to that observed in cell culture studies may provide insights into the optimal delivery and formulation of OL(1)p53 and other phosphorothioate oligonucleotides.

2. Preclinical Pharmacology and Toxicology

2.1. Pharmacokinetics of OL(1)p53 in the Mouse

Pharmacokinetic data in the mouse were restricted to the plasma kinetics, since the urine volume was to small to reliably recover and thus measurements were unreliable. Four mice were examined following a 0.1 mg/mouse dose of OL(1)p53 sc. Blood samples were recovered at 2, 4, 8, 24, 48, and 72 h postinfusion, and radioactivity in each sample determined. The concentration of OL(1)p53 was calculated from the specific activity of the injected material and reported as μg/mL. The radioactive material measured may be in the form of both fully intact oligonucleotide and/or degradation products.

The mean plasma concentrations were evaluated with PKCALC software developed by R. Shumaker release 9/28/87. The ESTRIP program was employed to calculate the pharmacokinetic values reported for a two-compartmental model in Table 1. The 12 mg/m² dose is equivalent to 4 mg/kg and is expressed in terms of surface area for ease of comparison to the dose in other species. Since the injection was via a subcutaneous route, plasma samples were taken beginning 2 h after the injection. Hence, only the elimination half-life of

Table 1
Summary of Mouse Pharmacokinetic Parameters for OL(1)p53

Dose, mg/m^2	C_p^0, µg/mL	$t_{1/2}$, h	V_d, L/kg	AUC, µg · h/mL	MRT, h	r^2
12	2.34	28.2	1.7	24.4	44.4	0.95

Table 2
Summary of Rat Pharmacokinetic Parameters for OL(1)p53

Dose, mg/m^2	C_p^0, µg/mL	$t_{1/2}$, h	V_d, L/kg	Excretion rate, µg/h	Amount excreted, µg	Cl_{renal}, mL/min
17.7	19.7	35.5	0.15	2.6	125	0.022
59.0	65.5	87.3	0.15	4.9	236	0.001
162.3	279.9	132.4	0.10	14.7	1408	0.001
590.0	358.9	167.0	0.28	17.8	1704	0.001
1770.0	962.1	191.0	0.31	130.6	3134	0.002

the single compartment model was observed. The volume of distribution of 1.71 L/kg is equivalent to 42.75 mL in a 25-g mouse, suggesting the OL(1)p53 is sequestered in addition to distribution throughout the body.

2.2. Pharmacokinetics of OL(1)p53 in the Rat

Studies involved three rats at each of five dose levels: 0.6, 2.0, 6.0, 20.0, and 60.0 mg/rat. The blood samples were withdrawn from the jugular vein via a surgically implanted catheter. The OL(1)p53 was injected under anesthesia into the right femoral vein. Plasma and urine kinetics are provided in Table 2. In this case, five different dose levels, 3.0, 10.0, 27.5, 100.0, and 300 mg/kg (equivalent to 17.7, 59.0, 162.3, 590.0, and 1770.0 mg/m^2, respectively), were administered iv as a bolus injection to 4 animals/dose groups.

The elimination half-life, concentration in plasma at time 0 (Cp0), area under the plasma concentration vs time curve (AUC), amount excreted in the urine, and excretion rates were increased as a result of increased dose. However, the volume of distribution (Vd) and renal clearance were not increased as the dose was increased.

The volume of distribution observed in the rat is substantially smaller than that observed in the mouse. This is possibly because of the route of administration and the fact that the OL(1)p53 is injected as a bolus injection in the rat, but in the mouse the release of OL(1)p53 is relatively slow from the subcutaneous site. This would only be true if the rate of distribution from the blood into tissue is relatively slow in comparison to the rate of loss owing to excretion.

Table 3
Summary of Monkey Pharmacokinetic Parameters
for OL(1)p53

Parameter	Bolus injection, $N = 3$	Continuous infusion, $N = 3$
Dose (mg/m^2)	149.0 ± 22.0	166.9 ± 6.2
Elimination half-life (h)	18.1 ± 1.5	33.3 ± 7.3
Volume of distribution (L/kg)	4.0 ± 0.9	4.9 ± 1.0
Plasma clearance (mL/min)	25.2 ± 4.7	3.5 ± 0.43
Maximal plasma concentration (µg/mL)	1.93 ± 0.18	9.2 ± 2.9

2.3. Pharmacokinetics of OL(1)p53 in the Rhesus Monkey

Single bolus injections of 149 mg/m^2 OL(1)p53 were injected intra-arterially into three Rhesus monkeys over a period of 4 h. A biphasic elimination from plasma was observed with an elimination half-life of 18.1 h. The volume of distribution was 4 L/kg, again suggesting the OL(1)p53 is widely distributed in the body and may be sequestered at some sites. The clearance rate is 25.2 mL/min, which is less than the glomerular filtration rate. The maximal plasma concentration achieved was 1.93 µg/mL, which was predominantly full-length material when evaluated by gel electrophoresis. From 13 to 27% of the dose was excreted into urine in 6 d. Finally the liver, kidney, heart, spleen, and pancreas were organs of greatest accumulation.

Three monkeys were administered continuous intravascular infusions with doses 169 mg/m^2/d over a period of 6–15 d. Steady-state plasma concentration was observed in 4–9 d reaching concentrations of 1.5–5.6 µ*M*. Plasma and urine kinetics are provided in Table 3. The volume of distribution is equal following continuous infusion or single 4-h injection. The maximal plasma concentrations following the infusion reached nearly five times that following the single injection. This is expected, since a new steady-state plasma concentration should be achieved in 3–5 half-lives.

The plasma half-life following the single injections was significantly shorter than that observed following continuous infusions. This is consistent with the higher plasma concentration and the smaller clearance. These data suggest a "deep compartment" where OL(1)p53 accumulates, which is characterized by slower accumulation and slower efflux into the blood where it is available for elimination. Further, the elimination half-life observed after the single injection is the average efflux out of a more shallow compartment and the deep compartment. The rate of efflux from the deep compartment is slow, and it is very difficult to observe following a single injection, since the OL(1)p53 is below the limits of detection 4–5 d postinjection. It is not possible to define this deep compartment at this time, since it is not clear if it is defined by a tissue or a property of OL(1)p53 with multiple subcellular compartments in cells within tissues. The most important observation pertaining to these data is that if a therapeutic response is related to accumulation of OL(1)p53 in the deep compartment, very little response would be expected from a single injection.

3. Phase I Clinical Trial

3.1. Primary Cells in Culture

Bone marrow aspirates recovered from normal donors were incubated in the presence of ^{35}S-labeled OL(1)p53 in culture. The cultures were harvested after 3, 6, 12, and 24 h of incubation with 1 μM OL(1)p53. Triplicate cultures were washed three times in cold PBS to remove nonspecific binding. Then cells were harvested and radioactivity measured by liquid scintillation counting. The average cellular uptake half-life was 19.9 ± 2.4 h ($N = 4$). Bone marrow aspirates from four patients with AML were also incubated with the radioactive OL(1)p53. The rate of cellular uptake was 58.2 h, which is significantly slower than that observed with the cell from normal donors. Although the rate of cellular absorbance was slower, the quantity associated with cells was somewhat greater than that observed in the cells from normal donors. These data indicate a single injection would not provide for cellular availability to the AML cells, which are the target for the OL(1)p53.

3.2. Pharmacokinetics of OL(1)p53 in the Human

OL(1)p53 was administered by 10 d of continuous iv infusion to 17 patients with either relapsed or refractory AML or myelodysplastic syndrome (MDS). Analysis of blood and urine by high-performance electrophoretic chromatography (HPEC) and high-performance liquid chromatography (HPLC) followed by postlabeling with ^{32}P-ATP and polynucleotide kinase revealed approx 36% of the recovered material in urine and >90% of the material recovered in plasma retain the identical electrophoretic mobility as OL(1)p53, and approx 53% of the recovered material in urine is equivalent to mononucleotide monophosphorothioate material. A single patient was administered uniformly ^{35}S-labeled OL(1)p53, and the analysis confirms approx 90% agreement with chromatographic methods with >90% total recovery of labeled material. The plasma concentration and area under the plasma concentration curve (AUC) were linearly proportional to the dose (mg/m^2) of OL(1)p53. The elimination half-life increased as a function of dose from 24.4 h at 0.05 mg/kg/h to 62.5 h at 0.25 mg/kg/h. The renal clearance did not increase with increasing dose remaining at 14.4 ± 2.56 mL/min. Hence, the renal clearance accounts for a smaller percentage of total clearance with increasing dose from 54.3 ± 6.6 at 0.05 mg/kg/h to 15.8 ± 3.3 at 0.25 mg/kg/h. This suggests a saturable renal clearance pathway, which suggests glomerular filtration may either be limited by the amount of free oligonucleotide or is not the major mechanism for renal clearance. Therefore, nonrenal clearance increases with dose, and the volume of distribution is increased from 34.1 L at 0.05 mg/kg/h to 454.1 L at 0.25 mg/kg/h. The half-life did change with increased dose in the rat, but the volume of distribution did not. This may have been owing to the differences in the injections, since the rat received a bolus injection and the humans received a continuous infusion.

OL(1)p53 reaches an apparent steady-state plasma concentration in <24 h. The expected time to steady state would be between 3 and 5 half-lives, which would be several days. The difference between the observed and expected time to reach steady-state plasma concentrations is probably owing to a high degree of plasma protein binding. We employed human serum albumin (HSA) protein bound to Sepharose beads to determine the HSA-protein dissociation constant *(32)*. These observations indicate OL(1)p53 is approx 98% protein bound in plasma. Hence, the infusion establishes equilibration of plasma protein binding. The small fraction of free OL(1)p53 in blood plasma is then available for renal clearance, distribution throughout the body, and uptake by target cells.

The rate of OL(1)p53 absorbed in peripheral blood mononuclear cells (PBMCs) was estimated from the blood plasma-to-PBMC ratio. The absorption half-life was calculated to be 27.7 h. This value is between the rate observed in vitro for normal (19.9 h) and AML samples (58.2 h). This is the first calculation of the rate of oligonucleotide uptake in vivo and is interesting in that it tends to confirm the rate observed in vitro with primary cells. The concentration of OL(1)p53 at the target cell site should be proportional to the potential therapeutic benefit derived from OL(1)p53. This concentration is estimated as the net difference of the amount absorbed minus the amount eliminated. Hence, if the rate of elimination is faster than the rate of absorption, then this rate of target cell uptake tends to support the need for continuous infusion of OL(1)p53 in favor of single iv bolus injections. Extravascular injection, such as subcutaneous or intramuscular, may provide options to continuous infusion but appear to be limited by the amount of compound that can be injected and a high potential for local irritation.

3.3. Toxicity and Efficacy

No specific toxicities were observed. However the following interesting observations were noted. First, urine samples were evaluated for thiobarbituric acid products indicative of lipid peroxidation in vivo. The data indicate a 180% increase (relative to d 0–6) in thiobarbituric acid products following 7 d of exposure to OL(1)p53 at a dose of 0.2 mg/ kg/h. This observation is consistent with increased lipid peroxidation in mice and rats treated with polyriboinosinic acid-

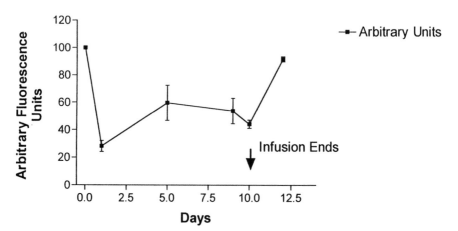

Fig. 1. The average fluorescence generated after monobromobimane incubation with patient urine samples. Data represent the average relative fluorescence from two patients on the ordinate plotted vs time relative to the 10-d continuous infusion on abscissa. The d 0 time-point represents urine samples recovered before administration of the OL(1)p53, and the d 12 sample comes from patients 2 d after the end of the continuous infusion.

polyribocytidylic acid (poly I. C.), a potent interferon inducer *(33)*. The mechanism may involve the activation of 5-lipoxygenase, which will result in lipid peroxidation and has been observed following exposure to phosphorothioate derivatives *(34)*. Our laboratory has also observed these responses, which were not sequence-dependent and could be elicited by phosphodiester DNA in vitro (unpublished results).

Second, an attempt to utilize monobromobimane (MBB) to detect the phosphorothioate oligonucleotide in the plasma and urine *(35–37)* reveals another mystery. MBB fluoresces as a result of covalent attachment to the thiol group in the phosphorothioate backbone *(35)*. This method produces reliable standard curves. However, in the presence of the host of biological substances of plasma and urine high background signal obscures the quantitative utility of this method. When urine samples were evaluated before administration of OL(1)p53, the MBB fluorescence was greater than that observed once the infusion of OL(1)p53 began (*see* Fig. 1). The observation that MBB fluorescence returns to pretreatment levels at the end of the infusion indicates the fluorescence quenching is reversible. This may be the result of:

1. Thiol-containing substances in the urine, which combine with the phosphorothioate oligonucleotide and "mask" the MBB binding sites,
2. A physiological response to the OL(1)p53, which results in less MBB reactive substances being excreted in the urine, or

3. Formation of an insoluble complex between MBB and biological substances, which is formed as a result of OL(1)p53 interaction.

Finally, 2 of the 17 patients experienced dermal features that may have been associated with the administration of OL(1)p53. In one case, this appeared to be exfoliative dermatitis, which occurred approx 7 d after the 10-d infusion was complete. The skin on the palms of the hands and soles of the feet sloughed off, but did not result in scar formation. In the other case the patient experienced skin peeling over a week after the completion of the 10-d infusion. The peeling resulted in the improvement of previously discolored skin regions, which were brought about following a severe sunburn from several months prior.

4. Conclusions

Once the promise of nuclease-resistant phosphorothioate oligonucleotides demonstrated in vivo efficacy *(38–40)* and feasible pharmacokinetic parameters *(41–44)*, the task of identification of target cell sites and oligonucleotide functional bioavailability had to be resolved. The most simple solution to this complex problem is to optimize the dose scheduling of the oligonucleotide. The data presented here address the problem of matching the rate of target cell uptake with the rate of loss from systemic circulation. In this case, analysis of OL(1)p53 pharmacokinetic behavior was established in the mouse, rat, and monkey. These observations support the conclusion that target cell uptake is slow relative to the rate of elimination from plasma following a bolus injection. Hence, continuous exposure of cells to oligonucleotide will optimize target cell accumulation of OL(1)p53. It is not possible to extrapolate these observations to other target cells or for other oligonucleotide sequences or oligonucleotide chemistries. However, these observations should provide a basis for comparison.

Another substantial task is to elucidate the patterns and mechanistic causes of oligonucleotide toxicity in vivo. We did not observe toxicity that could easily be linked to OL(1)p53 in either the preclinical pharmacology and toxicology studies or the phase I clinical trial. However, three interesting observations are described. The generation of lipid peroxidation and attenuation of monobromobimane reactive substances present in the urine appear to be independent of the oligonucleotide sequence. These observations will be clarified when similar clinical trials are conducted with oligonucleotide sequences other than OL(1)p53.

Acknowledgment

This work was supported in part by Lynx Therapeutics, Inc. Hayward CA, support from the University of Nebraska Medical Center Hospital, and by a grant from the Nebraska State Department of Health LB595. The authors would like to express gratitude to Jerry Zon, Tim Geiser, Sam Eletr, and the capable staff at Lynx Therapeutics, Inc.

References

1. Maltzman, W. and Czyzyk, L. (1984) UV irradiation stimulates levels of p53 cellular antigen in nontransformed mouse cells. *Mol. Cell Biol.* **4**, 1689–1694.
2. Kastan, M. B., Onyekwere, O., and Sidransky, D. (1991) Participation of p53 protein in the cellular response to DNA damage. *Cancer Res.* **51**, 6304–6311.
3. Kastan, M. B., Zhan, Q., and El-Deiry, W. S. (1992) A mammalian cell cycle checkpoint pathway utilizing p53 and GADD45 is defective in Ataxia-Telangiectasia. *Cell* **71**, 587–597.
4. Fritsche, M., Haessler, C., and Brandner, G. (1993) Induction of nuclear accumulation of the tumor-suppressor protein p53 by DNA-damaging agents. *Oncogene* **8**, 307–318.
5. Hall, P. A., McKee, P. H., and Menage, H. P. (1993) High levels of p53 protein in UV-irradiated normal human skin. *Oncogene* **8**, 203–207.
6. Tishler, R. B., Calderwood, S. K., and Coleman, C. N. (1993) Increases in sequence specific DNA binding by p53 following treatment with chemotherapeutic and DNA damaging agents. *Cancer Res.* **53**, 2212–2216.
7. Zhan, X., Carrier, F., and Fornace, A. J., Jr. (1993) Induction of cellular p53 activity by DNA-damaging agents and growth arrest. *Mol. Cell. Biol.* **13**, 4242–4250.
8. Zhan, Q., Bae, I., and Kastan, M. B. (1994) The p53-dependent gamma-ray response of GADD45. *Cancer Res.* **54**, 2755–2760.
9. Yonish-Rouach, E., Resnitzky, D., and Lotem, J. (1991) Wild-type p53 induces apoptosis of myeloid leukemic cells that is inhibited by interleukin 6. *Nature* **352**, 345–347.
10. Shaw, P., Bovey, R., and Tardy, S. (1992) Induction of apoptosis by wild-type p53 in a human colon tumor-derived cell line. *Proc. Natl. Acad. Sci. USA* **89**, 4495–4499.
11. Ryan, J. J., Danish, R., and Gottlieb, C. A. (1993) Cell cycle analysis of p53-induced cell death in murine erythroleukemia cells. *Mol. Cell Biol.* **13**, 711–719.
12. Hartwell, L. H. and Weinert, T. A. (1989) Checkpoints: controls that ensure the order of cell cycle events. *Science* **246**, 629–634.
13. Hartwell, L. (1992) Defects in a cell cycle checkpoint may be responsible for the genomic instability of cancer cells. *Cell* **71**, 543–546.
14. Kuerbitz, S. J., Plunkett, B. S., and Walsh, W. V. (1992) Wild-type p53 is a cell cycle checkpoint determinant following irradiation. *Proc. Natl. Acad. Sci. USA* **89**, 7491–7495.
15. Tsuji, K. and Ogawa, K. (1994) Recovery from ultraviolet-induced growth arrest of primary rat hepatocytes by p53 antisense oligonucleotide treatment. *Mol. Carcinog.* **9**, 167–174.
16. El-Deiry, W. S., Velculescu, T. T., and Levy, V. E. (1993) WAF-1, a potential mediator of p53 tumor suppression. *Cell* **75**, 817–825.
17. Weinert, T. A. and Hartwell, L. H. (1988) The RAD9 gene controls the cell cycle response to DNA damage in *Saccharomyces cerevisiae*. *Science* **241**, 317–322.

18. Weinert, T. A. and Hartwell, L. H. (1990) Characterization of RAD9 of *Saccharomyces cerevisiae* and evidence that its function acts posttranslationally in cell cycle arrest after DNA damage. *Mol Cell. Biol.* **10**, 6554–6564.

19. McKirmon, P. J. (1987) Ataxia-telangiectasia: an inherited disorder of ionizing-radiation sensitivity in man. *Hum. Genet.* **75**, 197–208.

20. Petty, R. D., Cree, I. A., and Sutherland. L. A. (1994) Expression of the p53 tumor suppressor gene product is a determinant of chemosensitivity. *Biochem. Biophys. Res. Comm.* **199**, 264–270.

21. Clarke, A. R., Purdie, C. A., and Harrison, D. J. (1993) Thymocyte apoptosis induced by p53-dependent and independent pathways. *Nature* **362**, 849–852.

22. Lotem, J. and Sachs, L. (1993) Hematopoietic cells from mice deficient in wild-type p53 are more resistant to induction of apoptosis by some agents. *Blood* **82**, 1092–1096.

23. Lowe, S. W., Schmitt, E. M., and Smith, S. W. (1993) p53 is required for radiation-induced apoptosis in mouse thymocytes. *Nature* **362**, 847–849.

24. O'Connor, P. M., Jackman, J., and Jondle, D. (1993) Role of p53 tumor suppressor gene in cell cycle arrest and radiosensitity of Burkitt's Lymphoma cell lines. *Cancer Res.* **53**, 4776–4780.

25. Bayever, E., Haines, K. M., and Iversen, P. L. (1994) Selective cytotoxicity to human leukemic myeloblasts produced by oligodeoxyribonucleotide phosphorothioates complementary to p53 nucleotide sequences. *Leukem. Lymph.* **12**, 223–231.

26. Koeffler, H. P., Miller, C., and Nicolson, M. A. (1986) Increased expression of p53 protein in human leukemia cells. *Proc. Natl. Acad. Sci. USA* **83**, 4035–4039.

27. Smith, L. J., McCulloch, E. A., and Benchimol, S. (1986) Expression of the p53 oncogene in acute myelogenous leukemia. *J. Exp. Med.* **164**, 751–761.

28. Sugimoto, K., Toyoshima, H., and Sakai, R. (1991) Mutations of the p53 gene in lymphoid leukemia. *Blood* **77**, 1153–1156.

29. Slingerland, J. M., Minden, M. D., and Benchimol, S. (1991) Mutation of the p53 gene in human acute myelogenous leukemia. *Blood* **77**, 1500–1507.

30. Fenaux, P., Jonveaux, P., and Quiquandon, I. (1991) p53 gene mutations in myeloid leukemia with 17p monosomy. *Blood* **78**, 1652–1657.

31. Stein, C. A., Iversen, P. L, Subasinghe, C., Cohen, J. S., Stec, W., and Zon, G. (1990) Preparation of ^{35}S-labelled polyphosphorothioate oligonucleotides by use of hydrogen phosphonate chemistry. *Anal. Biochemistry* **188**, 11–16.

32. Srinivasan, S. K., Tewary, H. K., and Iversen, P. L. (1995) Phosphorothioate interactions with albumin and implications for potential drug interactions. *Antisense Res. Dev.*, in press.

33. Koizumi, A., Walford, R. L., and Imamura, T. (1986) Treatment with poly I. C. enhances lipid peroxidation and the activity of xanthene oxidase, and decreases hepatic P-450 content and activities in mice and rats. *Biochem. Biophys. Res. Commun.* **134**, 632–637.

34. Denis, D., Choo, L. Y., and Riendeau, D. (1989) Activation of 5-lipoxygenase by guanine 5'-*O*-(3thiotriphosphate) and other nucleoside phosphorothioates: redox properties of thionucleotide analogs. *Arch. Biochem. Biophys.* **273**, 592–596.

35. Fidanza, J. A. and McLaughlin, L. W. (1989) Introduction of reporter groups at specific sites in DNA containing phosphorothioate diesters. *J. Am. Chem. Soc.* **111,** 9117–9119.
36. Hodges, R. R., Conway, N. E., and McLaughlin, L. W. (1989) Post-assay covalent labeling of phosphorothioate-containing nucleic acids with multiple fluorescent markers. *Biochemistry* **28,** 261–267.
37. Conway, N. E and McLaughlin, L. W. (1991) The covalent attachment of multiple fluorophores to DNA containing phosphorothioate diesters results in highly sensitive detection of single-stranded DNA. *Bioconjugate Chem.* **2,** 452–457.
38. Ratajczak, M. Z., Kant, J. A., Luger, S. M., Hijiya, N., Zhang, J., Zon, G., and Gewirtz, A. M. (1992) *In vivo* treatment of human leukemia in a scid mouse model with c-*myb* antisense oligodeoxynucleotides. *Proc. Natl. Acad. Sci. USA* **89,** 11,823–11,827.
39. Simmons, M., Edelman, E. R., Dekyser, J.-L., Langer, R., and Rosenberg, R. D. (1992) Antisense c-*myb* oligonucleotides inhibit intimal arterial smooth muscle accumulation *in vivo*. *Nature*. **359,** 67–70.
40. Skorski, T., Nieborowska-Skorska, M., Nicolaides, N. C., Szcylik, C., Iversen, P., Iozzo, R. V., Zon, G., and Calabretta, B. (1994) Suppression of Philadelphia leukemia cell growth in mice by *bcr-abl* antisense oligodeoxynucleotides. *Proc Natl. Acad. Sci. USA* **91,** 4504–4508.
41. Agrawal, S., Temsamani, J., and Tang, J. Y. (1991) Pharmacokinetics, biodistribution and stability of oligodeoxynucleotide phosphorothioates in mice. *Proc. Natl. Acad. Sci. USA* **88,** 7595–7599.
42. Goodarzi, G., Watabe, M., and Watabe, K. (1992) Organ distribution and stability of phosphorothioated oligodeoxyribonucleotides in mice. *Biopharm. Drug Dispos.* **13,** 221–227.
43. Cossum, P. A., Sasmor, H., Dellinger, D., Truong, L., Cummins, L., Owens, S. R., Markham, P. M., Shea, J. P., and Crooke, S. (1993) Disposition of the 14C-labeled phosphorothioate oligonucleotide ISIS 2105 after intravenous administration to rats. *J. Pharmacol. Exp. Ther.* **267,** 1181–1190.
44. Iversen, P. L., Mata, J., Tracewell, W. G., and Zon, G. (1994) Pharmacokinetics of an antisense phosphorothioate oligodeoxynucleotide against rev from human immunodeficiency virus type I in the adult male rat following a single injections and continuous infusion. *Antisense Res. Dev.* **4,** 43–52.

12

Protein Kinase-A Directed Antisense Therapy of Tumor Growth In Vivo

The Antisense Effects Outlast Antisense Survival

Yoon S. Cho-Chung and Maria Nesterova

1. Introduction

Standard chemotherapy for cancer usually is accompanied by systemic toxicity, reflecting the large number of cellular targets affected by the chemotherapeutic agent. In principle, an antisense oligonucleotide targeted at a gene essential for neoplastic cell growth should interfere with only that gene's expression, resulting in arrest of cancer cell growth.

Enhanced expression of the RI_α subunit of cAMP-dependent protein kinase (PKA) *(1)* has been shown in human cancer cell lines and in primary tumors, as compared with normal counterparts, in cells after transformation with the Ki-*ras* oncogene or transforming growth factor-α, and on stimulation of cell growth with granulocyte-macrophage colony-stimulating factor (GM–CSF) or phorbol esters *(2,3)*. Conversely, a decrease in the expression of RI_α correlates with growth inhibition induced by site-selective cAMP analogs in a broad spectrum of human cancer cell lines *(3)*.

It has been hypothesized that the RI_α is an ontogenic growth-inducing protein and that its constitutive expression disrupts normal ontogenic processes, resulting in a pathogenic outgrowth, such as malignancy *(3)*. We describe here the RI_α antisense phosphorothioate oligodeoxynucleotide (RI_α antisense) producing long-lasting inhibition of in vivo tumor growth. The results suggest that antisense like RI_α antisense, which produces a biochemical imprint for growth control, requires infrequent dosing to halt neoplastic growth in vivo.

2. cAMP-Dependent Protein Kinase Type I and Type II

The primary mediator of cAMP action in mammalian cells is PKA *(4,5)*. The PKA is uniquely composed of two genetically distinct catalytic (C) and

From: *Methods in Molecular Medicine: Antisense Therapeutics*
Edited by: S. Agrawal Humana Press Inc., Totowa, NJ

regulatory (R) subunits. The activating ligand, cAMP, by binding to the R subunit, induces conformational changes and dissociates the holoenzyme R_2C_2 into a R_2-$(cAMP)_4$ dimer and two free C subunits that are catalytically active (6,7).

There are two types of PKA, type I (PKA-I) and type II (PKA-II), which share a common C subunit but contain distinct R subunits, RI and RII, respectively (6). Through biochemical studies and gene cloning, four isoforms of the R subunits, RI_α, RI_β, RII_α, and RII_β, have been identified (8,9). Three distinct C subunits, C_α (10), C_β (11,12), and C_γ (13), also have been found; however, preferential coexpression of one of these C subunits with any of the R subunits has not been found (12,13). The two general classes of R subunits, RI and RII, share a conserved carboxyl terminus but differ significantly at the amino terminus. The R isoforms also differ in tissue distribution (14,15), biochemical properties (6,16), and their subcellular localization (2,17,18). Importantly, the expression of RI/PKA-I and RII/PKA-II has an inverse relationship during ontogenic development, cell differentiation, and malignant transformation (2,3). However, the significance of the presence of these two PKA isoforms in the biological functions of cAMP has not been determined.

3. RI Subunit of Protein Kinase as a Positive Regulator of Cell Growth

Evidence suggests that overexpression of RI/PKA-I is associated with cell proliferation and neoplastic transformation. An increase in RI_α or PKA-I is an early response to the mitogenic effects of growth factors, such as GM-CSF in human leukemic cells (19) and phytohemagglutinin stimulation of resting lymphocytes (20).

7,12-dimethylbenz[a]anthracene-induced mammary carcinogenesis in rats (21) and N-methyl-N'-nitro-N-nitrosoguanidine-induced gastric adenocarcinoma in rats (22) correlated with an increase in RI and PKA-I. SV40 viral transformation of Balb 3T3 fibroblasts and transformation of rat 3Y1 cells by human adenovirus type 12 were accompanied by an increase in RI and PKA-I (23,24). A marked increase in RI expression with a decrease in RII expression was detected in Ha-MuSV-transformed NIH3T3 clone 13-3B-4 cells (25–27), normal rat kidney cells transformed with TGFα or v-Ki-ras oncogene (28), TGFα-induced transformation of mouse mammary epithelial cells (29), and point-mutated c-Ha-ras and c-erb B-2 proto-oncogene-transformed human mammary epithelial cell line, MCF-10A HE cells (30).

RI is the major, or sole, R subunit of protein kinase detected in a variety of types of human cancer cell lines (3,31). The majority of primary human breast and colon carcinomas examined show an enhanced expression of RI and a higher ratio of PKA-I/PKA-II as compared with normal counterparts (32–35). The ratio of PKA-I/PKA-II in renal cell carcinomas was about twice that in renal cortex, although the total soluble PKA activity was similar in both tissues (36). In surgical specimens of Wilms' tumor, the ratio of PKA-I/PKA-II was twice that in normal kidney and the RI/RII ratio was 0.79 for normal and 2.95 for tumor ($P < 0.01$) *(37)*.

A retroviral vector *(17,38,39)* for human RI_α provided a tool to show direct evidence that RI_α plays a role in cell proliferation by regulating cell-cycle progression. The RI_α retroviral vector-infected FRTL5 rat thyroid cells, human mammary MCF-10A cells, and Chinese hamster ovary cells demonstrate independence from a hormone/serum requirement for cell proliferation and exhibit a cell cycle distribution similar to that of parental cells growing in the presence of hormone/serum *(40–42)*. These results provide evidence that RI/PKA-I may act as a mediator of various mitogenic stimuli and thus represents a potential target for the pharmacological control of cell proliferation.

4. 8-Cl-cAMP

A site-selective cAMP analog, 8-Cl-cAMP, modulates the intracellular concentration of PKA isozymes. PKA is usually present in tissue as a mixture of its isozymes, PKA-I and PKA-II *(43,44)*. At high millimolar concentrations, cAMP saturates both PKA-I and PKA-II maximally and equally without discrimination *(6)*; therefore, selective modulation of cAMP receptor isoforms has not been possible in previous studies where high concentrations of cAMP/cAMP analogs have been employed.

Unlike parental cAMP, site-selective cAMP analogs demonstrate selective binding for either one of the two known cAMP binding sites, site A (site 2) and site B (site 1) *(45,46)* in the R subunit, resulting in preferential binding and activation of either protein kinase isozyme. Significantly, this was the first demonstration that a cAMP analog can induce growth inhibition at micromolar concentrations (the physiological concentration of cAMP), as opposed to the millimolar pharmacological or cytotoxic concentrations of cAMP analogs reported in all previous literature. It was found that the potency of a cAMP analog in growth inhibition depends on the analog's ability to selectively modulate the RI and RII isoforms of the cAMP receptor protein. Precisely, the downregulation of RI/PKA-I with the upregulation of RII/PKA-II restores the normal balance of these cAMP-transducing proteins in cancer cells *(3,47)*.

It was discovered that site-selective cAMP analogs can act as novel biological agents capable of inducing growth inhibition and differentiation in a broad spectrum of human cancer cell lines, including carcinomas, sarcomas, and leukemias, without causing cytotoxicity *(31,48–50)*. These studies resulted in the selection of 8-Cl-cAMP, the most potent site-selective cAMP analog, as a preclinical phase I antineoplastic agent of the US National Cancer Institute (January 27, 1988). A phase I clinical study of 8-Cl-cAMP has been completed in Naples, Italy *(51)*. It is the first introduction of a cAMP analog into clinical testing in over 30 years of cAMP research.

5. RI$_\alpha$ Antisense Oligodeoxynucleotide Inhibition of Human Cancer Cells In Vitro

The possibility that the RI$_\alpha$ cAMP receptor is a positive regulator essential for cancer cell growth was explored with the use of antisense strategy. The experimental data provided the first direct evidence that the RI$_\alpha$ receptor is a positive effector of cancer cell growth. It was found that downregulation of RI$_\alpha$ by a 21-mer antisense oligodeoxynucleotide directed to codons 1–7 of human RI$_\alpha$ (15–30 μM) led to growth arrest and differentiation in HL-60 leukemia cells *(52)* and the inhibition of growth in human cancer cells of epithelial origin, including breast (MCF-7), colon (LS-174T), and gastric (TMK-1) carcinoma and neuroblastoma (SK-N-SH) cells, with no sign of cytotoxicity *(53)*.

The normal, unmodified β-oligodeoxynucleotides are far more sensitive to nuclease hydrolysis than are α-oligodeoxynucleotides, methylphosphonate, or phosphorothioate oligodeoxynucleotides *(54,55)*. It was found that the unmodified antisense oligomer, at 30 μM, produced a growth inhibitory effect similar to that caused by 6 μM phosphorothioate oligodeoxynucleotide, demonstrating a greater potency of phosphorothioate oligomer as compared with unmodified oligomer *(53)*. The growth inhibition accompanied changes in cell morphology. In HL-60 leukemia cells, the RI$_\alpha$ antisense oligodeoxynucleotide induced monocytic morphologic change that is indistinguishable from that induced by cAMP analogs or phorbol esters *(52)*.

The effect of RI$_\alpha$ antisense oligodeoxynucleotide correlated with a decrease in RI$_\alpha$ receptor and a concomitant increase in RII$_\beta$ receptor levels. Thus, suppression of RI$_\alpha$ by the antisense oligodeoxynucleotides brought about a compensatory increase in RII$_\beta$ level *(52,53)*. Such coordinated expression of RI and RII without changes in the amount of C subunit has been observed elsewhere *(44)*. The increase in RII$_\beta$ may, therefore, be responsible for the differentiation in these cells exposed to RI$_\alpha$ antisense oligodeoxynucleotide. In fact, exposure of HL-60 cells to a 21-mer RII$_\beta$ antisense oligodeoxynucleotide resulted in a blockage of cAMP-induced growth inhibition and differentiation without apparent effect on the differentiation induced by phorbol esters *(56)*. Thus, RII$_\beta$ cAMP receptor, but not RI$_\alpha$, is the mediator of cAMP-induced differentiation in HL-60 cells. The increase in RII$_\beta$ at mRNA and protein level has also been correlated with differentiation of K562 chronic myelocytic leukemia *(57)* and Friend erythrocytic leukemia cells *(58)*.

6. A Single Injection RI$_\alpha$ Antisense Treatment Inhibits In Vivo Tumor Growth

It was demonstrated that the sequence-specific inhibition of RI$_\alpha$ gene expression results in inhibition of in vivo tumor growth *(59)*. A single subcutaneous (sc) injection into nude mice bearing LS-174T human colon carcinoma

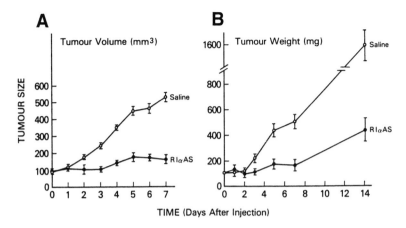

Fig. 1. Inhibition of in vivo tumor growth by a single dose of RI_α antisense. Reprinted with kind permission from ref. *59*. **(A)** Tumor volume obtained from daily measurement. **(B)** Tumor weight at the time mice were killed. LS-174T human colon carcinoma cells (1×10^6 cells) were inoculated sc into the left flank of athymic mice. The RI_α antisense phosphorothioate oligodeoxynucleotide (corresponding to the RI_α NH$_2$-terminus 8–13 codons [RI_α antisense] [5'-GCG-TGC-CTC-CTC-ACT-GGC-3']) and control antisense (the same base composition as the RI_α antisense with the sequence jumbled) were kindly provided by T. Geiser (Lynx Therapeutics, Hayward, CA). A single dose of RI_α antisense or control antisense (1 mg/0.1 mL saline/mouse) or saline (0.1 mL/mouse) was injected sc into the right flank of mice when tumor size reached 80–100 mg, 1 wk after cell inoculation. Tumor volumes were obtained from daily measurement of the longest and shortest diameters and calculation by the formula $4/3\pi r^3$ where r = (length + width)/4. At each indicated time, two animals from the control and antisense-treated groups were killed, and tumors were removed, weighed, immediately frozen in liquid N_2, and kept frozen at –80°C until used.

with RI_α antisense resulted in almost complete suppression of tumor growth for 7 d (Fig. 1). There was no apparent sign of toxicity. Even after 14 d, tumor growth was significantly inhibited in the antisense-treated animals (Fig. 1). In contrast, tumors in saline-treated animals showed continued growth (Fig. 1). Tumors in untreated or control antisense-treated animals grew at a rate similar to those in saline-treated animals. The results of tumor growth inhibition were confirmed with three additional RI_α antisense constructs (each of the 21-base polymers directed to codons 1–7, 14–20, and 94–100, respectively, of human RI_α), which previously have been shown to inhibit growth in a variety of human cancer cell lines *(53)*.

The RI_α antisense treatment resulted in a marked decrease in RI_α levels in tumors within 24 h and sustained the low RI_α levels (10–20% of control tumors)

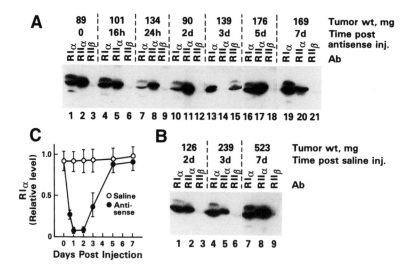

Fig. 2. Suppression of RI_α levels in tumors by a single dose of RI_α antisense. Reprinted with kind permission from ref. *59*. The R subunit levels in tumors were determined by photoaffinity labeling with $8\text{-}N_3\text{-}[^{32}P]$ cAMP followed by immunoprecipitation with R antibodies and SDS-PAGE (*56*). **(C)** Quantification of RI_α levels by densitometric tracings of autoradiographs **(A,B)**. The tumors were homogenized with a Teflon/glass homogenizer in ice-cold buffer 10 (Tris-HCl, pH 7.4, 20 mM; NaCl, 100 mM; NP-40, 1%; sodium deoxycholate, 0.5%; $MgCl_2$, 5 mM; pepstatin, 0.1 mM; antipain, 0.1 mM; chymostatin, 0.1 mM; leupeptin, 0.2 mM; aprotinin, 0.4 mg/mL; and soybean trypsin inhibitor, 0.5 mg/mL^4; filtered through a 0.45-μm pore size membrane), and centrifuged for 5 min in an Eppendorf microfuge at 4°C. The supernatants were used as tumor extracts.

for up to 2–3 d (Fig. 2). Specific targeting of RI_α by the antisense is evident since RII_α levels remained unchanged (Fig. 2). At 5–7 d after antisense treatment, the RI_α levels in tumors were elevated to levels similar to those in control tumors (Fig. 2). Three days after antisense treatment, tumors that contained unreduced amounts of RI_α contained a new species of R, RII_β, along with a lower level of RII_α (Fig. 2). RII_β appeared 24 h to 3 d after antisense treatment but was not detected in control tumors. These data show that the antisense-targeted suppression of RI_α brought about a compensatory increase in RII_β levels. Similar observations were made in cultured cancer cell lines on treatment with RI_α antisense (*53*).

The RI_α antisense effect on the PKA isozyme distribution in tumors was examined using DEAE ion-exchange chromatography. The antisense treatment completely eliminated PKA-I, the RI_α-containing holoenzyme, and the RI_α subunit from tumors within 24 h. Importantly, this downregulation of

PKA-I lasted for up to 5–7 d after antisense treatment even when the RI_α levels increased and reached the levels of control tumors. This indicates that the RI_α that increased subsequently to its initial suppression after antisense treatment was present mostly in its subunit form rather than in its holoenzyme form, PKA-I.

Concomitant with the suppression of PKA-I, the antisense brought about changes in the PKA-II profile of tumors. PKA-II in antisense-treated tumors contained PKA-II$_\beta$ (RII$_\beta$ containing PKA-II). These data suggest that the C subunits of PKA in tumors are in equilibrium between PKA-I and PKA-II and that downregulation of RI_α and PKA-I by the antisense led to an increase in PKA-II through the induction of PKA-II$_\beta$.

The in vivo pharmacokinetic studies in rodents showed that a single intravenous dose of phosphorothioate oligodeoxynucleotide leaves the vascular space after 2–3 h (phase α $t_{1/2}$ = 15–25 min), and its elimination from the body, which is almost completely urinary, requires 72 h (phase β $t_{1/2}$ = 20–40 h) *(60)*. In accordance with such pharmacokinetics of oligonucleotide, a single sc dose of RI_α antisense produced an acute reduction in RI_α ($t_{1/2}$ RI_α = 31 h *[61]*) content within 24 h, and thereafter for 2–3 d. This reduction in RI_α triggered a compensatory increase in RII$_\beta$ and an elimination of PKA-I activity.

The downregulation of PKA-I lasted for several days, even after the RI_α suppression ceased, suggesting that RI_α may be functionally different because it no longer formed the holoenzyme, PKA-I. This may be because:

1. Once RI_α is downregulated, the free C subunits complex with all the available RII$_\alpha$ subunits to form PKA-II$_\alpha$;
2. The remaining free C subunits trigger the synthesis of RII$_\beta$ and form PKA-II$_\beta$;
3. RII has a greater half-life than RI (RII, $t_{1/2}$ = 125 h; RI, $t_{1/2}$ = 31 h *[61]*); therefore, once RII is synthesized, it remains in the cell for a longer time and favors complex formation with the C subunit as compared with RI;
4. The subsequently formed RI_α, after its initial suppression (because of the antisense), can no longer form PKA-I holoenzyme in the presence of the increased amount of PKA-II, which is favored over PKA-I in its holoenzyme formation *(9)*; and
5. RI may be degraded and cannot form PKA-I.

Although the exact mechanisms of action await future studies, the results showed that the antisense produced a biochemical imprint in tumor cells. The cells behaved like untransformed cells by making less PKA-I than PKA-II. This may be the basis for the suppression of tumor growth.

7. Conclusion

A single injection of RI_α antisense triggered the suppression of RI_α and inhibition of tumor growth. Importantly, the growth inhibition persisted, even after RI_α suppression ceased, as long as PKA-I (the RI_α-containing holoenzyme) downregulation was present. The single-injection antisense treatment introduced a program for growth control in tumor cells, and was sufficient to produce a sustained inhibition of growth. Thus, the antisense effects outlasted antisense survival. This unexpected finding has a great impact on the application of antisense oligonucleotides as therapeutic agents, especially in terms of potency, targeting, and cost. These results suggest that an antisense like RI_α antisense, which is capable of producing a biochemical imprint for growth control, may require infrequent dosing to maintain its inhibitory effect toward tumor growth in vivo.

References

1. Krebs, E. G. (1972) Protein kinase. *Curr. Top. Cell Reg.* **5**, 99–133.
2. Lohmann, S. M. and Walter, U. (1984) Regulation of the cellular and subcellular concentrations and distribution of cyclic nucleotide-dependent protein kinases, in *Advances in Cyclic Nucleotide and Protein Phosphorylation Research*, vol. 18 (Greengard, P. and Robison, G. A., eds.), Raven, New York, pp. 63–117.
3. Cho-Chung, Y. S. (1990) Role of cyclic AMP receptor proteins in growth, differentiation and suppression of malignancy: new approaches to therapy (perspectives in cancer research). *Cancer Res.* **50**, 7093–7100.
4. Krebs, E. G. and Beavo, J. A. (1979) Phosphorylation-dephosphorylation of enzymes. *Ann. Rev. Biochem.* **48**, 923–939.
5. Kuo, J. F. and Greengard, P. (1969) Cyclic nucleotide-dependent protein kinases. IV. Widespread occurrence of adenosine 3',5'-monophosphate-dependent protein kinase in various tissues and phyla of the animal kingdom. *Proc. Natl. Acad. Sci. USA* **64**, 1349–1355.
6. Beebe, S. J. and Corbin, J. D. (1986) Cyclic nucleotide-dependent protein kinases, in *The Enzymes: Control by Phosphorylation*, part A, vol. 17 (Boyer, P. D. and Krebs, E. G., eds.), Academic, New York, pp. 43–111.
7. Bramson, H. N., Kaiser, E. T., and Mildvan, A. S. (1983) Mechanistic studies of cAMP-dependent protein kinase actions. *CRC Crit. Rev. Biochem.* **15**, 93–124.
8. Levy, F. O., Oyen, O., Sandberg, M., Tasken, K., Eskild, W., Hansson, V., and Jahnsen, T. (1988) Molecular cloning, complementary deoxyribonucleic acid structure and predicted full-length amino acid sequence of the hormone-inducible regulatory subunit of 3',5'-cyclic adenosine monophosphate-dependent protein kinase from human testis. *Mol. Endocrinol.* **2**, 1364–1373.
9. McKnight, G. S., Clegg, C. H., Uhler, M. D., Chrivia, J. C., Cadd, G. G., Correll, L. A., and Otten, A. D. (1988) Analysis of the cAMP-dependent protein kinase system using molecular genetic approaches. *Rec. Prog. Horm. Res.* **44**, 307–335.
10. Uhler, M. D., Carmichael, D. F., Lee, D. C., Chrivia, J. C., Krebs, E. G., and McKnight, G. S. (1986) Isolation of cDNA clones coding for the catalytic subunit of mouse cAMP-dependent protein kinase. *Proc. Natl. Acad. Sci. USA* **83**, 1300–1304.

11. Uhler, M. D., Chrivia, J. C., and McKnight, G. S. (1986) Evidence for a second isoform of the catalytic subunit of cAMP-dependent protein kinase. *J. Biol. Chem.* **261,** 15,360–15,363.

12. Showers, M. O. and Maurer, R. A. (1986) A cloned bovine cDNA encodes an alternate form of the catalytic subunit of cAMP-dependent protein kinase. *J. Biol. Chem.* **261,** 16,288–16,291.

13. Beebe, S. J., Oyen, O., Sandberg, M., Froysa, A., Hansson, V., and Jahnsen, T. (1990) Molecular cloning of a unique tissue-specific protein kinase (C_γ) from human testis—representing a third isoform for the catalytic subunit of the cAMP-dependent protein kinase. *Mol. Endocrinol.* **4,** 465–475.

14. Clegg, C. H., Cadd, G. G., and McKnight, G. S. (1988) Genetic characterization of a brain-specific form of the type I regulatory subunit of cAMP-dependent protein kinase. *Proc. Natl. Acad. Sci. USA* **85,** 3703–3707.

15. Øyen, O., Frøysa, A., Sandberg, M., Eskild, W., Joseph, D., Hansson, V., and Jahnsen, T. (1987) Cellular localization and age-dependent changes in mRNA for cyclic adenosine 3',5'-monophosphate-dependent protein kinase in rat testis. *Biol. Reprod.* **37,** 947–956.

16. Cadd, G. G., Uhler, M. D., and McKnight, G. S. (1990) Holoenzymes of cAMP-dependent protein kinase containing the neural form of type I regulatory subunit have an increased sensitivity to cyclic nucleotides. *J. Biol. Chem.* **265,** 19,502–19,506.

17. Kapoor, C. L. and Cho-Chung, Y. S. (1983) Compartmentalization of regulatory subunits of cyclic adenosine 3',5'-monophosphate-dependent protein kinases in MCF-7 human breast cancer cells. *Cancer Res.* **43,** 295–302.

18. Nigg, E. A., Schäfer, G., Hilz, H., and Eppenberger, H. M. (1985) Cyclic-AMP-dependent protein kinase type II is associated with the golgi complex and with centrosomes. *Cell* **41,** 1039–1051.

19. Tortora, G., Pepe, S., Yokozaki, H., Meissner, S., and Cho-Chung, Y. S. (1991) Cooperative effect of 8-Cl-cAMP and rhGM-CSF on the differentiation of HL-60 human leukemia cells. *Biochem. Biophys. Res. Commun.* **177,** 1133–1140.

20. Byus, C. V., Klimpel, G. R., Lucas, D. O., and Russell, D. H. (1977) Type I and type II cyclic AMP-dependent protein kinase as opposite effectors of lymphocyte mitogenesis. *Nature* **268,** 63,64.

21. Cho-Chung, Y. S., Clair, T., and Shepheard, C. (1983) Anticarcinogenic effect of N^6, $O^{2'}$-dibutyryl cyclic adenosine 3',5'-monophosphate on 7.12-dimethylbenz(α) anthracene mammary tumor induction in the rat and its relationship to cyclic adenosine 3',5'-monophosphate metabolism and protein kinase. *Cancer Res.* **43,** 2736–2740.

22. Yasui, W. and Tahara, E. (1985) Effect of gastrin on gastric mucosal cyclic adenosine 3',5'-monophosphate-dependent protein kinase activity in rat stomach carcinogenesis induced by N-methyl-N'-nitro-N-nitrosoguandine. *Cancer Res.* **45,** 4763–4767.

23. Gharrett, A. J., Malkinson, A. M., and Sheppard, J. R. (1976) Cyclic AMP-dependent protein kinases from normal and SV40-transformed 3T3 cells. *Nature* **264,** 673–675.

24. Ledinko, N. and Chan, I.-J. A. D. (1984) Increase in type I cyclic adenosine 3',5'-monophosphate-dependent protein kinase activity and specific accumulation of type I regulatory subunits in adenovirus type 12-transformed cells. *Cancer Res.* **44,** 2622–2627.

25. Tagliaferri, P., Clair, T., DeBortoli, M. E., and Cho-Chung, Y. S. (1985) Two classes of cAMP analogs synergistically inhibit p21 ras protein synthesis and phenotypic transformation of NIH/3T3 cells transfected with Ha-MuSV DNA. *Biochem. Biophys. Res. Commun.* **130,** 1193–1200.

26. Tagliaferri, P., Katsaros, D., Clair, T., Neckers, L., Robins, R. K., and Cho-Chung, Y. S. (1988) Reverse transformation of Harvey murine sarcoma virus-transformed NIH/3T3 cells by site-selective cyclic AMP analogs. *J. Biol. Chem.* **263,** 409–416.

27. Clair, T., Ally, S., Tagliaferri, P., Robins, R. K., and Cho-Chung, Y. S. (1987) Site-selective cAMP analogs induce nuclear translocation of the RII_β cAMP receptor protein in Ha-MuSV-transformed NIH/3T3 cells. *FEBS Lett.* **224,** 337–384.

28. Tortora, G., Ciardiello, F., Ally, S., Clair, T., Salomon, D. S., and Cho-Chung, Y. S. (1989) Site-selective 8-chloroadenosine 3',5'-cyclic monophosphate inhibits transformation and transforming growth factor a production in Ki-*ras*-transformed rat fibroblasts. *FEBS Lett.* **242,** 363–367.

29. Ciardiello, F., Tortora, G., Kim, N., Clair, T., Ally, S., Salomon, D. S., and Cho-Chung, Y. S. (1990) 8-Chloro-cAMP inhibits transforming growth factor a transformation of mammary epithelial cells by restoration of the normal mRNA patterns for cAMP-dependent protein kinase regulatory subunit isoforms which show disruption upon transformation. *J. Biol. Chem.* **265,** 1016–1020.

30. Ciardiello, F., Pepe, S., Bianco, C., Baldassarre, G., Ruggiero, A., Bianco, C., Selvam, M. P., Bianco, A. R., and Tortora, G. (1993) Down-regulation of RI_α subunit of cAMP-dependent protein kinase induces growth inhibition of human mammary epithelial cells transformed by c-Ha-*ras* and c-*erb* B-2 proto-oncogenes. *Int. J. Cancer* **53,** 438–443.

31. Cho-Chung, Y. S., Clair, T., Tagliaferri, P., Ally, S., Katsaros, D., Tortora, G., Neckers, L., Avery, T. L., Crabtree, G. W., and Robins, R. K. (1989) Basic science review: site-selective cyclic AMP analogs as new biological tools in growth control, differentiation and proto-oncogene regulation. *Cancer Invest.* **7,** 161–177.

32. Handschin, J. C. and Eppenberger, U. (1979) Altered cellular ratio of type I and type II cyclic AMP-dependent protein kinase in human mammary tumors. *FEBS Lett.* **106,** 301–304.

33. Weber, W., Schwoch, G., Schroder, H., and Hilz, H. (1981) Analysis of cAMP-dependent protein kinases by immunotitration: multiple forms—multiple functions? *Cold Spring Harbor Conf. Cell Prolif.* **8,** 125–140.

34. Watson, D. M. S., Hawkins, R. A., Bundred, N. J., Stewart, H. J., and Miller, W. R. (1987) Tumor cyclic AMP binding proteins and endocrine responsiveness in patients with inoperable breast cancer. *Br. J. Cancer* **56,** 141–142.

35. Bradbury, A. W., Miller, W. R., and Carter, D. C. (1991) Cyclic adenosine 3',5'-monophosphate binding proteins in human colorectal cancer and mucosa. *Br. J. Cancer* **63,** 201–204.

36. Fossberg, T. M., Døskeland, S. O., and Ueland, P. M. (1978) Protein kinases in human renal cell carcinoma and renal cortex. A comparison of isozyme distribution and of responsiveness to adenosine 3',5'-cyclic monophosphate. *Arch. Biochem. Biophys.* **189**, 372–381.

37. Nakajima, F., Imashuku, S., Wilmas, J., Champion, J. E., and Green, A. A. (1984) Distribution and properties of type I and type II binding proteins in the cyclic adenosine 3',5'-monophosphate-dependent protein kinase system in Wilms' tumor. *Cancer Res.* **44**, 5182–5187.

38. Cho-Chung, Y. S., Clair, T., Tortora, G., and Yokozaki, H. (1991) Role of site-selective cAMP analogs in the control and reversal of malignancy. *Pharmacol. Ther.* **50**, 1–33.

39. Cho-Chung, Y. S., Clair, T., Tortora, G., Yokozaki, H., and Pepe, S. (1991) Suppression of malignancy targeting the intracellular signal transducing proteins of cAMP: the use of site-selective cAMP analogs, antisense strategy and gene transfer. *Life Sci.* **48**, 1123–1132.

40. Tortora, G., Pepe, S., Cirafici, A. M., Ciardiello, F., Porcellini, A., Clair, T., Colletta, G., Cho-Chung, Y. S., and Bianco, A. R. (1993) Thyroid-stimulating hormone-regulated growth and cell cycle distribution of thyroid cells involve type I isozyme of cyclic AMP-dependent protein kinase. *Cell Growth Differen.* **4**, 359–365.

41. Tortora, G., Pepe, S., Bianco, C., Damiano, V., Ruggiero, A., Baldassarre, G., Corbo, C., Cho-Chung, Y. S., Bianco, A. R., and Ciardiello, F. (1994) Differential effects of protein kinase A sub-units on Chinese-hamster-ovary cell cycle and proliferation. *Int. J. Cancer* **59**, 712–716.

42. Tortora, G., Pepe, S., Bianco, C., Baldassarre, G., Budillon, A., Clair, T., Cho-Chung, Y. S., Bianco, A. R., and Ciardiello, F. (1994) The RI$_\alpha$ subunit of protein kinase A controls serum dependency and entry into cell cycle of human mammary epithelial cells. *Oncogene* **9**, 3233–3240.

43. Corbin, J. D., Keely, S. L., and Park, C. R. (1975) The distribution and dissociation of cyclic adenosine 3',5'-monophosphate-dependent protein kinases in adipose, cardiac, and other tissues. *J. Biol. Chem.* **250**, 218–225.

44. House, C. and Kemp, B. E. (1987) Protein kinase C contains a pseudosubstrate prototype in its regulatory domain. *Science* **238**, 1726–1728.

45. Døskeland, S. O. (1978) Evidence that rabbit muscle protein kinase has two kinetically distinct binding sites for adenosine 3',5'-cyclic monophosphate. *Biochem. Biophys. Res. Commun.* **83**, 542–549.

46. Rannels, S. R. and Corbin, J. D. (1980) Two different intrachain cAMP binding sites of cAMP-dependent protein kinases. *J. Biol. Chem.* **255**, 7085–7088.

47. Ally, S., Tortora, G., Clair, T., Grieco, D., Merlo, G., Katsaros, D., Ogried, D., Døskeland, S. O., Jahnsen, T., and Cho-Chung, Y. S. (1988) Selective modulation of protein kinase isozymes by the site-selective analog 8-chloradenosine 3',5'-cyclic monophosphate provides a biological means for control of human colon cancer cell growth. *Proc. Natl. Acad. Sci. USA* **85**, 6319–6322.

48. Katsaros, D., Tortora, G., Tagliaferri, P., Clair, T., Ally, S., Neckers, L., Robins, R. K., and Cho-Chung, Y. S. (1987) Site-selective cyclic AMP analogs provide a new approach in the control of cancer cell growth. *FEBS Lett.* **223,** 97–103.

49. Cho-Chung, Y. S. (1989) Commentary. Site-selective 8-chloro-cyclic adenosine 3',5'-monophosphate as a biologic modulator of cancer: restoration of normal control mechanisms. *J. Natl. Cancer Inst.* **81,** 982–987.

50. Ally, S., Clair, T., Katsaros, D., Tortora, G., Yokozaki, H., Finch, R. A., Avery, T. L., and Cho-Chung, Y. S. (1989) Inhibition of growth and modulation of gene expression in human lung carcinoma in athymic mice by site-selective 8-Cl-cyclic adenosine monophosphate. *Cancer Res.* **49,** 5650–5655.

51. Tortora, G., Ciardiello, F., Pepe, S., Tagliaferri, P., Ruggiero, A., Bianco, C., Guarrasi, R., Miki, K., and Bianco, A. R. (1995) Phase I clinical study with 8-chloro-cAMP and evaluation of immunological effects in cancer patients. *Clin. Cancer Res.* **1,** 377–384.

52. Tortora, G., Yokozaki, H., Pepe, S., Clair, T., and Cho-Chung, Y. S. (1991) Differentiation of HL-60 leukemia by type I regulatory subunit antisense oligodeoxynucleotide of cAMP-dependent protein kinase. *Proc. Natl. Acad. Sci. USA* **88,** 2011–2015.

53. Yokozaki, H., Budillon, A., Tortora, G., Meissner, S., Beaucage, S. L., Miki, K., and Cho-Chung, Y. S. (1993) An antisense oligodeoxynucleotide that depletes RI_α subunit of cyclic AMP-dependent protein kinase induces growth inhibition in human cancer cells. *Cancer Res.* **53,** 868–872.

54. Rothenberg, M., Johnson, G., Laughlin, C., Green, I., Craddock, J., Sarver, N., and Cohn, J. S. (1989) Commentary: oligodeoxynucleotides as anti-sense inhibitors of gene expression: therapeutic implications. *J. Natl. Cancer Inst.* **81,** 1539–1544.

55. Cho-Chung, Y. S. (1993) Antisense oligonucleotides for the treatment of cancer. *Curr. Opin. Ther. Pat.* **3,** 1737–1750.

56. Tortora, G., Clair, T., and Cho-Chung, Y. S. (1990) An antisense oligodeoxynucleotide targeted against the RIIβ regulatory subunit mRNA of protein kinase inhibits cAMP-induced differentiation in HL-60 leukemia cells without affecting phorbol ester effects. *Proc. Natl. Acad. Sci. USA* **87,** 705–708.

57. Tortora, G., Clair, T., Katsaros, D., Ally, S., Colamonici, O., Neckers, L. M., Tagliaferri, P., Jahnsen, T., Robins, R. K., and Cho-Chung, Y. S. (1989) Induction of megakaryocytic differentiation and modulation of protein kinase gene expression by site-selective cAMP analogs in K-562 human leukemic cells. *Proc. Natl. Acad. Sci. USA* **86,** 2849–2852.

58. Schwartz, D. A. and Rubin, C. S. (1985) Identification and differential expression of two forms of regulatory subunits (RII) of cAMP-dependent protein kinase II in Friend erythroleukemic cells. *J. Biol Chem.* **260,** 6296–6303.

59. Nesterova, M. and Cho-Chung, Y. S. (1995) A single-injection protein kinase A-directed antisense treatment to inhibit tumour growth. *Nat. Med.* **1,** 528–533.

60. Iversen, P. (1991) In vivo studies with phosphorothioate oligonucleotides: pharmacokinetics prologue. *Anti-Cancer Drug Des.* **6,** 531–538.

61. Weber, W. and Hilz, H. (1986) cAMP-dependent protein kinases I and II: divergent turnover of subunits. *Biochemistry* **25,** 5661–5667.

13

Antisense Therapeutics in the Central Nervous System

The Induction of c-fos

Bernard J. Chiasson, Murray Hong, Michele L. Hooper, John N. Armstrong, Paul R. Murphy, and Harold A. Robertson

1. Introduction

Immediate-early genes (IEGs) are members of a class of genes that respond, in many cell types, to a variety of stimuli by rapid, but transient expression *(1)*. Several of these IEGs code for transcription factors and include the widely studied activator protein-1 (AP-1) transcription factor complex believed to be homo- and heterodimeric assemblies of the Fos and Jun families *(1–3)*. IEGs are induced in the central nervous system (CNS) by diverse physiological and pharmacological stimuli, many of which, when presented once or on multiple occasions, can alter the "normal" functioning of the brain in a permanent or semipermanent fashion. Examples of pharmacological stimuli that lead to long-term changes are the highly addictive psychostimulant drugs, amphetamine and cocaine These drugs produce a robust activation of IEGs (e.g., *c-fos*, *jun*-B, *egr*-1) in areas of the brain that are believed to be part of the neural substrates of addiction *(4–8)*. In animal models of epileptogenesis or memory, such as kindling and long-term potentiation (LTP), respectively, electrical stimuli produce activation of IEGs within the brain structures thought to underlie the long-lasting changes associated with these experimental procedures *(9–16)*. IEGs can also be induced by noninvasive stimuli, such as a simple light pulse given to animals in a dark room. The circadian rhythms of animals that are housed in darkened conditions can be shifted by exposing them to a light during their subjective night. Activation of IEGs in such experiments are restricted to the suprachiasmatic nucleus (SCN), which is believed to be the

From: *Methods in Molecular Medicine: Antisense Therapeutics*
Edited by: S. Agrawal Humana Press Inc., Totowa, NJ

seat of the biological clock *(17)*. These are a few of many examples that illustrate the association between stimuli that lead to changes in brain function and the activation of IEGs in the CNS.

As a consequence of such observations, many investigators have been led to speculate that IEGs, such as *c-fos*, play a role in initiating the molecular events leading to medium- and long-term changes (hours to years) in brain function. However, until recently, the evidence for this has been either poor, circumstantial, or totally lacking *(15,16,18,19)*. Studies performed in vitro substantiate the notion that the Fos/Jun dimers regulate the expression of various neuropeptides and trophic molecules, such as nerve growth factor (NGF) *(20,21)*. However, the physiological consequences of either acutely or chronically altering the expression of IEGs within a freely behaving animal are less certain. Thus, at both the cellular and behavioral levels, the role of IEGs is largely unknown. In this chapter, we will describe experiments performed by us and others that attempt to address the role of *c-fos* at both the molecular level in individual neurons and in freely behaving animals. We demonstrate that IEGs, such as *c-fos*, are more than mere markers of cellular activity and that altering their expression even temporarily can have pronounced effects on cellular function and even on behavior.

2. Some Strategies Used to Study the Role of Specific Genes

There are at least three strategies currently used to alter mammalian gene expression. In transgenic animals, single-gene "knock-outs" eliminate a given gene early in embryonic development. However, it should be clear that in many cases, the genes being "knocked out" may not only play a role in adult animals, but also in development *(22)*. For example, mice carrying a Fos-LacZ fusion gene under the control of *c-fos* regulatory elements express a complex pattern of expression during development, suggesting an important role in ontogeny *(1,23)*. Other IEGs, such as *egr*-1 (also known as *zif/268*, *Krox24*, *Tis 8*, *ngfi-a*) are also expressed embryonically, but their role in development remains unclear *(24)*. Consequently, studying neurological phenomena as complex as memory or addiction must be interpreted cautiously when using animals with null mutations. Such animals are perhaps better suited to examining developmental phenomena as well as compensatory mechanisms, which undoubtedly operate under transgenic conditions. Finally, "knock-out" transgenic technology is largely restricted to application in rodents (mice and rats).

A second strategy employs viral vectors to introduce genes or antisense constructs to a restricted region of the brain by stereotaxic infusion. Since adult neurons are typically postmitotic, the vectors used must be capable of infecting such cells. Therefore, retroviral vectors that have proven so useful in lineage analysis are likely inappropriate *(25,26)*. Most laboratories have focused on developing adenovirus and herpes simplex virus-based vectors, which have the capacity to infect postmitotic neurons *(26,27)*. Although this strategy has advantages over the transgenic approach, there are also problems with this approach. Construct design is important to eliminate viral genes whose products are cytotoxic. Identification of the proper site in which to introduce the foreign gene cassette within the viral vector *(26)* and the design of a suitable promoter-regulator, which permits some control of latency or neuron-specific expression, are also important considerations. The delivery of genes via viral vectors is likely to provide considerable information about the function of various genes, but it is not currently the method of choice for most neuroscience laboratories.

The last approach and the one we shall discuss at length in this chapter is the antisense approach. Antisense technology is conceptually simple. It is based on the idea that a short oligodeoxyribonucleotide (ODN) sequence, which is complementary to a target mRNA, hybridizes according to Watson-Crick bonding and inactivates the mRNA *(28–30)*. Although simple in concept, the exact mechanism by which the normal function of mRNA is disrupted is currently not known. Furthermore, this mechanism probably differs from system to system, but the end point is a reduction in the protein product of the target mRNA. Typically, antisense technology results in a "knock-down" of the protein product, as opposed to the transgenic approach, which results in a complete "knock-out" of the gene product *(30)*. Thus, results obtained from antisense studies must be interpreted cautiously, keeping in mind that the observed response is potentially the result of a decrease in but not an elimination of the target protein(s). It should also be noted that antisense ODNs are purported to have a narrow dose response and limited duration of action following a single administration to the brain *(30–33)*. This brief period of antisense action is particularly appropriate for the study of IEGs, which are expressed rapidly and transiently. This possibility of having a partial and a time-limited effect may itself be of great use in our attempts to understand regulation of gene expression.

In our search for ways of understanding the function of *c-fos* and other immediate-early genes, it was apparent that antisense ODN knock-down of gene expression could be a powerful and relatively simple tool. We have thus adopted this technique and determined empirically some of the factors involved in the application of antisense ODNs to use in experimental neuroscience. On the other hand, it is now evident that the increase in immediate-early gene expression after various manipulations in brain provides an excellent system in which to study the effects of antisense ODNs in vivo. This is because, first, the constitutive expression of *c-fos* and some other immediate-early genes is essentially zero in the absence of stimulation, but increases rapidly to a high level on stimulation. Second, there are a number of paradigms where the brain structures to be targeted are bilateral, and antisense ODNs can be administered into one hemisphere of the brain and nontargeting ODNs (sense, random, or mismatch) can be used as a control in the other hemisphere. Finally, knocking down immediate-early gene expression in the brain often leads to a behavioral change that can be correlated with the changes in protein or mRNA expression. Areas of discussion include the behavioral consequences of attenuating *c-fos* expression in the striatum, the toxicity of antisense ODNs associated with their repeated use, and methods to reduce or eliminate this toxicity while retaining an effective antisense compound.

3. Activation of c-*fos* in the Striatum: The Effects of Dopaminergic Drugs on c-*fos* Expression and on Behavior

It has long been known that dopamine plays an important part in the regulation of fine motor control by the striatum. Disruptions of the dopaminergic system (i.e., death of the dopaminergic neurons in the substantia nigra that innervate the striatum) lead to Parkinson's disease. The effects of selective lesions of the substantia nigra have been well-characterized by the studies of Ungerstedt *(34)*. In this system (Fig. 1), rats with a lesion to the dopamine neurons on the right side of the brain rotate to the left (contraversive rotation) when given a dopaminergic agonist, such as apomorphine, and to the right (ipsiversive rotation) when given a dopamine-releasing drug, such as amphetamine or cocaine. Thus, rotation occurs away from the side exhibiting the greatest dopaminergic receptor activation. The contraversive rotation observed after an agonist is the result of direct activation by agonists of supersensitive dopamine receptors on the side of the brain that is depleted of dopamine. Indirectly acting drugs, such as amphetamine and cocaine, can only act by releasing dopamine (amphetamine) or increasing dopamine levels (cocaine) on the side that has an intact dopamine innervation.

Rotation in this model can be induced by drugs active at either the D_1 dopamine family of receptors or by drugs active at receptors of the D_2 dopamine receptor family *(35–38)*. Activation of rotation by D_1 dopamine receptor agonists, L-dopa or amphetamine, induces expression of IEGs, including *c-fos* on the side of the brain opposite to the direction of rotation *(39,40)*. In naive unlesioned animals, amphetamine and cocaine induce the expression of *c-fos*, *egr*-1, and other IEGs equally in both striata through a D_1 receptor mechanism *(4–7,40)* (Fig. 1). In the absence of a lesion, both sides of the brain theoretically receive equal dopaminergic stimulation, and animals, as a group, do not demonstrate any preference to rotate in one direction more than the other Taken together, this information suggests that *c-fos*, *egr*-1, and perhaps other IEGs may play a role in the rotational behavior seen in both the lesioned and naive animals. In lesioned animals, D_1 receptor stimulation results in rotation away

Measuring Rotation Behavior

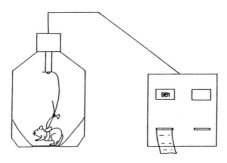

D-Amphetamine Induced Fos-LI in the Striatum and Rotation

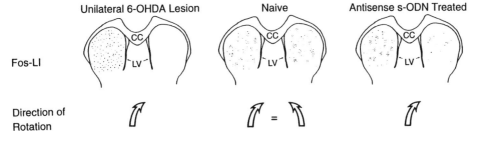

Fig. 1. Schematic representation of the apparatus used to measure rotation behavior following systemic D-amphetamine administration under lesioned, naive, and antisense S-ODN conditions. Rotometer (top) consists of a rounded bowl (approx 35 cm in diameter) in which the animal can easily move. The animal is harnessed to a transducing mechanism that translates each complete rotation into a digital display on the memory module (measuring both directions independently). Hard copy printout allows for either sequential or total counts of the rotational behavior. Shown below are representations of coronal sections through the striatum under three different conditions where Fos-LI is induced along with rotational behavior. Animals whose right nigrostriatal pathway has been interrupted by a 6-OHDA lesion (usually produced 3 wk before testing dopaminergic drugs) show ipsiversive rotation when challenged with amphetamine. That is, they rotate toward the lesioned side (to the right) or away from the side expressing Fos-LI. Naive animals show a balanced bilateral Fos-LI in the striata and rotate in both directions equally. Animals that receive infusions of an antisense S-ODN to c-*fos* mRNA into the right striatum and a sense (or other control) S-ODN into the left striatum show rotation to the right, away from the side expressing the greatest number of Fos-positive cells. CC= corpus callosum, LV = lateral ventricle.

from the side expressing *c-fos*, and in naive animals, no overall preference in rotation is seen when *c-fos* expression is balanced between sides of the brain (striata). However, the mechanism by which Fos would regulate the neurophysiological output of the striata (basal ganglia) is not known. We do know, however, that several of the key neuropeptides (e.g., neurotensin, dynorphin, enkephalin, substance P) located within the basal ganglia have AP-1 consensus sequences in their regulatory regions. Furthermore, Fos expression in striatum is colocalized with substance P and with dynorphin in neurons that project to the substantia nigra pars reticulata *(41)* and with neurotensin containing neurons of the dorsolateral striatum *(42)*. Thus, Fos expression may play a key role in regulating neuropeptide levels To begin to answer some of the questions about the role of *c-fos* and other IEGs in the brain, we have studied the effects of antisense ODNs targeting the *c-fos* gene in the striatum. These studies take advantage of the fact that the nigrostriatal system is bilateral, and that dopaminergic drugs, such as amphetamine, increase the expression of *c-fos*, *egr*-1, and a number of other IEGs.

4. Infusion of Antisense Oligodeoxynucleotides into the Striatum

Using fully substituted pentadecamers (15-mers) phosphorothioate oligodeoxynucleotides (S-ODNs), we first examined whether we could alter D-amphetamine induced striatal *c-fos* expression between hemispheres. This was accomplished by infusing an antisense S-ODN to *c-fos* directly into one striatum and a sense control S-ODN into the opposing striatum of the same animal. At different times (10 and 22 h) following the ODN treatment, Fos production was stimulated by the systemic administration of D-amphetamine (5 mg/kg, ip). We then examined the effects of the antisense and sense S-ODNs on Fos-like immunoreactivity (Fos-LI) in brain sections and in adjacent sections examined Egr-1-like immunoreactivity (Egr-1-LI). Egr-1-LI was studied in order to determine if other IEGs that are activated through the same D_1 receptor mechanism would also be affected by the S-ODN treatment. In other words, Egr-1 expression served as a measure of selectivity of the antisense S-ODN. The immunohistochemical procedures revealed that animals in the 12-h group (10 h post S-ODN treatment + 2 h amphetamine) showed a remarkable attenuation of Fos-LI in approx 20–50% of the striatum treated with the antisense S-ODN compared to the sense-treated side (Fig. 2A,B) *(31)*. However, animals in the 24 h group (22 h post S-ODN treatment + 2 h amphetamine) showed no difference in Fos-LI between the antisense- or sense-treated striata (data not shown; *see* ref. *33*). Egr-1-LI remained unaffected in either time group, suggesting that the D_1 receptor mechanism remained intact and that the effects seen on Fos-LI were selective for *c-fos* (Fig. 2C,D).

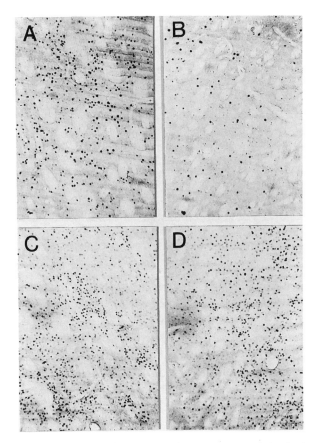

Fig. 2. Amphetamine-induced Fos and Egr-1 immunoreactivity in the striatum following c-*fos* antisense and control S-ODN infusions (12-h group). **(A)** Fos-LI in the striatum on the sense S-ODN-treated side. **(B)** Fos-LI in the striatum in the antisense S-ODN-treated side. **(C)** Egr-1-LI (sense-treated side) in an adjacent section from the same animal as in (A) and (B). **(D)** Egr-1-LI (antisense-treated side). An obvious knockdown of Fos immunoreactive nuclei can be discerned between the antisense- and sense-treated sides. However, Egr-1-LI remains intact, demonstrating specificity of action of the antisense treatment. Photographs were taken in opposing striata at the same anatomical region. (*see also* Table 1 for quantification of the knock-down).

When animals were placed in a rotometer (as described in Fig. 1) following an antisense and sense S-ODN infusion (as described above) and given amphetamine 10 h later, they demonstrated a preference to rotate in the direction away from the sense S-ODN-treated side. That is, animals turned (within 10–15 min following the administration of amphetamine) in the direction away from the side expressing the greater Fos-LI. However, at 22 h following the ODN treatment,

Table 1
Quantitative Analysis of Fos and Egr-1 Immunoreactive Nuclei in Animals Receiving the Antisense and Sense c-*fos* S-ODNs (15-mers) into Opposing Striata

Group	(*n*)	IEG	Mean ratio of positive nuclei on antisense side compared to control side		
4 h	(6)	Fos	0.94	±	0.18
	(3)	Egr-1	1.15	±	0.10
12 h	(7)	Fos	0.35	±	0.04[a]
	(7)	Egr-1	0.98	±	0.13
24 h	(10)	Fos	0.89	±	0.06
	(6)	Egr-1	0.93	±	0.04

Values are means ± SEM.
[a]Indicates significant difference from control side, $p < 0.001$.

animals no longer demonstrated a preference to rotate in either direction *(32)*. A more comprehensive study revealed time and dose dependency to both the knockdown of Fos-LI and the rotation behavior *(33)*. This study revealed that as long as a significant difference in Fos-LI existed between the striata, animals would rotate away from the side expressing the most Fos expression *(see* Table 1 and Fig. 3A). These behavioral findings have also been observed in other laboratories *(43,44)*. Subsequent studies have demonstrated that the antisense sequence used originally *(31,32,43–46)* and sequences that are extensions of this original sequence are also effective antisense ODNs against *c-fos (47,48)*. However, we have also demonstrated that the knock-down of *c-fos* expression and rotation is sequence-specific, since not all sequences targeting the *c-fos* mRNA were effective at attenuating expression (Fig. 3B) *(33)*. Thus, our experience with antisense ODNs to *c-fos* reveals a general phenomenon in the field of antisense technology, which is that not all antisense sequences work well or at all *(see* ref. *49* for an example).

Fig. 3. *(opposite page)* Effect of antisense S-ODN infusion on D-amphetamine-induced rotation behavior. Total number of rotations were registered in a 2-h period following D-amphetamine injection (5 mg/kg, ip). All numbers shown are means ± SEM for animals in that group. Animals received oligonucleotide infusions into the striata (antisense into one side and control ODN into the other side) and were challenged at varying times with D-amphetamine. The times indicated for each group include the 2 h following the D-amphetamine injection. Rotations in the direction toward the antisense-infused side are termed ipsiversive (I) and those in the opposite direction contraversive (C). Animals in the control group received bilateral infusions of vehicle into opposing striata prior to amphetamine challenge. Control animals that rotated in the counterclockwise direction were arbitrarily considered to have turned

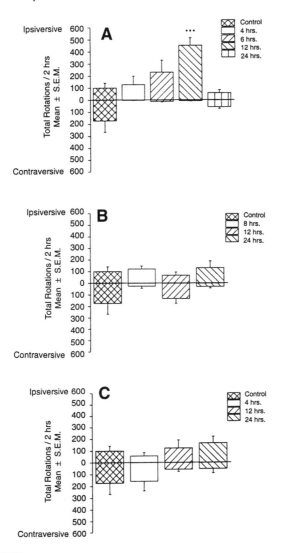

ipsiversively, and those that rotated clockwise contraversively. In order to achieve a statistically significant difference from the control, the group under investigation had to show a statistical difference from both the clockwise and counterclockwise directed rotations of the control. **(A)** Rotational response in animals treated with an effective fully substituted antisense S-ODN (15-mer); $N = 6, 6, 6, 7,$ and 10 for the control, 4-, 6-, 12-, and 24-h groups, respectively. **(B)** Rotational response in animals treated with a fully substituted S-ODN, which is not effective at knocking down Fos expression at any time indicated; $N = 6, 6, 7,$ and 6 for the control, 8-, 12-, and 24-h groups, respectively. **(C)** Rotational response in animals treated with an ineffective phosphodiester ODN; $N = 6$ in each group. *** Indicates a level of significance of $p < 0.001$.

In addition to demonstrating a sequence-specific effect of the antisense S-ODN to *c-fos*, we examined the role of phosphorothioate substitution to the backbone of the ODN. Reports have suggested that chimeric ODNs (ODNs whose backbones possess various types of modifications, such as a mixture of phosphorothioate and phosphodiester internucleoside linkages) may demonstrate more rapid uptake into cells and diminished toxicity while retaining nuclease resistance *(50–52)*. We compared the ability of partially substituted S-ODNs and nonmodified phosphodiester ODNs to attenuate Fos expression and induce the direction-specific rotation behavior. Using the same doses of phosphodiester ODNs, which are effective when fully phosphorothioated, we were unable either to attenuate Fos-LI or affect behavior *(33)* (Fig. 3C). It is possible that higher doses of these nonmodified ODNs would be effective, since others have shown effects in vivo with repeated doses that are 5–10 times higher than used in our experiments. However, in addition to the higher doses, these studies using nonmodified ODNs involved icv administration *(53,54)*, as opposed to our studies where the ODNs were delivered directly into the tissue *(33)*.

Initial results using end-capped S-ODNs (ODNs with a partially modified phosphorothioate backbone containing sulfur substitutions only at the 5' and 3' ends) were slightly more promising than our results with the phosphodiester ODNs. In 50% of the animals treated with end-capped S-ODNs and then challenged 2 h later with amphetamine, we observed rotation and later demonstrated that these animals showed a knock-down of Fos expression. In the other animals treated with end-capped S-ODNs, no rotation was observed, and no difference in Fos-LI between the antisense and control ODN side could be discerned *(33)*. Thus, our initial studies using end-capped S-ODNs suggested that in some instances, antisense ODNs could be effective in vivo with only minor modification of the backbone (*see* Section 6.).

Taken together, our initial results using the amphetamine/striatal model of IEG expression demonstrated that some, but not all fully substituted S-ODNs targeting *c-fos* could effectively attenuate Fos-LI. It also appears that this D_1-mediated Fos expression was important in the functional output of the striatum (basal ganglia), since only animals in which Fos-LI differed significantly between striata demonstrated a significant directional preference in rotational behavior. This behavior was also consistent with what has been observed in the lesioned model where animals stimulated with D_1 agonist rotate away from the side of the brain expressing the most Fos-positive nuclei *(37,39,40,55)*. These results indicate that the IEG *c-fos* may play an important role in mediating the physiology of the basal ganglia.

The mechanism by which *c-fos* alters the physiology of the basal ganglia is currently unknown, but several possibilities exist. It has been suggested that *c-fos* expression is ongoing at very low levels (*1,56,* personal communication

J. Babity). Thus, the antisense ODN may exert its action by attenuating ongoing expression of *c-fos*, and genes under the control of this low-level expression of Fos may be affected accordingly. Candidate genes would include proenkephalin, prodynorphin, and proneurotensin, all of which possess AP-1 regulatory elements and show alteration in gene expression in conditions where *c-fos* levels are changed *(20,57–59)*. In fact, in recent studies using antisense S-ODN in a D2 receptor model of Fos expression, two groups have shown independently that neurotensin levels are regulated in vivo by *c-fos (45,47)*. However, in these studies proenkephalin expression remained unaffected by the antisense treatment, suggesting that enkephalin expression is regulated differently than neurotensin. In fact, this observation is consistent with a report by Konradi et al., which demonstrated that cAMP-response element-binding protein (CREB), and not Fos, mediates enkephalin levels *(60)*. These findings strongly suggest that neuropeptide gene expression is regulated by *c-fos* and other transcription factors in vivo. The potential asymmetry in neuropeptide levels could account for the directional preference in rotation observed once animals are stimulated with amphetamine *(33)*. Amphetamine-induced Fos protein, along with other immediate-early gene proteins, may function as transcriptional regulators of genes whose responses occur rapidly and have immediate consequences on striatal function. These changes may involve fluctuations in neurotransmission, ion channel permeability, homeostasis, and other aspects of cellular function capable of altering the output of the basal ganglia in a short period of time. As noted earlier, the rotational behavior emerges within 10–15 min following the injection of amphetamine. Thus, the genes whose expression could be altered within this time frame are limited, but probably include other IEGs, like jun-B and *egr*-1 *(see ref. 43)*. It is also possible that *c-fos* possesses a role as a second messenger within the cellular environment that is capable of influencing neuronal function *(33)*. This role could be played by either the mRNA or the protein. What is certain is that changes brought about by altering Fos expression do not directly change the function of the D_1 receptor system, since the response of *egr*-1 expression via the D_1 system is still intact following the *c-fos* antisense treatment *(see* Table 1 and Fig. 2C,D). Although several explanations are possible, it is clear that we do not currently understand the mechanism by which *c-fos* may alter the physiology of the basal ganglia.

The work performed by Heilig and colleagues is consistent with the observation that rotational behavior is associated with the asymmetry in Fos expression between hemispheres. They demonstrated that bilateral attenuation of cocaine-induced Fos expression in the nucleus accumbens was associated with a decrease in cocaine-induced locomotory activity *(46)*. This finding suggests that cocaine-induced *c*-fos expression is also associated with the functional output of the nucleus accumbens. This is particularly interesting in view of the potential role of this mesolimbic structure in the addictive process *(8)*. This would suggest that *c-fos* may play a role in coupling the short-term exposure to a drug like cocaine to longer-lasting events underlying addiction. Further studies examining this relationship are warranted.

5. Repeated Infusion of Antisense S-ODNs into the Amygdala in Studies on Kindling

By applying the antisense approach, we and others have been able to implicate *c-fos* in certain short-term biological phenomena in vivo. However, our primary interests are oriented at gaining an understanding of the mechanisms that lead to long-lasting changes in brain function in conditions such as addiction and kindling. These phenomena typically require repeated stimuli resulting in repeated activation of IEGs. Therefore, in order to knock down IEG expression under these conditions, either repeated or continuous delivery of antisense ODNs is required.

One of the most robust models involving a long-lasting change in brain function is kindling. Animals can be kindled using either electrical or chemical stimuli. Kindling is the progressive development of behavioral convulsions and epileptiform activity following the repeated administration of an initially subconvulsant electrical or chemical stimulation to the brain *(61,62)*. Although many areas of the brain can be kindled, most researchers use electrical kindling of the amygdala as the model of choice *(63)*, but one of the first examples of kindling used repeated exposure to cocaine *(64)*.

In 1987, Dragunow and Robertson demonstrated that a kindling stimulation to the hippocampus activated *c-fos* within the hippocampal formation, and thus established the first link between kindling and IEG expression *(9)*. Since that time, numerous studies have demonstrated IEG activation following kindling *(10,12,65,66)*, but few have conclusively demonstrated that the crucial electrographic event necessary for kindling, namely the afterdischarge (AD) is tied to the activation of IEGs. It is important to establish this link, since in order to consider IEGs as candidate genes involved in the development of kindling, it would seem that the minimal requirement for kindling, the AD, be able to induce their expression. Recently, using stimulation parameters that deliver the same overall electrical charge to the amygdala, but do not lead to an AD, we have established a strong link between the AD and the activation of IEGs in the amygdala/piriform area in amygdala kindling *(11;* Note: Hippocampal kindling differs, but is beyond the scope of this chapter).

We first had to establish whether the kindling stimulus-induced Fos-LI could be knocked down by a single antisense treatment. We demonstrated that a single infusion of *c-fos* antisense S-ODNs, but not control S-ODN, could attenuate Fos-LI when infused 10 h prior to the amygdala kindling stimulus *(67)*. As before, Egr-1-LI remained unaffected. However, since kindling requires daily stimulation, and the S-ODNs' effectiveness dissipates by 24 h *(31–33)*, repeated applications of S-ODN were required to knock down the transient expression in *c-fos* induced by each kindling stimulus. Under this protocol, "normal" AD signals recorded from the amygdala were absent after three to four stimulations, suggesting that recording problems or tissue damage had arisen. Histological examination of the brain tissue (cresyl violet and glial fibrillary acidic protein immunohistochemistry) revealed lesions created by the ODNs (both control and antisense, but not vehicle) (Fig. 4A,B).

Extending the interinfusion interval to either 3 or 5 d decreased the damage, but did not eliminate the lesions (Fig. 4C,D). Thus, it appears that single infusions of fully phosphorothioated S-ODN are well tolerated, but that multiple infusions lead to deleterious effects on tissue extending from severe gliosis to large lesions *(67)*. Although others have also reported toxicity problems *(51;* Heilig, personal communication), several studies using repeated or continuous infusions of S-ODNs did not reveal any noticeable damage *(68,69)*. Many possible factors could contribute to these disparate observations, including oligonucleotide purity, vehicle solution, rate of infusion, location of infusion, and many others. These are important factors for future investigations using S-ODNs. Thus, our experience to date has been that single infusions of S-ODNs do not lead to any detectable damage to brain tissues when infused directly into parenchyma, but that repeated infusions of S-ODNs cause considerable neurotoxicity when delivered in this way *(67)*.

6. Alternative Strategies Used to Circumvent Toxicity Problems

In order to reduce the toxicity seen with repeated dosing of S-ODNs, we have developed an approach using chimeric end-capped S-ODNs. Based on our earlier findings *(33)*, which demonstrated that some animals treated with end-capped S-ODNs showed a knock-down of Fos-LI and rotation behavior, we investigated further the potential of these compounds as less toxic alternatives to fully substituted phosphorothioate S-ODNs. As has been reported by us and others, toxicity and nonselective effects of S-ODNs are potentially related to the overall sulfur substitutions within the backbone of the ODN *(51,67,70,71)*.

We first tested single and double end-capped phosphorothioate S-ODNs (one or two sulfur atoms substituted at both the 5' and 3' ends of the ODN, 20-mers in this case) in the striatal/ amphetamine model as described above. When animals were infused with a *c-fos* antisense and control (sense or random) end-capped S-ODNs into opposing striata and then 2.5 h later stimulated with D-amphetamine (5 mg/kg, ip), we observed rotational behavior similar to that

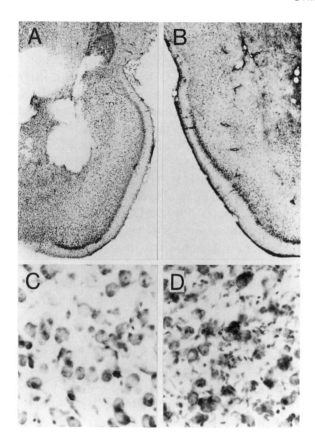

Fig. 4. Cresyl violet-stained sections from animals infused repeatedly into the amygdala region with fully substituted S-ODNs either daily **(A)** or every third day **(B)** (total of four infusions for each). (A) Repeated infusions on a daily basis of either sense or antisense S-ODNs lead to neurotoxic damage. (B) Extending the interinfusion interval to 3 d reduces, but does not eliminate the damage. **(C)** High magnification of the noninfused side. **(D)** High magnification of the infused side (3-d interval). The gliosis present on the infused side is considerable when compared to the noninfused side of the brain.

previously described and a knock-down of Fos-LI. The single and double end-capped S-ODNs did not appear to differ in their ability to cause the behavior nor did they differ in the knock-down of Fos-LI (Hong, Chiasson, and Robertson, in preparation). Thus, this antisense end-capped S-ODN, which is an extension of the original 15-mer, appears to have effective antisense action at doses similar to those effective with the fully substituted 15-mer S-ODN *(48,72)*. Additionally, we have examined the cellular uptake and biodegradation rates of the end-capped S-ODNs, and have demonstrated that they follow a pattern that is consistent with their time-course of action in vivo *(73)*.

Having determined that the end-capped S-ODNs can be effective antisense molecules in vivo, we examined the toxicity of these compounds following repeated injections into brain parenchyma. The histological effects of multiple infusions of end-capped S-ODNs into the amygdala were examined using cresyl violet to detect any gross morphological abnormalities. Animals were infused daily at a dose two to three times that necessary for effective knockdown. Histological examinations revealed local damage that did not extend far (within a millimeter) from the site of infusion. Animals that received the end-capped S-ODNs every third day showed little or no damage when compared to vehicle-infused animals *(73;* Chiasson, Hong, and Roberston, in preparation). Additionally, animals treated with antisense or control end-capped S-ODNs demonstrated normal ADs to kindling stimulations even after multiple infusions. These observations suggest that the basic electrophysiological properties (capability to generate action potentials and thus generate an AD) remain intact, and thus end-capped S-ODN provide a tool by which to dissect out the role of Fos in a normal physiological response to a kindling stimulus (Chiasson and Robertson, in preparation). In fact, preliminary results suggest that antisense end-capped S-ODNs can alter the rate at which animals kindle *(73)*. These electrophysiological results substantiate the histological observations showing a significantly reduced toxicity associated with the end-capped S-ODNs over the fully substituted S-ODNs.

Thus, our work to date suggests that end-capped S-ODNs are appropriate antisense compounds for studies requiring repeated applications, since they are both effective at attenuating the target gene and produce little toxicity. Although these compounds appear much less toxic than the fully substituted compounds when injected repeatedly, the reason for this reduced toxicity remains unclear. The most obvious reason is simply related to the reduced sulfur content. Recent work has suggested that fully substituted S-ODNs bind with a very high affinity, in a sequence-independent fashion, to heparin binding growth factors and members of the fibroblast growth factor family, and inhibit their biological activities *(70)*. Thus, repeatedly inactivating a biologically important group of molecules such as these might be expected to cause general neurotoxic trauma, since many such factors are essential for cellular subsistence. Further evidence suggesting that the backbone modification is associated with toxicity comes from studies that have used high doses (5–10 times higher than the ones used in our studies) of unmodified ODNs administered via icv or it routes *(53,54,74)* in which no toxicity is reported. Although toxicity is clearly an important issue, it appears that there are strategies that circumvent this problem and thus allow for the application of antisense ODNs under a variety of circumstances.

In conclusion, antisense ODNs have proven useful in identifying putative roles for the IEG *c-fos* in vivo. The expression of *c-fos* plays an important role in regulating the output of the basal ganglia and mesolimbic systems in animals challenged with psychostimulant drugs *(32,33,43,44,46)*. These observations suggest a coupling between the activation of this gene and the longer-term consequences of taking drugs with abuse potential. Others have demonstrated that *c-fos* may regulate neuropeptide levels in vivo following antipsychotic treatment *(45,47)*. Taken together, these findings indicate an important biological role for this gene in the brain of adult animals. Other IEGs currently under investigation may also display important roles in neurophysiology.

7. Summary

Research using antisense ODNs in vivo has provided valuable information regarding the putative function of IEGs in the brain. The development of ODNs as research tools will likely lead to their wide applicability in both basic research and therapeutics *(29,75,76)*. However, it is clear that we must be cautious when designing and performing experiments using these compounds *(29,67)*. Specific and nonspecific effects must be clearly delineated that emphasize the importance of proper controls. Although there is no general consensus on the issue of controls, it is clear that at least two different nontargeting ODNs should be used in addition to vehicle controls in any experiment. Experiments using antisense ODNs should also monitor gene products other than those targeted in order to demonstrate some degree of specificity. It has been suggested that isoforms of the same family as the target gene be monitored as well as other genes regulated by the experimental manipulation.

Although antisense ODNs are clearly not without problems, they currently appear to offer the best approach to examining the role of genes in adult animals. In fact, with the growing use of antisense technology, considerable improvements are likely to present themselves. Perhaps antisense ODNs work because they exploit the selectivity and specificity of Watson-Crick base pairing, a clearly proven strategy used by all living organisms.

Acknowledgments

We would like to extend thanks to K. M. A. Murphy and B. Ross for excellent technical assistance in various aspects of this work, as well as G. Evan (Imperial Cancer Research Fund, London, UK) and R. Bravo (Bristol-Myers Squibb, Princeton, NJ) for supplying antibodies to *egr*-1 (Krox 24) and K. G. Baimbridge (University of British Columbia, Vancouver, Canada) for the antibody to Calbindin-D_{28K} used in earlier work (discussed in Hooper et al. *[33]*) not described here. Financial support was provided by the Savoy Foundation for Epilepsy (B. J. C.), The Parkinson Foundation of Canada (M. H.), the Medical Research Council of Canada, and SmithKline Beecham Pharma. Inc. (Canada) (H. A. R.). P. R. M. is an M. R. C. scholar.

References

1. Curran, T. and Morgan, J. I. (1995) Fos: an immediate-early transcription factor in neurons. *J. Neurobiol.* **26,** 403–412.
2. Curran, T. and Franza, B. R., Jr. (1988) Fos and Jun: the AP-1 connection. *Cell* **55,** 395–397.
3. Morgan, J. I. and Curran, T. (1991) Stimulus-transcription coupling in the nervous system: involvement of the inducible proto-oncogenes *fos* and *jun*. *Annu. Rev. Neurosci.* **14,** 421–451.
4. Graybiel, A. M., Moratalla, R., and Robertson, H. A. (1990) Amphetamine and cocaine induce drug-specific activation of the *c-fos* gene in striosome-matrix compartments and limbic subdivisions of the striatum. *Proc. Natl. Acad. Sci. USA* **87,** 6912–6916.
5. Moratalla, R., Robertson, H. A., and Graybiel, A. M. (1992) Dynamic regulation of *NGFI-A* (zif268, egr 1) gene expression in the striatum. *J. Neurosci.* **12,** 2609–2622.
6. Moratalla, R., Vickers, E. A., Robertson, H. A., Cochran, B. H., and Graybiel, A. M. (1993) Coordinate expression of *c-fos* and *junB* is induced in the striatum by cocaine. *J. Neurosci.* **13,** 423–433.
7. Beretta, S., Robertson, H. A., and Graybiel, A. M. (1993) Neurochemically specialized projection neurons of the striatum respond differently to psychomotor stimulants. *Prog. Brain Res.* **99,** 201–205.
8. Nestler, E. J., Hope, B. T., and Widnell, K. L. (1993) Drug addiction: a model for the molecular basis of neural plasticity. *Neuron* **11,** 995–1006.
9. Dragunow, M. and Robertson, H. A (1987) Kindling stimulation induces *c-fos* protein(s) in granule cells of the rat dentate gyrus. *Nature* **329,** 441,442.
10. Dragunow, M., Robertson, H. A., and Robertson, G. S. (1988) Effects of kindled seizures on the induction of c-fos protein(s) in mammalian neurons. *Exp. Neurol.* **102,** 261–263.
11. Chiasson, B. J, Dennison, Z., and Robertson, H. A. (1995) Amygdala kindling and immediate-early genes. *Mol. Brain Res.* **29,** 191–199.
12. Sinomato, M., Hosford, D. A., Labiner, D. M., Shin, C., Mansbach, H. H., and McNamara, J. O. (1991) Differential expression of immediate early genes in the hippocampus in the kindling model of epilepsy. *Mol. Brain Res.* **11,** 115–124.
13. Cole, A. J., Saffen, D. W., Baraban, J. M., and Worley, P. F. (1989) Rapid increase of an immediate early gene messenger RNA in hippocampal neurons by synaptic NMDA receptor activation. *Nature* **340,** 474–476.
14. Wisden, W., Errington, M. L., Williams, S., Dunnett, S. B., Waters, C., Hitchcock, D., Evan, G., Bliss, T. V. P., and Hunt, S. P. (1990) Differential expression of immediate-early genes in the hippocampus and spinal cord. *Neuron* **4,** 603–614.
15. Dragunow, M., Currie, R. W., Faull, R. L. M., Robertson, H. A., and Jansen, K. (1989) Immediate-early genes, kindling and long-term potentiation. *Neurosci. Biobehav. Rev.* **13,** 301–313.

16. Robertson, H. A. (1992) Immediate-early genes, neuronal plasticity, and memory. *Biochem. Cell Biol.* **70,** 729–737.

17. Rusak, B., Robertson, H. A, Wisden, W., and Hunt, S. P. (1990) Light pulses that shift rhythms induce gene expression in the suprachiasmatic nucleus. *Science* **248,** 1237–1240.

18. Sheng, M. and Greenberg, M. E. (1990) The regulation and function of *c-fos* and other immediate-early genes in the nervous system. *Neuron* **4,** 477–485.

19. Robertson, H. A. and Dragunow, M. (1990) From synapse to genome: the role of immediate-early genes in permanent alterations in the central nervous system, in *Current Aspects of the Neurosciences* (Osborne, N. N., eds.), Macmillan, London, pp. 143–157.

20. Sonnenberg, J. L., Rauscher, J. R., III, Morgan, J. I., and Curran, T. (1989) Regulation of proenkephalin by Fos and Jun. *Science* **246,** 1622–1625.

21. Hengerer, B., Lindholm, D., Heumann, R., Ruther, U., Wagner, E. F., and Thoenen, H. (1990) Lesion-induced increase in nerve growth factor mRNA is mediated by *c-fos. Proc. Natl. Acad Sci. USA* **87,** 3899–3903.

22. Routtenberg, A. (1995) Knockout mouse fault lines. *Nature* **374,** 314,315.

23. Smeyne, R. J., Curran, T., and Morgan, J. I. (1992) Temporal and spatial expression of a *fos-lacZ* transgene in the developing nervous system. *Mol. Brain Res.* **16,** 158–162.

24. Herms, J., Zurmohle, U., Schlingensiepen, R., Brysch, W., and Schlingensiepen, K. H. (1994) Developmental expression of transcription factor zif268 in the rat brain. *Neurosci. Lett.* **165,** 171–174.

25. Gray, G. E. and Sanes, J. E. (1992) Lineage of radial glia in the chicken optic tectum. *Development* **114,** 271–283.

26. Glorioso, J. C., Goins, W. F., Meaney, C. A., Fink, D. J., and DeLuca, N. A. (1994) Gene transfer to brain using herpes simplex virus vectors. *Ann. Neurol.* **35,** S28–S34.

27. Akli, S., Caillaud, C., Vigne, E., Stratford-Perricaudet, L. D., Poenaru, L., Perricaudet, M., Kahn, A., and Peschanski, M. R. (1993) Transfer of a foreign gene into brain using adenovirus vectors. *Nature Genet.* **3,** 219–223.

28. Stein, C. A. and Cohen, J. S. (1988) Oligonucleotides as inhibitors of gene expression: a review. *Cancer Res.* **48,** 2659–2668.

29. Wagner, R. W. (1994) Gene inhibition using antisense oligodeoxynucleotides. *Nature* **372,** 333–335.

30. Wahlestedt, C. (1994) Antisense oligonucleotide strategies in neuropharmacology. *Trends Pharmacol. Sci.* **15,** 42–46.

31. Chiasson, B. J., Hooper, M. L., Murphy, P. R., and Robertson, H. A. (1992) Antisense oligonucleotide eliminates in vivo expression of *c-fos* in mammalian brain. *Eur. J. Pharmacol. Mol. Pharmacol.* **227,** 451–453.

32. Chiasson, B. J., Hooper, M. L., and Robertson, H. A. (1992) Amphetamine induced rotational behavior in non-lesioned rats: a role for *c-fos* expression in the striatum. *Soc. Neurosci. Abst.* **18,** 562.

33. Hooper, M. L., Chiasson, B. J., and Robertson, H. A. (1994) Infusion into the brain of an antisense oligonucleotide to the immediate-early gene c-fos suppresses production of Fos and produces a behavioral effect. *Neuroscience* **63,** 917–924.

34. Ungerstedt, U. (1971) Striatal doparnine release after amphetamine or nerve degeneration revealed by rotational behaviour. *Acta Physiol. Scand. Suppl.* **367,** 49–68.
35. Robertson, G. S. and Robertson, H. A. (1987) D1 and D2 dopamine agonist synergism: separate sites of action. *Trends Pharmacol. Sci.* **8,** 295–299.
36. Robertson, G. S. and Robertson, H. A. (1989) Evidence that L-dopa-induced rotational behavior is dependent on both striatal and nigral mechanisms. *J. Neurosci.* **9,** 3326–3331.
37. Paul, M. L., Graybiel, A. M., David, J.-C., and Robertson, H. A. (1992) D_1-like and D_2-like dopamine receptors synergistically activate rotation and *c-fos* expression in the dopamine-depleted striatum in a rat model of Parkinson's disease. *J. Neurosci.* **12,** 3729–3742.
38. Robertson, H. A. (1992) Dopamine receptor interactions: some implications for the treatment of Parkinson's disease. *Trends Neurosci.* **15,** 201–206.
39. Robertson, H. A., Peterson, M. R., Murphy, K., and Robertson, G. S. (1989) D_1-dopamine receptor agonists selectively activate striatal *c-fos* independent of rotational behaviour. *Brain Res.* **503,** 346–349.
40. Robertson, G. S., Herrera, D. G., Dragunow, M., and Robertson, H. A. (1989) L-Dopa activates *c-fos* expression in the striatum of 6-hydroxydopamine-lesioned rats. *Eur. J. Pharmacol.* **159,** 99,100.
41. Robertson, G. S., Vincent, S. R., and Fibiger, H. C. (1992) D1 and D2 dopamine receptors differentially regulate c-fos expression in striatonigral and striatopallidal neurons. *Neuroscience* **49,** 285–296.
42. Merchant, K. M. and Miller, M. A. (1994) Coexpression of neurotensin and c-fos mRNAs in rat neostriatal neurons following acute haloperidol. *Brain Res. Mol. Brain. Res.* **23,** 271–277.
43. Dragunow, M., Lawlor, P. A., Chiasson, B. J., and Robertson, H. A. (1993) Antisense to *c-fos* suppresses both Fos and Jun B expression in rat striatum and generates apomorphine- and amphetamine-induced rotation. *Neuroreport* **5,** 305,306.
44. Sommer, W., Bjelke, B., Ganten, D., and Fuxe, K. (1993) Antisense oligonucleotide to *c-fos* induces ipsilateral rotational behavior to *d*-amphetarnine. *Neuroreport* **5,** 277–280.
45. Merchant, K. M. (1994) *c-fos* antisense oligonucleotide specifically attenuates haloperidolinduced increases in neurotensin/neuromedin N mRNA expression in rat dorsal striatum. *Mol. Cell. Neuroscience* **5,** 336–344.
46. Heilig, M., Engel, J. A., and Söderpalm, B. (1993) C-*fos* antisense in the nucleus accumbens blocks the locomotor stimulant action of cocaine. *Eur. J. Pharmacol.* **236,** 339,340.
47. Robertson, G. S., Tetzlaff, W., Bedard, A., St-Jean, M., and Wigle, N. (1995) *c-fos* mediates antipsychotic-induced neurotensin gene expression in the rodent striatum. *Neuroscience* **67,** 325–344.
48. Gillardon, F., Beck, H., Uhlmann, E., Herdegen, T., Sandkuler, J., Peyman, A., and Zimmermann, M. (1994) Inhibition of c-fos protein expression in rat spinal cord by antisense oligodeoxynucleotide superfusion. *Eur. J. Neurosci.* **6,** 880–884.

49. Chiang, M.-Y., Chan, H., Zounes, M. A., Freier, S. M., Lima, W. F., and Bennett, C. F. (1991) Antisense oligonucleotides inhibit intercellular adhesion molecule 1 expression by two distinct mechanisms. *J. Biol. Chem.* **266,** 18,162–18,171.

50. Carter, G. and Lemoine, N. R. (1993) Antisense technology for cancer therapy does it make sense? *Br. J. Cancer* **67,** 869–876.

51. Woolf, T. M., Jennings, G. B., Rebagliati, M., and Melton, D. A. (1990) The stability, toxicity and effectiveness of unmodified and phosphorothioate antisense oligodeoxynucleotides in Xenopus oöcytes and embryos. *Nucleic Acids Res.* **18,** 1763–1769.

52. Krieg, A. M. (1993) Uptake and efficacy of phosphodiester and modified antisense oligonucleotides in primary cell cultures. *Clin. Chem.* **39,** 710–712.

53. Wahlestedt, C., Golanov, E., Yamamoto, S., Yee, F., Ericson, H., Yoo, H., Inturrisi, C. E., and Reis, D. J. (1993) Antisense oligonucleotides to NMDA-R1 receptor channel protect cortical neurons from excitotoxicity and reduce focal ischaemic infarctions. *Nature* **363,** 260–263.

54. Wahlestedt, C., Pich, E. M., Koob, G. F., Yee, F., and Heilig, M. (1993) Modulation of anxiety and neuropeptide Y-Y1 receptors by antisense oligodeoxynucleotides. *Science* **259,** 528–531.

55. Paul, M. L., Currie, R. W., and Robertson, H. A. (1995) Priming of a D1 dopamine receptor behavioural response is dissociated from striatal immediate-early gene activity *Neuroscience* **66,** 347–359.

56. Teskey, C. G., Atkinson, B. G., and Cain, D. P. (1991) Expression of the proto-oncogene *c-fos* following electrical kindling in the rat. *Mol. Brain. Res.* **11,** 1–10.

57. Naranjo, J. R., Mellstrom, B., Achaval, M., and Sassone-Corsi, P. (1991) Molecular pathways of pain: fos/jun-mediated activation of noncanonical AP-1 site in the prodynorphin gene. *Neuron* **6,** 606.

58. Rossier, J. (1993) Biosynthesis of enkephalin-derived peptides, in *Handbook of Experimental Pharmacology: Opioids I* (Herz, A., ed.), Springer-Verlag, Germany, pp. 423–447.

59. Höllt, V. (1993) Regulation of opioid peptide gene expression, in *Handbook of Experimental Pharmacology: Opioids I* (Herz, A., ed.), Springer-Verlag, Germany, pp. 307–346.

60. Konradi, C., Kobierski, L. A., Nguyen, T. V., Heckers, S., and Hyman, S. E. (1993) The c-AMP-response-element-binding-protein interacts but Fos protein does not interact, with the proenkephalin enhancer in rat striatum. *Proc. Natl. Acad. Sci. USA* **90,** 7005–7009.

61. Goddard, G. V., McIntyre, D., and Leech, C. (1969) A permanent change in brain function resulting from daily electrical stimulation. *Exp. Neurol.* **25,** 295–330.

62. Racine, R. J. (1972) Modification of seizure activity by electrical stimulation: motor seizure. *Electroencephalogr. Clin. Neurophysiol.* **38,** 281–294.

63. Cain, D. P. (1993) Kindling and the amygdala, in *The Amygdala: Neurobiological Aspects of Emotion, Memory, and Mental Dysfunction* (Aggleton, J. P., ed.), Wiley-Liss, New York, pp. 539–560.

64. Downs, A. W. and Eddy, N. B. (1932) The effect of repeated doses of cocaine on the rat. *J Pharmacol. Exp. Therap.* **46,** 199,200.
65. Shin, C., McNamara, J. O., Morgan, J. I., Curran, T., and Cohen, D. R. (1990) Induction of c-fos mRNA expression by afterdischarge in the hippocampus of naive and kindled rats. *J. Neurochem.* **55,** 1050–1055.
66. Labiner, D. M., Butler, L. S., Cao, Z., Hosford, D. A., Shin, C., and McNamara, J. O. (1993) Induction of c-fos mRNA by kindled seizures: complex relationship with neuronal burst firing. *J. Neurosci.* **13,** 744–751.
67. Chiasson, B. J., Armstrong, J. N., Hooper, M. L., Murphy, P. R., and Robertson, H. A. (1994) The application of antisense oligonucleotide technology to the brain: some pitfalls. *Cell. Mol. Neurobiol.* **14,** 507–521.
68. Zhang, M. and Creese, I. (1993) Antisense oligonucleotide reduces brain dopamine D2 receptor: behavioral correlates. *Neurosci. Lett.* **161,** 223–226.
69. Zhou, L. M., Zhang, S. P. Qin, Z. H., and Weiss, B. (1994) In vivo administration of an oligonucleotide antisense to the D2 dopamine receptor messenger RNA inhibits D2 dopamine receptor mediated behavior and the expression of D2 dopamine receptors in mouse striatum. *J. Pharmacol. Exp. Ther.* **268,** 1015–1023.
70. Guvakova, M. A., Yakubov, L. A., Vlodavsky, I., Tonkinson, J. L., and Stein, C. A. (1995) Phosphorothioate oligodeoxynucleotides bind to basic fibroblast growth factor, inhibit its binding to cell surface receptors, and remove it from low affinity binding sites on extracellular matrix. *J. Biol. Chem.* **270,** 2620–2627.
71. Perez, J. R., Li, Y., Stein, C. A., Majumder, S., van Oorschot, A., and Narayanan, R. (1994) Sequence-independent induction of Sp1 transcription factor activity by phosphorothioate oligonucleotides. *Proc. Natl. Acad Sci. USA* **91,** 5957–5961.
72. Hong, M., Chiasson, B. J., Murphy, K. M. A., and Robertson, H. A. (1995) Effects of thio end-capped oligonucleotides on rat striatal c-fos expression and behaviour following amphetamine challenge. *International Symposium on the Regulation of Gene Expression: Practical Approaches.* Oxford, UK, March 26–28.
73. Chiasson, B. J., Hong, M., Armstrong, J. N., and Robertson, H. A. (1995) Comparative effects of fully thio substituted and end-capped oligonucleotides directed at *c-fos* in amygdala kindling. *International Symposium on the Regulation of Gene Expression: Practical Approaches.* Oxford, UK, March 26–28.
74. Standifer, K. M., Chien, C.-C., Wahlestedt, C., Brown, G. P., and Pasternak, G. W. (1994) Selective loss of delta opioid analgesia and binding by antisense oligodeoxynucleotides to a delta opioid receptor. *Neuron* **12,** 805–810.
75. Crooke, S. T. (1995) Progress in antisense technology. *International Symposium on the Regulation of Gene Expression: Practical Approaches.* Oxford, UK, March 26–28.
76. Stein, C. A. and Cheng, Y.-C. (1993) Antisense oligonucleotides as therapeutic agents—is the bullet really magical? *Science* **261,** 1004–1012.

14

Comparative Pharmacokinetics of Antisense Oligonucleotides

Sudhir Agrawal and Jamal Temsamani

1. Introduction

Antisense oligonucleotides have attracted special interest as a novel class of therapeutic agents for the treatment of viral infection, cancers, and genetic disorders because of their ability to inhibit expression of a disease-associated gene in a sequence-specific manner. Gene expression is inhibited by hybridization of the oligonucleotide to sequences in the gene or the messenger RNA (mRNA) target by Watson-Crick base pairing. The first example of specific inhibition of gene expression by an oligonucleotide was reported by Zamecnik and Stephenson *(1)*, who demonstrated that a short oligonucleotide inhibited Rous sarcoma virus replication in cell culture. Since then, the field has progressed enormously. Numerous studies have demonstrated the ability of antisense oligonucleotides to modulate gene expression *(2–4)*. Accompanying chapters in this volume describe the use of antisense oligonucleotides for various disease targets.

The potential utility of antisense oligonucleotides for the treatment of viral infections and other diseases will depend on whether effective concentrations of oligonucleotide reach the target site of action in vivo. Although the naturally occurring phosphodiester oligonucleotides have displayed activities against some targets in vitro *(2,5)*, they are degraded rapidly in animals, which limits their use in vivo *(6–9)*. The key to the future value of oligonucleotide therapeutics resides in the development of modified oligonucleotides. Several modified versions of the oligonucleotide have been synthesized *(10)*. The major modifications have been carried out on the phosphodiester backbone. Another interesting strategy is to prepare mixed-backbone oligonucleotides, capitalizing on the ideal aspects of each type of nucleotide analog. Numerous studies have

From: *Methods in Molecular Medicine: Antisense Therapeutics*
Edited by: S. Agrawal Humana Press Inc., Totowa, NJ

identified the biochemical and physical characteristics of modified oligonucleotides. However, in vivo data regarding the pharmacokinetics, biodistribution, and stability of these modified oligonucleotides have been few. In this chapter, we describe the pharmacokinetics, tissue disposition, in vivo stability, and excretion of a phosphorothioate oligodeoxynucleotide (GEM 91) in rats after a single bolus iv injection. GEM 91 is a 25-mer oligonucleotide complementary to the *gag* region of HIV-1 and has been extensively studied for its anti-HIV activity *(11,12)*. Further, to understand the effect of chemical or structural modification of phosphorothioate oligodeoxynucleotide on pharmacokinetics parameters, we studied three modified analogs called chimeric, hybrid, and self-stabilized oligonucleotides.

2. Properties of Oligonucleotides

2.1. Phosphorothioates

Phosphorothioate oligonucleotides (PS-oligonucleotides) are among the most obvious and are likely one of the first analogs of naturally occurring phosphates used, and are still the most widely applied analogs of oligonucleotides for antisense application *(3)*. PS-oligonucleotides have a sulfur for oxygen substitution of one of the nonbridged oxygen atoms of internucleotide linkages (Fig. 1). PS-oligonucleotides are negatively charged, taken up in the cells by endocytosis, activate RNase H, have strong interaction with plasma proteins, and show sequence-specific and nonsequence-specific biological activity *(13–17)*. In in vivo studies, PS-oligonucleotides are widely distributed to various tissues with slow elimination, and are metabolized primarily from the 3' end *(8,18–22)*.

2.2. Chimeric Oligonucleotides

The rationale for designing chimeric oligonucleotides is based on biophysical, biochemical, and biological properties of various phosphate-modified oligonucleotides, including phosphorothioates *(23)*. Some of the oligonucleotide analogs, e.g., methylphosphonate, have properties that are different than those of phosphorothioates. Methylphosphonate oligonucleotides are nonionic chemical analogs in which the negatively charged phosphate oxygen is replaced by a neutral methyl group *(24)*. Methylphosphonate oligonucleotides are highly lipophilic, resistant to nucleases, have weak interaction with plasma proteins, and are taken up by cells by a combination of endocytosis and passive diffusion *(24)*. The major drawback of these analogs is that they do not activate RNase H. Pharmacokinetic studies show that they are widely distributed in various tissues and are eliminated rapidly in urine *(25)*. In chimeric oligonucleotides, we have incorporated methylphosphonate internucleotide linkages at both the 3' and the 5' ends of the oligonucleotide, whereas the central core is of phosphorothioate

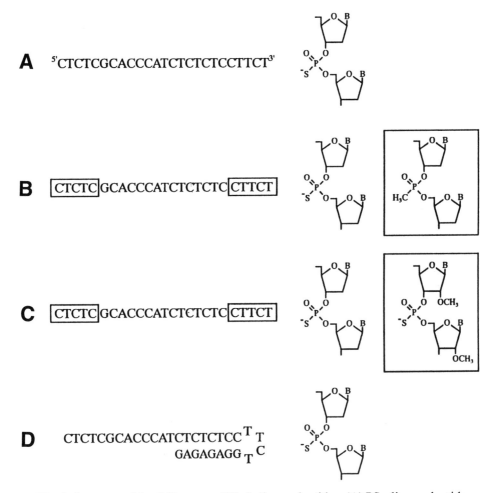

Fig. 1. Structure of the different modified oligonucleotides: **(A)** PS-oligonucleotide; **(B)** Chimeric oligonucleotide; **(C)** Hybrid oligonucleotide; **(D)** Self-stabilized oligonucleotide.

composition (Fig. 1) *(23)*. The methylphosphonate linkages increase the uptake and the stability of the oligonucleotide, whereas the central core enhances the hybridization and activation by RNase H *(23)*. The overall negative charge of the oligonucleotide as well as binding to plasma proteins will be reduced.

2.3. Hybrid Oligonucleotides

The rationale of designing hybrid oligonucleotides is quite similar to the chimeric oligonucleotide. In hybrid oligonucleotides, the 3' and 5' ends of the oligonucleotide are of 2'-*O*-methylribonucleoside, whereas the central core is of phosphorothioate composition *(26)* (Fig. 1). The incorporation of 2'-*O*-methyl-ribonucleoside residues provides increased nuclease resistance and higher affinity and specificity for the target RNA, whereas the central core enhances the hybridization and activation by RNase H.

2.4. Self-Stabilized Oligonucleotides

Another design of oligonucleotides is to self-stabilize them against nucleases, which is accomplished by structural modification rather than chemical modification (Fig. 1). Self-stabilized oligonucleotides are of phosphorothioate composition and have two domains: a single-stranded antisense sequence and a hairpin loop at the 3' end, which reduces accessibility of the oligonucleotide to nucleases *(27)*. In the presence of a complementary target nucleic acid, the hairpin loop is destabilized and the antisense sequence hybridizes to the target sequence.

3. Synthesis and Labeling of Oligonucleotides

3.1. ^{35}S-Labeled Phosphorothioate

To obtain ^{35}S-label PS-oligonucleotide, synthesis was carried out in two steps. The first 19 nucleotides of the sequence of PS-oligonucleotide (from the 3' end) were assembled using the β-cyanoethylphosphoramidite approach *(28)*, and the last six nucleotides were assembled using the H-phosphonate approach *(29)*. CPG support-bound oligodeoxynucleotide (30 mg of CPG; approx 1 μ*M*) containing five H-phosphonate linkages was oxidized with unlabeled sulfur ($^{35}S_8$) (4 mCi, 1 Ci/mg, Amersham [Arlington Heights, IL]; 1 Ci = 37 gBq) in 60 mL carbon disulfide:pyridine:triethylamine (10:10:1). The oxidation reaction was performed at room temperature for 1 h with occasional shaking. Then, 2, 5, and 200 μL of 5% $^{35}S_8$ in the same solvent mixture were added every 30 min to complete the oxidation. The solution was removed, and the CPG support was washed with carbon disulfide:pyridine:triethylamine (10:10:1) (3 × 5 μL) and with acetonitrile (3 × 700 μL). The product was deprotected in concentrated ammonium hydroxide (55°C, 14 h) and evaporated. The resultant product was purified by polyacrylamide gel electrophoresis (PAGE) (20% polyacrylamide containing 7*M* urea). The desired band was excised under UV shadowing, and the PS-oligonucleotide was extracted from the gel and desalted with a Sep-Pak C 18 cartridge (Waters, Milford, MA) and a Sephadex G-18 column. The yield was 20 A_{260} U (600 μg; SA 1 μCi/μg).

3.2. ^{35}S-Labeled Hybrid Oligonucleotide

To prepare the ^{35}S-labeled hybrid oligonucleotide, synthesis was carried out in a similar way as described above, except that the last four couplings were carried out using 2'-*O*-methylribonucleoside H-phosphonate. 2'-*O*-methylribonucleoside H-phosphonates, U and C, were synthesized by following the PCl$_3$/ triazole method *(29)*, starting from the appropriate 2'-*O*-methylribonucleoside and isolated as triethylammonium salts. The isolated 2'-*O*-methylribonucleoside H-phosphonates, U and C, were analyzed by ^{31}P- and ^{1}H-NMR. Prior to use in the oligonucleotide synthesis, the nucleoside H-phosphonates were evaporated to dryness twice with anhydrous pyridine and dissolved into anhydrous pyridine/CH$_3$CN (1/1) to a concentration of 40 m*M*. After the assembly, the CPG-bound oligonucleotide containing four H-phosphonate linkages was oxidized with [35]*S* elemental sulfur (Amersham, 0.5–2.5 Ci/milliatom) and deprotected by the same procedure as reported earlier *(18)*. Purification of the [35]*S*-labeled hybrid oligonucleotide was carried out on 20% PAGE (containing 7*M* urea). The bands under UV light were excised, extracted in 100 m*M* ammonium acetate, and desalted using Sep-Pak C18 Column (Waters). The specific activity of hybrid oligonucleotide obtained was 0.25 µCi/µg.

3.3. ^{35}S-Labeled Chimeric Oligonucleotide

Synthesis of [35]*S*-labeled chimeric oligonucleotide was carried out on a 10-µmol scale using an automated DNA synthesizer (Expedite 8909, Biosearch). The first four couplings were carried out using nucleoside methylphosphonamidites (Glen Research, Sterling, VA), followed by oxidation with iodine reagent. The next six couplings were then carried out using nucleoside β-cyanoethyl phosphoramidite, followed by oxidation with ^{3}H-benzodithiol-3-one-1,1-dioxide reagent. The next coupling was carried out using nucleoside β-cyanoethyl phosphoramidite, and the CPG-bound oligonucleotide was oxidized with a mixture of [35]*S* (4.5 µCi, 1 Ci/milliatom in 50 µL toluene, Amersham) in a solution of CS$_2$/pyridine/triethylamine (200/200/4 µL) at 25°C, for 1 h. After 1 h ^{3}H-1,2-benzodithiole-^{3}H-one-1,1 dioxide (1 mL, 2% in CH$_3$CN) was added, and the reaction mixture was allowed to remain at 25°C for 10 min. The supernatants were removed and the CPG-bound material was washed with CH$_3$CN (10 × 1 mL). After capping with acetic anhydride (300 µL, Tetrahydrofuran/lutidine/ascetic anhydride, 8:1:1) and Dimethylaminopyridine (300 µL, 0.625% in pyridine), the [35]*S*-CPG material was washed with CH$_3$CN (10 × 1 mL). The remaining 13-mer synthesis was continued as normal. The crude CPG-bound 25-mer chimeric oligonucleotide was treated with NH$_4$OH (28%, 3 mL, 25°C, 2 h). Evaporation on a Speed-Vac yielded a dried yellow pellet, which was immediately incubated with a solution of ethylenediamine/ethanol/water (50/45/5, v/v/v, 4 mL) for 4.5 h at 25°C. Evaporation gave the crude [[35]*S*] chimeric oligonucleotide as a yellow pellet. PAGE purification (20%, 7*M* urea) gave [[35]*S*] chimeric oligonucleotide as a white pellet (A$_{260}$ = 194, 155 µCi, 180 µCi/µmol).

3.4. *^{35}S-Labeled Self-Stabilized Oligonucleotide*

^{35}S-labeled self-stabilized oligonucleotide was synthesized, purified, and analyzed by the methods previously described *(18,27)*. In short, synthesis was carried out on a 1-μ*M* scale using an automated synthesizer (Millipore 8801, Milford, MA) The first 28 nucleotides were coupled using the β-cyanoethyl phosphoramidite approach followed by oxidation with ^{3}H-1,2-benzodithiole-3-one-1,1-dioxide *(28)*. The last five couplings were carried out using H-phosphonate approach. After the assembly of the required sequence (33-mer), CPG-bound oligonucleotide containing five H-phosphonate linkages was oxidized with ^{35}S elemental sulfur (Amersham, 0.5–2.5 Ci/milliatom) in the presence of carbon disulfide and pyridine. After the oxidation, CPG-bound oligonucleotide was deprotected in concentrated ammonium hydroxide and purified by PAGE (20% gel containing 7*M* urea). The bands under UV light were excised and extracted in 100 m*M*/L ammonium acetate buffer. The product was desalted using Sep-Pak C18 (Waters) cartridge. The specific activity of the product obtained was 0.21 μCi·μg (Total yield 1.2 mg).

4. Protocols for Oligonucleotide Administration, Tissue Distribution, and In Vivo Stability

Male Sprague-Dawley rats (100–120 g, Harlan Laboratories, Indianapolis, IN) were used in these studies. The animals were fed commercial diet and water ad lib for 1 wk prior to the studies. Animals were dosed via a single bolus iv injection into a tail vein at a dose of 30 mg/kg. Unlabeled and ^{35}S-labeled oligonucleotides were dissolved in physiological saline (0.9% NaCl) in a concentration of 10 mg/mL. The final specific activity for PS-oliognucleotide was 27.8 μCi/mL; chimeric 15 μCi/mL; hybrid 20 μCi/mL; and self-stabilized 20 μCi/mL. After injection, each animal was placed in a metabolism cage and fed with commercial diet and water ad lib. Total voided urine was collected, and each metabolism cage was then washed following the collection intervals. Total excreted feces was collected from each animal at various time-points, and feces samples were homogenized in a ninefold volume of 0.9% NaCl saline prior to quantitation of radioactivity. Blood samples were collected in heparinized tubes from animals at various time-points, and plasma was separated by centrifugation. At the indicated time-points, animals were sacrificed, and tissues were collected. All tissues were weighed and homogenized in 0.9% NaCl saline (3–5 mL/g of tissue). The total radioactivities in tissue and body fluids were determined by liquid scintillation spectrometry as described previously *(30)*. To assess the in vivo stability of the oligonucleotide, plasma and homogenized tissues were incubated with proteinase K (2 mg/mL) in extraction buffer (0.5% SDS, 10 m*M* NaCl, 20 m*M* Tris-HCl, pH 7.6, 10 m*M* EDTA) for 2 h at 37°C. The samples were then extracted twice with phenol/chloroform (1:1, v/v) and once with

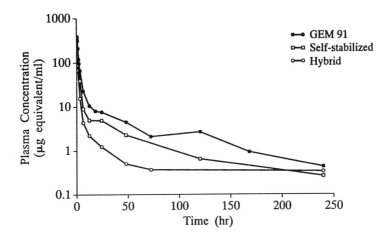

Fig. 2. Plasma concentration time-course of oligonucleotide-derived radioactivity. Plasma concentrations were expressed as micrograms of oligonucleotide Eq/mL after iv bolus administration of ^{35}S-labeled oligonucleotides into rats at a dose of 30 mg/kg. Plasma concentration was based on the quantitation of radioactivity (based on the data from refs. *30–32*).

chloroform. After ethanol precipitation, the oligonucleotides were analyzed by electrophoresis on 20% polyacrylamide gels containing 7M urea. The gels were fixed in 10% acetic acid/10% methanol solution and then dried before autoradiography *(18)*.

5. Pharmacokinetics of PS-Oligonucleotide (GEM 91)
5.1. Kinetics and Tissue Distribution

The mean plasma concentrations over time of PS-oligonucleotide equivalents following iv bolus administration of ^{35}S-labeled GEM 91 are illustrated in Fig. 2. After iv administration of the 25-mer PS-oligonucleotide, a peak plasma radioactivity concentration of 418 µg/mL (52 µm) equivalents was achieved *(30)*. Pharmacokinetic analysis revealed that plasma disappearance curves for PS-oligonucleotide-derived radioactivity could be described by the sum of two exponentials with half-lives of 0.95 and 48 h (*see* Table 1). The chemical form of radioactivity in plasma was further evaluated by gel electrophoresis, demonstrating the presence of both intact PS-oligonucleotide and metabolites with smaller molecular weights. Urinary excretion represented the major pathway of elimination of PS-oligonucleotide. Rapid excretion of radioactivity was observed in urine for the first 48 h following administration with 27% of the administered dose excreted within 24 h, 40% within 48 h, and 58% over 240 h (Fig. 3). The radioactivity in urine was present as degraded products with

Table 1
Pharmacokinetic Parameters for GEM 91 in Various Tissues (Two-Compartmental iv Bolus Model)[a]

Tissue	C_{max}, µg/mL or µg/g	$T_{1/2}\alpha$, h	$T_{1/2}\beta$, h	AUC, (µg/mL) × h	MRT, h	CL, mL/(kg × h)
Plasma	418.40	0.95 ± 0.07	47.57 ± 14.48	1432.51	42.59	20.94
Liver	78.19	24.48 ± 19.28	97.16 ± 75.61	6538.18	116.36	4.59
Spleen	67.35	2.32 ± 0.54	136.72 ± 66.5	2389.02	182.00	12.56
Lung	48.36	4.36 ± 1.62	101.69 ± 42.43	2178.61	132.53	13.77
Adrenal	66.24	2.34 ± 0.06	76.98 ± 0.10	1879.03	101.24	15.97
Heart	44.48	5.17 ± 1.26	119.32 ± 59.74	1591.96	143.76	18.84
Small intestine	39.73	14.21 ± 3.13	118.44 ± 57.75	2263.93	129.93	13.25
Large intestine	35.90	11.48 ± 5.39	112.56 ± 54.08	2129.02	133.61	14.09
Pancreas	38.34	3.29 ± 1.57	97.35 ± 41.24	1445.03	127.55	20.76
Stomach	40.67	2.401 ± 0.74	96.62 ± 25.46	1913.95	132.63	15.67
Thymus	11.27	4.33 ± 0.76	198.05 ± 113.14	1188.42	275.09	25.24
Eye	7.24	22.47 ± 16.94	319.46 ± 572.73	1430.24	417.72	20.98
Skeletal muscle	9.31	8.50 ± 2.92	245.12 ± 150.27	1244.45	333.51	24.11
Stomach contents	3.67	126.63 ± 22.91	234.97 ± 1540.00	1091.05	313.45	27.50

[a]Values are means (±SEM) based on the experimental data from 51 rats following administration of ^{35}S-labeled GEM 91. Concentrations in various tissues were based on the quantitation of radioactivity (reproduced from ref. 30).
[b]MRT = mean residue time.

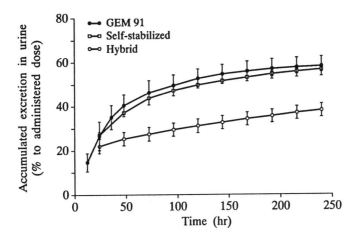

Fig. 3. Cumulative urinary excretion of oligonucleotide-derived radioactivity. Urinary excretion of oligonucleotide-derived radioactivity was expressed as mean ± SD of the cumulative percentage of administered dose excreted over time. Total excretion was based on the quantitation of radioactivity in urine (based on the data from refs. *30–32*).

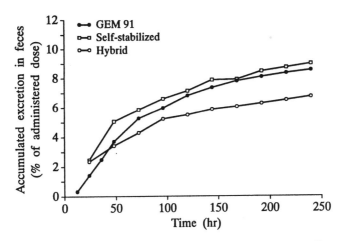

Fig. 4. Cumulative feces excretion of oligonucleotide-derived radioactivity. Feces excretion of oligonucleotide-derived radioactivity was expressed as the cumulative percentage of administered dose excreted over time. Total excretion was based on the quantitation of radioactivity in feces (based on the data from refs. *30–32*).

smaller molecular weights. Fecal excretion was a minor pathway of elimination of the PS-oligonucleotide GEM 91 with only 8% excreted over 240 h (Fig. 4).

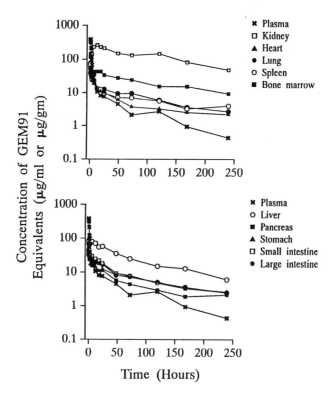

Fig. 5. Tissue concentration time-course of GEM 91-derived radioactivity. Tissue concentrations were expressed as micrograms of GEM 91 Eq/g of organ tissue. Tissue concentration was based on the quantitation of radioactivity (based on the data from ref. *30*).

Following administration of GEM 91, there was wide distribution of radioactivity in various tissues *(30)* (Fig. 5; *see also* Table 1). The accumulation of GEM 91 in tissues occurs when plasma concentrations have diminished. Initially, GEM 91 was distributed mainly in highly perfused tissues, including the liver, kidneys, heart, lungs, and spleen. GEM 91-derived radioactivity in the kidneys was consistently higher than the corresponding plasma level over the experimental period. Relatively high concentrations of GEM 91-derived radioactivity in bone marrow, heart, lung, spleen, adrenal, and thyroid were observed. In particular, consistently higher levels were observed in these tissues compared with the corresponding plasma during the elimination phase of GEM 91. The highest tissue concentration of GEM 91 was detected in most of the organs within 15–30 min following administration of GEM 91. Throughout the experiment period, a high concentration of radioactivity persisted in the liver and the kidney. A high concentration of radioactivity was also observed in the intestinal tract, especially in the small intestine. Limited amounts of radioactivity were observed in the brain.

5.2. In Vivo Stability

To determine the molecular nature of the radioactivity present in plasma and tissues, plasma and tissues were homogenized and analyzed by PAGE and HPLC. In plasma, both intact GEM 91 and its metabolites with smaller molecular weights were detected. Intact GEM 91 was still detectable in plasma up to 24 h, indicating that it is relatively stable in vivo. In kidney and liver, both GEM 91 and metabolic forms of GEM 91 were present (Fig. 6). The radioactivity in urine and feces was present as degraded products of GEM 91 with smaller molecular weight, indicating breakdown of the oligonucleotide.

6. Pharmacokinetics of Hybrid Oligonucleotide

6.1. Kinetics and Tissue Distribution

The mean plasma concentrations over time of the hybrid oligonucleotide equivalents following iv bolus administration of ^{35}S-labeled hybrid oligonucleotide are illustrated in Fig. 2. Pharmacokinetic analysis revealed that plasma disappearance curves for hybrid oligonucleotide-derived radioactivity could be described by the sum of two exponential with half-lives of 0.34 and 52 h *(31)* (Fig. 2; *see also* Table 2). Compared to GEM 91, the hybrid oligonucleotide had a shorter distribution half-life and a longer elimination half-life, suggesting that the hybrid oligonucleotide was taken up rapidly and retained in various tissues. Urinary excretion represented the major pathway of elimination with 22, 25, and 38% of the administered dose excreted within 24, 48, and 240 h, respectively (Fig. 3). Fecal excretion was a minor pathway of elimination of the hybrid oligonucleotide with 2 and 7% excreted over a period of 24 and 240 h, respectively (Fig. 4).

The hybrid oligonucleotide had a wide tissue distribution. In the first 30 min, the oligonucleotide was distributed in kidneys, liver, spleen, bone marrow, and intestine. Most tissues examined had significantly higher concentrations compared to those observed in plasma 3 h and longer after dosing. The kidneys had the highest concentration of radioactivity. Throughout the experiment, a high concentration of radioactivity persisted in kidneys, liver, heart, lungs, bone marrow, and the intestinal tract. Low concentrations of about 2 μg/ g were observed in the brain *(31)*.

6.2. In Vivo Stability

Analysis of the hybrid oligonucleotide by PAGE showed the presence of mainly intact hybrid oligonucleotide in plasma up to 12 h, as well as minor metabolites with lower molecular weights (Fig. 7). In urine, although the majority

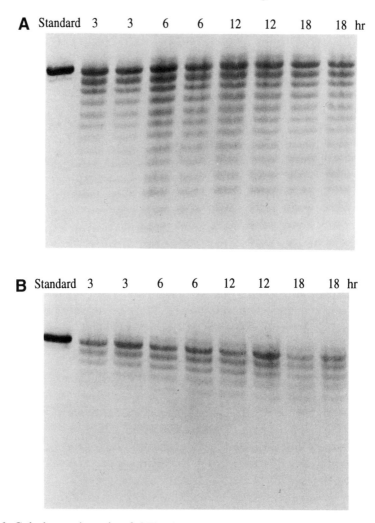

Fig. 6. Gel electrophoresis of GEM 91-derived radioactivity in kidneys **(A)** and liver **(B)**. Tissue samples were extracted and analyzed by PAGE. Standard represents GEM 91 before injection. Two animals were analyzed for each time-point (reproduced from ref. *30*).

of the radioactivity was present as degraded products of hybrid oligonucleotide, a trace of intact oligonucleotide was also detected. In most tissues analyzed (kidneys, liver, and intestine), mainly the intact hybrid oligonucleotide was present (Fig. 8). Greater biostability was observed for the hybrid oligonucleotide compared to the PS-oligonucleotide. The degradation pattern of the hybrid oligonucleotide suggests that the 3'-end segments of 2'-*O*-methylribonucleotide of hybrid were cleaved off

Table 2
Pharmacokinetic Parameters for Hybrid Oligonucleotide in Various Tissues (Two-Compartmental iv Bolus Model)[a]

Tissue	C_{max}, µg/mL or µg/g	$T_{1/2}\alpha$, h	$T_{1/2}\beta$, h	AUC, (µg/mL) × h	MRT, h	CL, mL/(kg × h)
Plasma	339.73	0.34	52.02	704.09	57.53	42.61
Adrenal	47.49	0.96	163.32	5745.29	234.36	5.21
Brain	4.82	0.16	172.41	384.69	248.30	77.99
Fat	8.45	7.48	287.43	2440.64	410.06	12.29
Heart	29.29	0.86	252.19	4016.00	361.85	7.47
Large intestine	41.22	5.57	181.49	8154.99	259.29	3.68
Pancreas	28.14	0.53	314.83	4795.25	453.01	6.26
Skeletal muscle	9.84	1.02	255.70	1768.30	367.43	16.97
Skin	44.29	16.76	113.95	4724.98	151.33	6.35
Small intestine	52.20	1.26	157.65	4725.93	224.27	7.02
Spleen	56.29	0.55	265.75	7233.67	381.89	4.15
Stomach	54.03	0.32	115.76	2919.24	166.07	10.28
Testes	8.38	2.45	252.57	2858.54	364.22	10.49
Thyroid	36.61	0.16	242.27	8789.45	349.49	3.41

[a]Values are means based on the experimental data from 30 rats following administration of ^{35}S-labeled hybrid oligonucleotide. Concentrations in various tissues were based on the quantitation of radioactivity (reproduced from ref. 31).

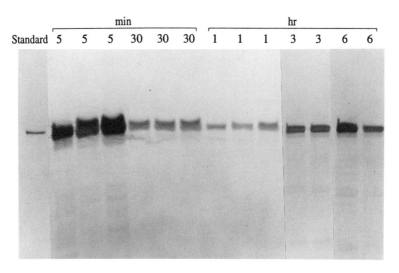

Fig. 7. Gel electrophoresis of hybrid-derived radioactivity in plasma. Plasma samples were extracted and analyzed by PAGE. Standard represents hybrid oligonucleotide before injection. Plasma from two or three animals was analyzed for each time-point (from ref. *31*).

by enzymes (possibly by endonuclease), thereby generating oligonucleotide population of 18- and 19-mers.

7. Pharmacokinetics of Self-Stabilized Oligonucleotide

7.1. Kinetics and Tissue Distribution

After iv bolus administration of ^{35}S-labeled self-stabilized oligonucleotide to rats, the pharmacokinetic analysis revealed that plasma disappearance curves for self-stabilized oligonucleotide-derived radioactivity could be described by the sum of two exponentials, with half-lives of 0.54 and 41 h *(32)* (Fig. 2, *see also* Table 3). Urinary excretion represented the major pathway of elimination of the oligonucleotide, similar to that observed for GEM 91 (Fig. 3). Urinary excretion of self-stabilized oligonucleotide-derived radioactivity mainly occurred in the first 48 h (approx 40%). Over the subsequent 8-d period, 3–5% of the administered dose was excreted in urine in each 24-h period. Fecal excretion was a minor pathway of elimination of self-stabilized oligomer with 9% excreted over a period of 240 h (Fig. 4). A wide tissue distribution of self-stabilized oligonucleotide was observed. Initially, the oligonucleotide was distributed mainly in highly perfused tissues, including the liver, kidneys, heart, lungs, and spleen (Table 3). In kidneys, the oligonucleotide-derived radioactivity was detected within 5 min following administration, peaked between 6 and 18 h, and high concentrations remained thereafter being 10- to 100-fold

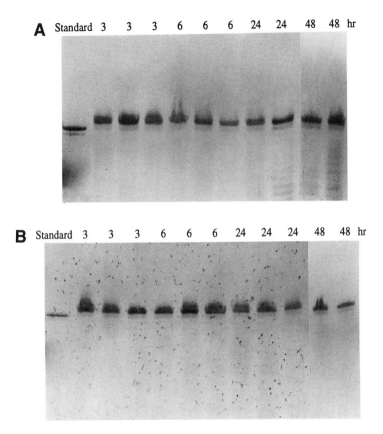

Fig. 8. Gel electrophoresis of hybrid-derived radioactivity in kidneys **(A)** and liver **(B)**. Tissue samples were extracted and analyzed by PAGE. Standard represents hybrid oligonucleotide before injection (from ref. *31*).

higher than that of plasma throughout the rest of the experimental period. In the liver, the highest concentration occurred between 3 and 6 h. A large amount of the self-stabilized oligonucleotide-derived radioactivity was also observed in lungs, bone marrow, spleen, and heart. In most tissues analyzed, the concentrations detected were higher than those observed in plasma *(32)*.

7.2. In Vivo Stability

The self-stabilized oligonucleotide was found to be more stable in various tissues compared to the linear PS-oligonucleotide. Both intact self-stabilized oligonucleotide and its metabolites were detected in plasma, with the presence of mainly the intact form up to 6 h after dosing. Intact oligomer was still detectable in plasma up to 12 h, indicating the relative stability of this com-

Table 3

Pharmacokinetic Parameters for Self-Stabilized Oligonucleotide Phsophorothioate in Various Tissues in Rats Following iv Administration of ^{35}S-Labeled Compound at a Dose of 30 mg/kg (Two-Compartmental iv Bolus Model)[a]

Tissue	C_{max}, mg/L, or mg/kg	$T_{1/2}\alpha$, h	$T_{1/2}\beta$, h	AUC, (mg/mL) × h	MRT, h	CL, mL/(kg × h)
Plasma	343.2	0.54	41.44	1039.1	45.2	28.9
Bone marrow	69.1	0.04	101.93	5182.2	147.0	
Brain	3.5	1.72	907.76	1632.2	1305.3	
Spleen	52.9	2.38	95.77	2043.0	129.3	
Lung	35.5	14.29	174.38	2109.0	185.0	
Heart	24.5	13.62	190.17	1348.8	195.9	
Stomach	49.8	2.56	63.41	2422.2	88.2	
Small intestine	41.7	18.19	138.11	1930.5	112.6	
Large intestine	38.4	1.12	54.89	2081.1	78.5	
Pancreas	29.8	1.58	69.11	1967.5	98.6	
Adrenal	61.9	6.60	88.74	1752.4	94.5	
Thymus	7.1	16.79	464.20	1645.8	624.0	
Skeletal muscle	6.5	78.91	500.48	737.3	113.9	

[a]Values are means based on the experimental data from 30 rats following administration of ^{35}S-labeled hybrid oligonucleotide. Concentrations in various tissues were based on the quantitation of radioactivity (reproduced from ref. 32).

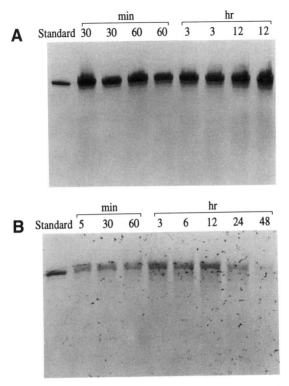

Fig. 9. Gel electrophoresis of self-stabilized-derived radioactivity in kidneys (**A**) and liver (**B**). Tissue samples were extracted and analyzed by PAGE. Standard represents self-stabilized oligonucleotide before injection (reproduced from ref. *32*).

pound. Intact self-stabilized oligonucleotide was also detectable in several tissues for a longer period, including kidney, liver, and intestine (Fig. 9). However, some 32-mer, 31-mer, and shorter lengths can also be seen. One advantage of the self-stabilized oligonucleotide over the linear PS-oligonucleotide is that after degradation, it generates metabolites in which the antisense sequence (25-mer) remains intact at the 5' end. Analysis of the urine samples by PAGE and HPLC indicated that the degradation of this oligomer is a slow process.

8. Pharmacokinetics of Chimeric Oligonucleotide
8.1. Kinetics and Tissue Distribution

The plasma concentration time-course of chimeric oligonucleotide equivalents after iv bolus administration showed a disappearance curve that could be described by the sum of two exponentials, with half-lives of 0.38 and 52 h *(33)* (Fig. 2, *see also* Table 4). The majority of the chimeric oligonucleotide was excreted in urine, with 21 and 36% of the dose excreted within 24 and 240 h, respectively. In feces, about 3% was excreted over a period of 240 h. Following administration of the chimeric oligonucleotide, there was a wide tissue distribution of radioactivity in various tissues, especially in the kidney, liver, bone marrow, and intestinal tract. The highest concentration of chimeric-derived radioactivity was found in the kidneys. Radioactivity was detected within 5 min following administration of the oligonucleotide, peaked at 18 h, and high concentrations remained thereafter, being 25- to 600-fold that of plasma throughout the rest of the experiment. High concentrations of the chimeric oligonucleotide were also observed in the liver. The highest concentration occurred between 6 and 12 h and remained 10- to 150-fold higher than the corresponding plasma levels. A relatively significant concentration of radioactivity was also observed in lymph nodes, bone marrow, spleen, heart, lungs, and gastrointestinal tract. Limited amounts of radioactivity were observed in the brain. In most of the tissues analyzed, the concentrations were higher than the corresponding plasma levels *(33)*.

8.2. In Vivo Stability

The chemical forms of radioactivity in plasma, urine, and various tissues were examined by PAGE and HPLC *(33)*. Both intact chimeric oligonucleotide and its metabolites were detected in plasma, with the presence of mainly the intact form up to 6 h after dosing. However, the radioactivity in the plasma was present as degraded products thereafter. In urine, the majority of the radioactivity was present as intact chimeric oligonucleotide, up to 6 h postdosing, and as mainly degraded products thereafter. The majority of the radioactivity in tissues was associated with intact chimeric oligonucleotide, indicating that the chimeric oligonucleotide is stable in vivo.

9. Comparative Studies

9.1. Plasma

Comparison of plasma concentrations vs time profiles after iv bolus injection of PS-oligonucleotide, hybrid, self-stabilized, and chimeric oligonucleotides are shown in Fig. 2. After administration, the concentrations in plasma for the four different oligonucleotides were similar, ranging from 300 to 400 µg/mL, corresponding to approx 35–45 µM. Plasma disappearances for the four oligonucleotides' radioactivity could be described by a two-compartment model with a short α-phase and a long β-phase. The hybrid and chimeric oligonucleotides had a relatively short mean distribution half-life (0.34 and 0.38 h, respectively) and a prolonged mean elimination half-life (52 h for both) compared to PS-oligonucleotide and self-stabilized oligonucleotide (α-phase: 0.95 h and β-phase: 46 h). This prolonged elimination phase reflected retention of

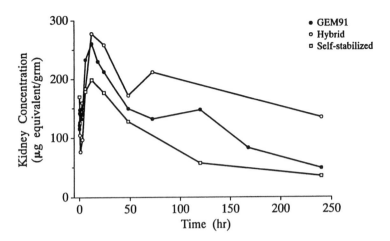

Fig. 10. Tissue concentration time-course of oligonucleotide-derived radioactivity in kidneys. Tissue concentrations were expressed as micrograms of oligonucleotides equivalents/g of kidneys. Tissue concentration was based on the quantitation of radioactivity (based on the data from refs. *30–32*).

hybrid and chimeric oligonucleotides (and possibly their metabolites) in several tissues. Analysis of the molecular nature of the radioactivity present in plasma for the four oligonucleotides by PAGE and HPLC showed that intact oligonucleotide could be detected in plasma for more than 6 h. Self-stabilized and hybrid oligonucleotides were significantly more stable than PS-oligonucleotide.

9.2. Liver and Kidney

Comparison of kidney and liver concentrations vs time profiles after iv bolus injection of PS-oligonucleotide, hybrid, self-stabilized, and chimeric oligonucleotides are shown in Figs. 10 and 11. The highest concentrations of the four oligonucleotides were detected in kidneys within 15 min following administration, peaked between 6 and 18 h, and decreased thereafter. The highest concentrations were between 200 and 270 µg/g. In the liver, the radioactivity was almost constant 12 h after administration and decreased thereafter. The highest concentrations were detected between 4 and 6 h after administration and were in the range of 75–110 µg/g. Analysis of the radioactivity present in the liver and kidney by PAGE and HPLC indicated that intact oligonucleotides could be detected throughout the experiment. Metabolic forms of oligonucleotides were also present. Greater stability was observed for hybrid and self-stabilized oligonucleotides. The degradation pattern for hybrid oligonucleotide suggests that the segments of 2'-*O*-methylribonucleotide are removed at once rather than nucleotide by nucleotide.

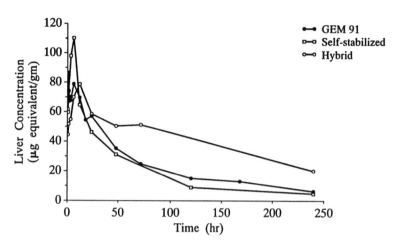

Fig. 11. Tissue concentration time-course of oligonucleotide-derived radioactivity in liver. Tissue concentrations were expressed as micrograms of oligonucleotides equivalents/g of liver. Tissue concentration was based on the quantitation of radioactivity (based on the data from refs. *30–32*).

9.3. Excretion

Urinary excretion represented the major pathway of elimination for all four modified oligonucleotides (Fig. 3). In the first 24 h, the excretion rate was similar: between 21 and 27%. Subsequently, the elimination rate for the hybrid and chimeric oligonucleotides was slower than PS-oligonucleotide and self-stabilized oligonucleotides. About 36–38% of the chimeric and hybrid oligonucleotides were excreted in urine after 240 h, whereas in the case of PS-oligonucleotide and self-stabilized oligonucleotides, the rate was between 56 and 58%. Fecal excretion was a minor pathway of elimination for the oligonucleotides. This suggests that the hybrid and chimeric oligonucleotides are retained more in the tissues. Fecal excretion was a minor pathway of elimination. The excretion rate was similar for all four modified oligonucleotides. At 240 h 7–9% of the oligonucleotides were excreted in feces (Fig. 4).

10. Pharmacokinetics of PS-Oligonucleotide in Humans

We have also evaluated the pharmacokinetics of the PS-oligonucleotide in humans. We administered a single dose of the 25-mer ^{35}S-labeled PS-oligonucleotide (GEM 91) into six HIV-1-infected individuals by 2-h iv infusion at a dose of 0.1 mg/kg to assess the plasma clearance profile and urinary excretion. The plasma disappearance curve was described by the sum of two exponentials with mean half-lives of 0.18 and 26.7 h based on radioactivity levels *(34)*. Urinary excretion represented the major pathway of elimination of oligonucleotide. On the basis of radioactivity levels, about 49% of the administered dose was excreted within 24 h and about 70% within 96 h of oligonucleotide administration. Analysis of the extracted radioactivity in plasma showed both intact and degraded forms of PS-oligonucleotide. In urine, however, all of the radioactivity was associated with the degraded form of the oligonucleotide.

11. Conclusion

Although antisense oligonucleotides have displayed activities against a large number of targets in vitro, only during the past 4 yr have in vivo studies been reported. In order for antisense oligonucleotides to have potential as effective therapeutic agents, appropriate drug concentrations must be attained and maintained in vivo at the site of drug action. The studies published to date on the pharmacokinetics of PS-oligonucleotides indicate that PS-oligonucleotides have a long half-life in plasma of about 40–50 h, which suggests that they may be given on either a daily or an alternating day basis. They have a wide tissue distribution, mainly in highly perfused tissues, including the liver, kidneys, heart, lungs, and spleen. The liver and kidneys seem to be the organs where most of the oligonucleotides are retained, and the brain had the lowest concentration. Urinary excretion represented the major pathway of elimination of PS-oligonucleotides. These studies suggest that the oligonucleotides have favorable absorption, distribution kinetics, and sufficient in vivo stability to be used as therapeutic agents.

Although the most active first-generation class of oligonucleotides, the phosphorothioates, are being evaluated in humans, the progress made in the medicinal chemistry of oligonucleotides suggests that in the near future oligonucleotides with significantly improved activity and properties will likely be evaluated in humans. Recent progress demonstrates ample opportunity to improve the properties of oligonucleotides via rational medicinal chemical approaches. In this chapter, we describe the pharmacokinetics of some of these modified oligonucleotides. Our results show that hybrid, chimeric, and self-stabilized oligonucleotides do not display significant changes in the tissue distribution in rats. However, these modifications provide significant increase in in vivo stability of the oligonucleotide. Clearly medicinal chemistry of antisense oligonucleotides will become increasingly useful as the pharmacokinetics and metabolism issues are elucidated.

References

1. Zamecnik, P. C. and Stephenson, M. L. (1978) Inhibition of Rous sarcoma virus replication and cell transformation by a specific oligonucleotide. *Proc. Natl. Acad. Sci. USA* **75**, 280–284.

2. Agrawal, S. (1992) Antisense oligonucleotides as antiviral agents. *TIBTECH* **10,** 152–157.
3. Stein, C. A. and Cheng, Y. C. (1993) Antisense oligonucleotides as therapeutic agents—Is the bullet really magical? *Science* **261,** 1004–1012.
4. Temsamani, J. and Agrawal, S. (1995) Antisense oligonucleotides as antiviral agents, in *Advances in Antiviral Drug Design,* vol. 2 (de Clercq, E., ed.), JAI, in press.
5. Uhlmann, E. and Peyman, A. (1990) Antisense oligonucleotides. A new therapeutic principle. *Chem. Rev.* **90,** 543–584.
6. Inagaki, M., Togawa, K., Carr, B. I., Ghosh, K., and Cohen, J. S. (1992) Antisense oligonucleotides: inhibition of liver cell proliferation and in vivo disposition. *Transplant. Proc.* **26,** 2971,2972.
7. Zendegui, J., Vasquez, K., Tinsley, J., Kessler, D. J., and Hogan, M. E. (1992) In vivo stability and kinetics of absorption and disposition of 3' phosphopropyl amine oligonucleotides. *Nucleic Acids Res.* **20,** 307–314.
8. Sands, H., Gorey-Feret, L. J., and Cocuzza, A. J. (1994) Biodistribution and metabolism of internally ^{3}H-labeled oligonucleotides. I. Comparison of a phosphodiester and a phosphorothioate. *Mol. Pharmacol.* **45,** 932–943.
9. Agrawal, S., Temsamani, J., Galbraith, W., and Tang, J.-Y. (1995) Pharmacokinetics of antisense oligonucleotides. *Clin. Pharmacokinet.* **28,** 7–16.
10 Agrawal, S., ed. (1993) *Protocols for Oligonucleotides and Analogs: Synthesis and Properties.* Humana, Totowa, NJ.
11. Agrawal, S. and Lisziewicz, J. (1994) Potential for HIV-1 treatment with antisense oligonucleotides. *J. Biotechnol. Health Care* **1,** 167–182.
12. Lisziewicz, J., Sun, D., Weichold, F. F., Thierry, A., Lusso, P., Tang, J. Y., Gallo, R. C., and Agrawal, S. (1994) Antisense oligodeoxynucleotide phosphorothioate complementary to gag mRNA blocks replication of human immunodeficiency virus type 1 in human peripheral blood cells. *Proc. Natl. Acad. Sci. USA* **91,** 7942–7946.
13. Cohen, J. S. (1993) Phosphorothioate oligonucleotides, in *Antisense Research and Applications* (Crooke, S. T. and Lebleu, B., eds.), CRC, Boca Raton, FL, pp. 205–222.
14. Stein, C. A. and Cohen, J. S. (1989) Phopshorothioate oligodeoxynucleotide analogs, in *Oligodeoxynucleotides: Antisense Inhibitors of Gene Expression* (Cohen, J. S., ed.), CRC, Boca Raton, FL, pp. 97–120.
15. Akhtar, S. and Juliano, R. L. (1992) Cellular uptake and intracellular fate of antisense oligonucleotides. *Trends Cell Biol.* **2,** 139–144.
16. Temsamani, J., Kubert, M., Tang, J.-Y., Padmapriya, A. A., and Agrawal, S. (1994) Cellular uptake of oligodeoxynucleotide phosphorothioates and their analogs. *Antisense Res. Dev.* **4,** 35–42.
17. Agrawal S. (1991) Antisense oligonucleotides: a possible approach for chemotherapy of AIDS, in *Prospects for Antisense Nucleic Acid Therapy of Cancer and AIDS* (Wickstrom, E., ed.), Wiley-Liss, New York, pp. 143–158.
18. Agrawal, S., Temsamani, J., and Tang, J.-Y. (1991) Pharmacokinetics biodistribution, and stability of oligodeoxynucleotide phosphorothioates in mice. *Proc. Natl. Acad. Sci. USA* **88,** 7595–7599.

19. Temsamani, J., Tang, J.-Y., Padmapriya, A. A., Kubert, M., and Agrawal, S. (1993) Pharmacokinetics, biodistribution and stability of capped oligodeoxynucleotide phosphorothioates in mice. *Antisense Res. Dev.* **3**, 277–284.
20. Temsamani, J., Tang, J.-Y., and Agrawal, S. (1992) Capped oligodeoxynucleotide phosphorothioates: pharmacokinetics and stability in mice. *Ann. NY Acad. Sci.* **660**, 318–320.
21. Iversen, P. L., Mata, J., Tracewell, W. G., and Zon, G. (1994) Pharmacokinetics of an antisense phosphorothioate oligodeoxynucleotide against *rev* from human immunodeficiency virus type 1 in adult male rat following single injections and continuous infusion. *Antisense Res. Dev.* **4**, 43–52.
22. Cossum, P. A., Sasmor, H., Dellinger, D., Truong, L., Cummins, L., Owens, S. R., Markham, P. M., Shea, J. P., and Crooke, S. (1993) Disposition of the ^{14}C-labeled phosphorothioate oligonucleotide ISIS 2105 after intravenous administration to rats. *J. Pharmacol. Exp. Ther. USA* **267**, 1181–1190.
23. Agrawal, S., Mayrand, S. M., Zamecnik, P. C., and Pederson, T. (1990) Site specific excision from RNA by RNase H and mixed phosphate backbone oligodeoxynucleotides. *Proc. Natl. Acad. Sci. USA* **87**, 1401–1405.
24. Miller, P. S., Ts'o, P. O. P., Hogrefe, R. I., Reynolds, M. A., and Arnold, L. J. (1993) J. Anticode oligonucleoside methylphosphonates and their Psoralen derivatives, in *Antisense Research and Applications* (Crooke, S. T. and Lebleu, B., eds.), CRC, Boca Raton, FL, pp. 189–204.
25. Chem, T. L., Miller, P., Ts'o, P., and Colvin, O. M. (1990) Disposition and metabolism of oligodeoxynucleoside methylphosphonate following a single iv injection in mice. *Drug Metab. Dispos.* **18**, 815–818.
26. Metelev, V., Lisziewicz, J., and Agrawal, S. (1994) Study of antisense oligonucleotide phosphorothioates containing segments of oligodeoxynucleotides and 2'-O-methyloligoribonucleotides. *Bioorg. Med. Chem. Lett.* **4**, 2929–2934.
27. Tang, J.-Y., Temsamani, J., and Agrawal, S. (1993) Self-stabilized antisense oligonucleotide phosphorothioates: properties and anti-HIV activity. *Nucleic Acids Res.* **21**, 2729–2735.
28. Beaucage, S. L. (1993) Oligodeoxyribonucleotide synthesis-phosphorothioate approach, in *Protocols for Oligonucleotides and Analogs* (Agrawal, S., ed.), Humana, Totowa, NJ, pp. 33–61.
29. Froehler, B. C. (1993) Oligodeoxyribonucleotide synthesis-H phosphonate approach, in *Protocols for Oligonucleotides and Analogs* (Agrawal, S., ed.), Humana, Totowa, NJ, pp. 63–80.
30. Zhang, R., Diasio, R. B., Lu, Z., Liu, T., Jiang, Z., Galbraith, W. M., and Agrawal, S. (1995) Pharmacokinetics and tissue distribution in rats of an oligodeoxynucleotide phosphorothioate (GEM 91) developed as a therapeutic agent for human immunodeficiency virus type-1. *Biochem. Pharmacol.* **49**, 929–939.
31. Zhang, R., Lu, Z., Zhang, H., Diasio, R. B, Habus, I., Jiang, Z., Iyer, R. P., Yu, D., and Agrawal, S. (1995) In Vivo stability and metabolism of a "hybrid" oligonucleoside phosphorothioate in rats. *Biochem. Pharmacol.* **50**, 545–556.

32. Zhang, R., Lu, Z., Zhang, X., Diasio, R., Liu, T., Jiang, Z., and Agrawal, S. (1995) In vivo stability and disposition of a self-stabilized oligodeoxynucleotide phosphorothioate in rats. *Clin. Chem.* **41,** 836–843.

33. Zhang, R., Iyer, R. P., Weitan, T., Yu, D., Zhang, X., Lu, Z., Zhao, H., and Agrawal, S. (1995) Pharmacokinetics and tissue distribution of a chimeric oligodeoxynucleotide phosphorothioate in rats following intravenous administration (unpublished results).

34. Zhang, R., Yan, J., Shahinian, H., Yan, J., Amin, G., Lu, Z., Jiang, Z., Temsamani, J., Saag, M. S., Schechter, P. S., Agrawal, S., and Diasio, R. B. (1995) Human pharmacokinetics of an anti-HIV antisense oligodeoxynucleotide phosphorothioate (GEM 91) in HIV-infected individuals. *Clin. Pharmacokinet. Ther.* **58,** 44–53.

Index

Methods in Molecular Biology™

Methods in Molecular Biology™ manuals are available at all medical bookstores. You may also order copies directly from Humana by filling in and mailing or faxing this form to: Humana Press, 999 Riverview Drive, Suite 208, Totowa, NJ 07512 USA, Phone: 201-256-1699/Fax: 201-256-8341.

☐ 60. **Protein NMR Protocols,** edited by *David G. Reid, 1996* • 0-89603-309-0 • Comb $69.50 (T)

☐ 59. **Protein Purification Protocols,** edited by *Shawn Doonan, 1996* • 0-89603-336-8 • Comb $64.50 (T)

☐ 58. **Basic DNA and RNA Protocols,** edited by *Adrian J. Harwood, 1996* • 0-89603-331-7 • Comb $69.50 • 0-89603-402-X • Hardcover $99.50

☐ 57. **In Vitro Mutagenesis Protocols,** edited by *Michael K. Trower, 1996* • 0-89603-332-5 • Comb $69.50

☐ 56. **Crystallographic Methods and Protocols,** edited by *Christopher Jones, Barbara Mulloy, and Mark Sanderson, 1996* • 0-89603-259-0 • Comb $69.50 (T)

☐ 55. **Plant Cell Electroporation and Electrofusion Protocols,** edited by *Jac A. Nickoloff, 1995* • 0-89603-328-7 • Comb $49.50

☐ 54. **YAC Protocols,** edited by *David Markie, 1995* • 0-89603-313-9 • Comb $69.50

☐ 53. **Yeast Protocols:** *Methods in Cell and Molecular Biology,* edited by *Ivor H. Evans, 1996* • 0-89603-319-8 • Comb $74.50 (T)

☐ 52. **Capillary Electrophoresis:** *Principles, Instrumentation, and Applications,* edited by *Kevin D. Altria, 1996* • 0-89603-315-5 • Comb $74.50

☐ 51. **Antibody Engineering Protocols,** edited by *Sudhir Paul, 1995* • 0-89603-275-2 • Comb $69.50

☐ 50. **Species Diagnostics Protocols:** *PCR and Other Nucleic Acid Methods,* edited by *Justin P. Clapp, 1996* • 0-89603-323-6 • Comb $69.50

☐ 49. **Plant Gene Transfer and Expression Protocols,** edited by *Heddwyn Jones, 1995* • 0-89603-321-X • Comb $69.50

☐ 48. **Animal Cell Electroporation and Electrofusion Protocols,** edited by *Jac A. Nickoloff, 1995* • 0-89603-304-X • Comb $64.50

☐ 47. **Electroporation Protocols for Microorganisms,** edited by *Jac A. Nickoloff, 1995* • 0-89603-310-4 • Comb $69.50

☐ 46. **Diagnostic Bacteriology Protocols,** edited by *Jenny Howard and David M. Whitcombe, 1995* • 0-89603-297-3 • Comb $69.50

☐ 45. **Monoclonal Antibody Protocols,** edited by *William C. Davis, 1995* • 0-89603-308-2 • Comb $64.50

☐ 44. **Agrobacterium Protocols,** edited by *Kevan M. A. Gartland and Michael R. Davey, 1995* • 0-89603-302-3 • Comb $69.50

☐ 43. **In Vitro Toxicity Testing Protocols,** edited by *Sheila O'Hare and Chris K. Atterwill, 1995* • 0-89603-282-5 • Comb $69.50

☐ 42. **ELISA:** *Theory and Practice,* by *John R. Crowther, 1995* • 0-89603-279-5 • Comb $59.50

☐ 41. **Signal Transduction Protocols,** edited by *David A. Kendall and Stephen J. Hill, 1995* • 0-89603-298-1 • Comb $64.50

☐ 40. **Protein Stability and Folding:** *Theory and Practice,* edited by *Bret A. Shirley, 1995* • 0-89603-301-5 • Comb $69.50

☐ 39. **Baculovirus Expression Protocols,** edited by *Christopher D. Richardson, 1995* • 0-89603-272-8 • Comb $64.50

☐ 38. **Cryopreservation and Freeze-Drying Protocols,** edited by *John G. Day and Mark R. McLellan, 1995* • 0-89603-296-5 • Comb $79.50

☐ 37. **In Vitro Transcription and Translation Protocols,** edited by *Martin J. Tymms, 1995* • 0-89603-288-4 • Comb $69.50

☐ 36. **Peptide Analysis Protocols,** edited by *Ben M. Dunn and Michael W. Pennington, 1994* • 0-89603-274-4 • Comb $64.50

☐ 35. **Peptide Synthesis Protocols,** edited by *Michael W. Pennington and Ben M. Dunn, 1994* • 0-89603-273-6 • Comb $64.50

☐ 34. **Immunocytochemical Methods and Protocols,** edited by *Lorette C. Javois, 1994* • 0-89603-285-X • Comb $64.50

☐ 33. **In Situ Hybridization Protocols,** edited by *K. H. Andy Choo, 1994* • 0-89603-280-9 • Comb $69.50

☐ 32. **Basic Protein and Peptide Protocols,** edited by *John M. Walker, 1994* • 0-89603-269-8 • Comb $59.50 • 0-89603-268-X • Hardcover $89.50

☐ 31. **Protocols for Gene Analysis,** edited by *Adrian J. Harwood, 1994* • 0-89603-258-2 • Comb $69.50

☐ 30. **DNA–Protein Interactions,** edited by *G. Geoff Kneale, 1994* • 0-89603-256-6 • Paper $64.50

☐ 29. **Chromosome Analysis Protocols,** edited by *John R. Gosden, 1994* • 0-89603-243-4 • Comb $69.50 • 0-89603-289-2 • Hardcover $94.50

☐ 28. **Protocols for Nucleic Acid Analysis by Nonradioactive Probes,** edited by *Peter G. Isaac, 1994* • 0-89603-254-X • Comb $59.50

☐ 27. **Biomembrane Protocols:** *II. Architecture and Function,* edited by *John M. Graham and Joan A. Higgins, 1994* • 0-89603-250-7 • Comb $64.50

☐ 26. **Protocols for Oligonucleotide Conjugates:** *Synthesis and Analytical Techniques,* edited by *Sudhir Agrawal, 1994* • 0-89603-252-3 • Comb $64.50

☐ 25. **Computer Analysis of Sequence Data:** *Part II,* edited by *Annette M. Griffin and Hugh G. Griffin, 1994* • 0-89603-276-0 • Comb $59.50

☐ 24. **Computer Analysis of Sequence Data:** *Part I,* edited by *Annette M. Griffin and Hugh G. Griffin, 1994* • 0-89603-246-9 • Comb $59.50

☐ 23. **DNA Sequencing Protocols,** edited by *Hugh G. Griffin and Annette M. Griffin, 1993* • 0-89603-248-5 • Comb $59.50

☐ 22. **Microscopy, Optical Spectroscopy, and Macroscopic Techniques,** edited by *Christopher Jones, Barbara Mulloy, and Adrian H. Thomas, 1993* • 0-89603-232-9 • Comb $69.50

☐ 21. **Protocols in Molecular Parasitology,** edited by *John E. Hyde, 1993* • 0-89603-239-6 • Comb $69.50

☐ 20. **Protocols for Oligonucleotides and Analogs:** *Synthesis and Properties,* edited by *Sudhir Agrawal, 1993* • 0-89603-247-7 • Comb $69.50 • 0-89603-281-7 • Hardcover $89.50

☐ 19. **Biomembrane Protocols:** *I. Isolation and Analysis,* edited by *John M. Graham and Joan A. Higgins, 1993* • 0-89603-236-1 • Comb $64.50

☐ 18. **Transgenesis Techniques:** *Principles and Protocols,* edited by *David Murphy and David A. Carter, 1993* • 0-89603-245-0 • Comb $69.50

☐ 17. **Spectroscopic Methods and Analyses:** *NMR, Mass Spectrometry, and Metalloprotein Techniques,* edited by *Christopher Jones, Barbara Mulloy, and Adrian H. Thomas, 1993* • 0-89603-215-9 • Comb $69.50

☐ 16. **Enzymes of Molecular Biology,** edited by *Michael M. Burrell, 1993* • 0-89603-322-8 • Paper $59.50

☐ 15. **PCR Protocols:** *Current Methods and Applications,* edited by *Bruce A. White, 1993* • 0-89603-244-2 • Paper $54.50

☐ 14. **Glycoprotein Analysis in Biomedicine,** edited by *Elizabeth F. Hounsell, 1993* • 0-89603-226-4 • Comb $69.50

☐ 13. **Protocols in Molecular Neurobiology,** edited by *Alan Longstaff and Patricia Revest, 1992* • 0-89603-199-3 • Comb $59.50

☐ 12. **Pulsed-Field Gel Electrophoresis:** *Protocols, Methods, and Theories,* edited by *Margit Burmeister and Levy Ulanovsky, 1992* • 0-89603-229-9 • Hardcover $69.50

☐ 11. **Practical Protein Chromatography,** edited by *Andrew Kenney and Susan Fowell, 1992* • 0-89603-213-2 • Hardcover $59.50

☐ 10. **Immunochemical Protocols,** edited by *Margaret M. Manson, 1992* • 0-89603-270-1 • Comb $69.50

☐ 9. **Protocols in Human Molecular Genetics,** edited by *Christopher G. Mathew, 1991* • 0-89603-205-1 • Hardcover $69.50

☐ 8. **Practical Molecular Virology:** *Viral Vectors for Gene Expression,* edited by *Mary K. L. Collins, 1991* • 0-89603-191-8 • Paper $54.50

☐ 7. **Gene Transfer and Expression Protocols,** edited by *Edward J. Murray, 1991* • 0-89603-178-0 • Hardcover $79.50

☐ 6. **Plant Cell and Tissue Culture,** edited by *Jeffrey W. Pollard and John M. Walker, 1990* • 0-89603-161-6 • Comb $69.50

☐ 5. **Animal Cell Culture,** edited by *Jeffrey W. Pollard and John M. Walker, 1990* • 0-89603-150-0 • Comb $69.50

Name _____

Department _____

Institution _____

Address _____

City/State/Zip _____

Country _____

Phone # _____ Fax # _____

"T" denotes a tentative price. Prices listed are Humana Press prices, current as of October 1995, and do not reflect the prices at which books will be sold to you by suppliers other than Humana Press. All prices subject to change without notice.
UK, Europe, Middle East, and Africa: Order directly from Chapman & Hall by faxing to: +44-171-522-9623.

Postage & Handling: *USA Prepaid (UPS):* Add $4.00 for the first book and $1.00 for each additional book. *Outside USA* (Surface): Add $5.00 for the first book and $1.50 for each additional book.

☐ **My check for $_____ is enclosed**
  (Drawn on US funds from a US bank).

☐ Visa   ☐ MasterCard   ☐ American Express

Card # _____

Exp. date _____

Signature _____